Plant Hormone Receptors

NATO ASI Series

Advanced Science Institutes Series

A series presenting the results of activities sponsored by the NATO Science Committee, which aims at the dissemination of advanced scientific and technological knowledge, with a view to strengthening links between scientific communities.

The Series is published by an international board of publishers in conjunction with the NATO Scientific Affairs Division

A Life Sciences B Physics	Plenum Publishing Corporation London and New York
C Mathematical and Physical Sciences	D. Reidel Publishing Company Dordrecht, Boston, Lancaster and Tokyo
D Behavioural and Social Sciences E Applied Sciences	Martinus Nijhoff Publishers Boston, The Hague, Dordrecht and Lancaster
F Computer and Systems Sciences G Ecological Sciences H Cell Biology	Springer-Verlag Berlin Heidelberg New York London Paris Tokyo

Series H: Cell Biology Vol. 10

Plant Hormone Receptors

Edited by

Dieter Klämbt

Botanisches Institut der Universität
Meckenheimer Allee 170
5300 Bonn 1, FRG

Springer-Verlag
Berlin Heidelberg New York London Paris Tokyo
Published in cooperation with NATO Scientific Affairs Divison

Proceedings of the NATO Advanced Research Workshop on Plant Hormone Receptors held at Bad Honnef (Bonn), FRG, August 18–22, 1986

ISBN 3-540-17981-X Springer-Verlag Berlin Heidelberg New York
ISBN 0-387-17981-X Springer-Verlag New York Berlin Heidelberg

Library of Congress Cataloging in Publication Data. NATO Advanced Research Workshop on Plant Hormone Receptors (1986 : Bad Honnef, Germany) Plant hormone receptors. (NATO ASI series. Series H, Cell biology ; vol. 10) "Proceedings of the NATO Advanced Research Workshop on Plant Hormone Receptors held at Bad Honnef (Bonn), FRG, August 18–22, 1986"—T.p. verso. 1. Plant hormones—Receptors—Congresses. I. Klämbt, Dieter, 1930- . II. Title. III. Series. QK731.N22 1986 581.19'27
87-13120
ISBN 0-387-17981-X (U.S.)

Printed in Germany

Printing: Druckhaus Beltz, Hemsbach; Bookbinding: J. Schäffer GmbH & Co. KG, Grünstadt
2131/3140-543210

PREFACE

The Nato Advanced Research Workshop on Plant Hormone Receptors was held at the Physik Zentrum in Bad Honnef near Bonn, August 18-22, 1986. This workshop was mainly supported by the Nato Scientific Affairs Division and additionally cosponsered by Hoechst AG, Frankfurt and BASF AG, Ludwigshafen.

The workshop aimed at focusing research on plant hormone receptors. It should provide an opportunity to all who work in this field to report on their very recent data and to discuss their results with the most competent colleagues. The total number of participants was limited to 30 to ensure personal contact and intensive discussions. Everyone had to either give a lecture or practical course. One half of the participants were invited, the other was selected by applications.

Plant hormone receptors are assumed to exist but clear results are still rare. Nevertheless encouraging results have been published over the last years. Receptors for animal hormones and neuronal transmitters are well characterized, both structurally and functionally. Therefore scientists dealing with receptors for steroid hormones - Prof. E.E. Baulieu, Paris and Prof. J. Å. Gustafsson, Huddinge - and for acetylcholine - Prof. A. Maelicke, Dortmund - were invited to participate in the workshop. Their expertise in the field and their contributions were expected to greatly stimulate our discussion. Unfortunately both steroid hormonologists canceled their participation due to illness. A. Maelicke, however, reported on "Structure and function of the nicotinic acetylcholine receptor as probed by monoclonal antibodies". His lecture pinpointed several of the most controversial issues that had arisen during the whole meeting.

We decided to integrate some experimental courses and demonstrations in the workshop. Therefore two half days were reserved for these courses, which aimed at demonstrating special extraction procedures and hormone binding assays currently used.

All presentations of the workshop are published in these proceedings. The annex contains the experimental procedures demonstrated during the workshop and those used by other participants in their laboratories. Therefore the book not only summarizes the recent knowledge about plant hormone receptors, but also presents the latest procedures for the extraction, analysis, and characterization of plant hormone binding proteins and receptors.

Finally I want to thank all, who were involved in the organization and realization of the workshop, including the staff of the Physik Zentrum, Bad Honnef and of course every participant, who had part on the success of this workshop.

Dieter Klämbt

CONTENTS

APPENDIX

TECHNIQUES IN PLANT HORMONE RECEPTOR RESEARCH

ORGANIZING COMMITTEE

KLÄMBT, D.	Botanisches Institut, University of Bonn, Meckenheimer Allee 170, D53 Bonn 1, FRG
HALL, M.A.	Department of Botany and Microbiology, University College of Wales, Aberystwyth, SY23 3DA, UK
HERTEL,R.	Institut für Biologie III,University of Freiburg, Schänzlestr. 1, D78 Freiburg,FRG
LINDE, P.van der	Bulb Research Centre, POB 85, 2160 AB Lisse, NL
VENIS, M.A.	Department of Plant Physiology, East Malling Research Station, East Malling, Maidstone, Kent ME19 6BJ, UK

CONTRIBUTORS

ADUCCI, P.	Universita' degli Studi di Roma "La Sapienza;Istituto di Chimica Biologica Città Universitatia - 00185 Roma, I
BALLIO,A.	Università degli Studi di Roma, Istituto di Chimica Biologica, Città Universitaria 00185 Roma, I
CLARK, K.	Department of Biology, Kline Biology Tower, P.O.Box 6666, New Haven, Connecticut 06511-8112, USA
FIRN, R.	University of York, Department of Biology, York, YO1 5DD, UK
FOX, E.	ARCO Plant Cell Research Institute, 6560 Trinity Court, Dublin, California 94568, USA
HAJEK, K.	Genetisches Institut, University of Bonn, Kirschallee 1, D5300 Bonn 1
HERTEL, R.	Institut für Biologie III, University of Freiburg, Schänzlestr. 1,D78 Freiburg,FRG
HORTON, R.	University of Guelph, College of Biological Science, Department of Botany, Guelph, Ontario, Canada - N1G 2W1
HUANG, H.	Shanghai Institute of Plant Physiology, Academia Sinica, 300 Fonglin Road, Shanghai, China 200032
JACOBS, M.	Department of Biology, Edward Martin Biological Laboratory, Swarthmore College, Swarthmore, Pennsylvania 19081, USA

JACOBSEN, H.J.	Botanisches Institut, University of Bonn, Kirschallee 1, 53 Bonn 1, FRG
KANG, B.G.	Department of Biology Yonsei University Seoul 120, Korea
KATEKAR, G.F.	CSIRO, Division of Plant Industry, Black Mountain, GPO Box 1600. Canberra, ACT 2601, Australia
LESHEM, Y.Y.	Department of Life Sciences, Bar-Ilan University, 52 100 Ramat Gan, Israel
LÖBLER, M.	Botanisches Institut, University of Bonn, Meckenheimer Allee 170, 53 Bonn 1, FRG
MAELICKE, A.	MPI für Ernährungsphysiologie, Rheinlanddamm 201, 46 Dortmund 1, FRG
MENNES, B.	Department of Plant Molecular Biology, Botanical Laboratory, Nonnensteeg 3, 2311VJ - Leiden, NL
RÜDELSHEIM, P	Departement Biologie, University of Antwerpen, Universiteitsplein 1, B-2610 Antwerpen (Wilrijk) B
SANDERS, I.O.	Department of Botany and Microbiology, University College of Wales, Aberystwyth, SY23 3DA, UK
SCHERER, G.	Botanisches Institut, University of Bonn, Venusbergweg 22, 53 Bonn 1, FRG
SISLER, E.C.	North Carolina State University, School of Agriculture and Life Sciences, School of Physical and Mathematical Sciences, Department of Biochemistry, Box 7622 Raleigh, N.C. 27695-7622, USA
SMITH, A.R.	The University College of Wales, Department of Botany and Microbiology, School of Biological Sciences, Aberystwyth SY23 3DA, UK
SRIVASTAVA, L.M.	Simon Fraser University, Department of Biological Sciences, Burnaby, British Columbia V5A 1S6, Canada
TELGEN, I.H.J.	Department of Plant Molecular Biology, Botanical Laboratory, Nonnensteeg 3, 2311VJ - Leiden, NL
VENIS, M.A.	Department of Plant Physiology, East Malling Research Station, East Malling, Maidstone, Kent ME 19 6BJ, UK
ZBELL, B.	Botanisches Institut, University of Heidelberg, Im Neuenheimer Feld 360, 69 Heidelberg 1

PARTICIPANTS

EHRSFELD, K.	Botanisches Institut, University of Bonn, Meckenheimer Allee 170, 53 Bonn 1, FRG
HESS, D.	Genetisches Institut, University of Bonn, Kirschallee 1, 53 Bonn 1, FRG
LALLU, N.	Plant Physiology, Division of Horticulture and Processing, DSIR, Private Bag, Auckland, New Zealand
LIBBENGA, K.	Department of Plant Molecular Biology, Botanical Laboratory, Nonnensteeg 3, 2311VJ - Leiden, NL
MARTINY, G.	Botanisches Institut, University of Bonn, Venusbergweg 22, 53 Bonn 1, FRG
NAPIER, R.	Department of Plant Physiology, East Malling Research Station, East Malling, Maidstone, Kent ME 19 6BJ, Uk
PALME, K.	MPI für Züchtungsforschung, 5 Köln-Vogelsang, FRG
RADERMACHER, E.	Botanisches Institut, University of Bonn, Meckenheimer Allee 170, 53 Bonn 1, FRG
SCHMIDT, O.,	Landwirtschaftliche Versuchsstation, BASF, Box 220, 6703 Limburger Hof, FRG
VERDENHALVEN, R.	Pflanzen Biochemie, Höchst, Box 800320, 6230 Frankfurt/M., FRG
WILLIAMS, R.A.N.	The University College of Wales, Department of Botany and Microbiology, School of Biolgical Sciences, Aberystwyth SY23 3DA, UK
ZAAL van der, E.J.	Department of Plant Molecular Biology, Botanical Laborytory, Nonnensteeg 3, 2311VJ - Leiden, NL

TOO MANY BINDING PROTEINS, NOT ENOUGH RECEPTORS?

Richard D. Firn
Department of Biology
University of York
England YO1 5DD

Why we need to find receptors

Those who originally sought plant hormone receptors in the late 1960's were largely concerned with understanding the mechanism of action of the regulators. At that time, rapid progress was being made in discovering the basic mechanisms by which animal hormones acted and receptors for the major groups of animal hormones were being found. In the case of animal hormones, evidence as to the mechanism of action of the major groups of hormone was available and the receptors, as they were found, provided key pieces of a jigsaw which was fitting together nicely. Those working on the mechanism of action of plant hormones were having less success and the search for receptors was begun while their jigsaw pieces were still being roughly arranged. Researchers seeking plant hormone receptors reasoned that the isolation of plant hormone receptors would provide the key to the puzzle and expected that information about the receptors would lead to a breakthrough in discovering the mechanism of action of the individual hormones. This dream remained unfulfilled. Indeed it now seems possible that, rather than the isolation of plant hormone receptors leading us to the mechanism of action of the hormones, an increased understanding of the mechanism of action of the hormones may be needed to lead us to the receptors.

More recently, another group of plant hormonologists have begun to talk about receptors. These are workers who are interested in the physiological role of hormones. Like their colleagues working on the mechanism of action of hormones, they found that the 1970's were a somewhat difficult time. It was becoming apparent that simply measuring hormone concentrations in extracts of organs was not providing the desired correlations between hormone concentration and physiological response. This problem was in fact an old one, overlooked for many years, and the two postulated ways

NATO ASI Series, Vol. H10
Plant Hormone Receptors. Edited by D. Klämbt

of resolving the problem were likewise of considerable lineage. Firstly it was suggested that tissue or cell "compartmentation" was the underlying cause, it being argued that hormones were found in many compartments but only acted in some. The need to identify the compartment containing the site of action became apparent. It was obvious that if one could locate the site of the hormone receptors, the compartment of interest would be identified. The second reason why these physiologists became interested in receptors was the effective advocacy by Trewavas (1983) that cells underwent developmental changes in "sensitivity" to hormones. He argued that such sensitivity changes were caused by changes in the number of receptors in target cells and that these changes were more important than changes in hormone concentration. Although it is clear that cells could change their sensitivity to a hormone by means other than changing the number of receptors (Firn, 1986), it is certainly true that an ability to measure the receptivity of cells would be very useful to physiologists.

Those working on plant hormone receptors can obviously take great comfort in this interest, from so many sides, in their work. Once a convincing receptor for any plant hormone has been found, answers about the mechanism of action of the hormone and its physiological role should follow. Advances made in other areas over the last decade have presented researchers interested in proteins (e.g. receptors) with some beautiful tools. Immunological and immunocytochemical techniques should allow the location and quantitation of the receptor at the cell level. Consequently, it will be possible to measure developmental changes in receptor concentration and questions about changes in hormone sensitivity will be answerable. Molecular biological techniques will also enable researchers to study the control of genes coding for the various receptors and it may be possible eventually to place receptors in receptorless or receptor deficient cells in order to probe the physiological consequences. With such exciting opportunities ahead, the need to isolate and characterise plant hormone receptors is clearly evident. However, a note of caution may be in order. Before this new phase of work begins we should really be very sure that the binding proteins we decide to work with are physiologically important.

Why have so many plant hormone binding proteins been found?

If one considers the literature accumulated on plant hormone receptors over the last 15 years (comprehensively and critically discussed by Venis (1985) in a recent monograph) it is apparent that progress has not been dramatic. The rate of publication in the area, for instance, is fairly constant and there is no sign of the exponential rise in publication which so often characterises fruitful endeavour (Fig 1). The effort expended on the search for

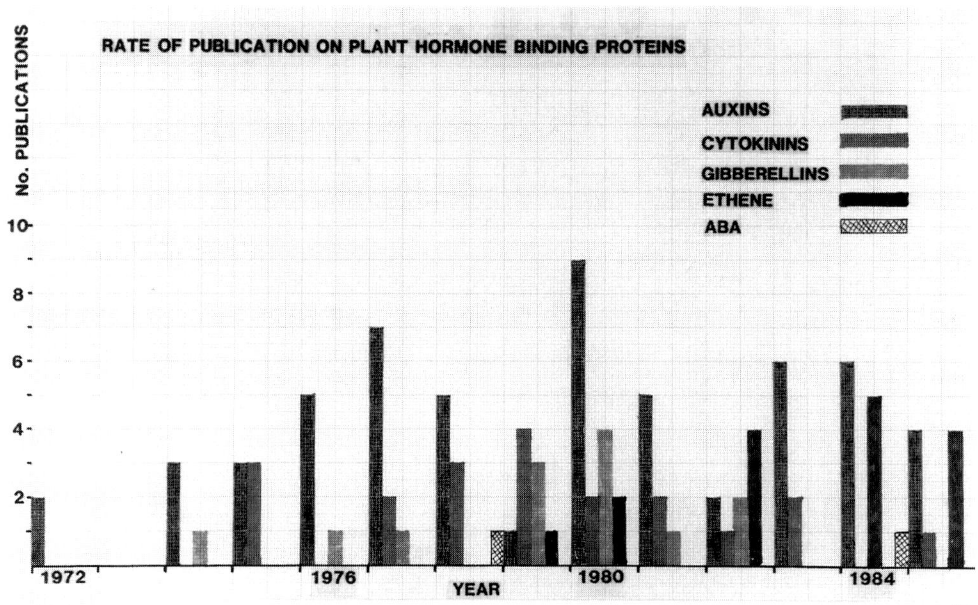

receptors for the various hormones has been very unequal (Table 1) and the majority of the papers published come from a few groups following up the binding of a hormone to a particular kind of binding site. In nearly all cases, such groups work with binding proteins in systems which are somewhat atypical in that they contain large amounts of binding. Quite a large proportion of the remaining papers are lone reports which have not led to further published work. Missing from any such survey of published work is the search for receptors which has been conducted but which did not lead to any formal publication. To the authors knowledge there has been at least several man/years unreported effort in this area.

Table 1. The relative effort expended in the search for various plant hormone binding proteins (1971 - 1985)

	Number of authors	Number of papers
Auxin		
Maize membrane	33	28
Others	53	36
Cytokinins	35	22
Ethylene	18	16
Gibberellins	18	12
Abscisic Acid	5	2

Considering all the work to date, it is obvious that no binding protein has a monopoly for one hormone. Admittedly it is difficult to compare the different reports of binding proteins for any one hormone, but it would seem that for all the hormones a number of different types of binding site have been found. Another striking feature of all the plant hormone binding proteins detected so far, is the the very limited range of species in which each has been found. It is usually very difficult to ascertain whether this limited distribution is because the search has been restricted to only a few favourite species or whether unsuccessful attempts to find binding proteins in other species have not generally been reported. These two factors, the apparent limited distribution of any one type of binding protein and the diversity of binding protein types for even a single hormone, are surely unexpected. Some workers obviously disagree and they point out that all plant hormones produce a diverse range of biochemical or physiological responses [the term "pleiotypic" has been used to describe this diversity of hormone action in animals (Michell & Houslay, 1986)]. It seems timely to consider this disagreement in more detail and to note the consequences of the various arguments to future work on plant hormone receptors.

Pleiotypic effects of plant hormone. It is very easy to draw up lists of the range of responses of plants to individual hormones and these show a striking diversity of response (Table 2).

Table 2. Examples of the diversity of effects of applied plant hormones

AUXINS

Promotion of cell elongation
Inhibition of root growth
Promotion of cell division in cell cultures
Stimulation of root formation

GIBBERELLINS

Promotion of cell elongation
Stimulates germination of some dormant seeds
Promotes synthesis of some hydrolytic enzymes in cereals
Induces parthenocarpic fruit set in some plants

CYTOKININS

Retards senescence of some cells
Influences differentiation in mosses
Promotes cell division in cell cultures
Promotes cell enlargement of some cells

ETHYLENE

Promotes abscission
Influences rate and direction of cell expansion
Stimulates rate of expansion of some submerged shoots
Promotes fruit ripening

ABSCISIC ACID

Promotes stomatal closure
Inhibits seed germination
Inhibits cell elongation
Stimulates turion formation in some pondweeds

The responses usually vary from rapid, specific effects in isolated cell types, to slow, complex, poorly understood developmental changes in whole organisms. This diversity of response has led to proposals that each hormone might have more than one fundamental mechanism of action. The possibility that a number of receptors exists for each hormone follows as a consequence. However, while this argument might explain the existence of two or more receptors for any one hormone it would not explain the limited distribution of any receptor because several responses can often be evoked in a single species by a single hormone. If each response were produced by the action of the hormone at a different receptor, each species capable of showing a range of pleiotypic effects to one hormone would have to contain several different receptors. Likewise, it would surely be expected that all species showing the equivalent response to a specific hormone would contain a similar receptor. These predictions do not seem to be borne out by the work on receptors to date, hence it seems unlikely to me that one can explain the diversity of binding proteins found for some plant hormones so far on the basis of pleiotypic effects.

A further point might also be made about the diversity of hormone effects in relation to the primary action of hormones. It must be remembered that the lists of hormone responses for each hormone, so often taught at an elementary level, are somewhat misleading. Very often some of these responses are very specific for one type of plant and there is little good evidence that such responses are common. Furthermore, although a single plant can show several effects when treated with one hormone, it would seem to me that each effect is often the consequence of hormone action in quite distinct cell types. Is there any good evidence that any hormone responsive cell can show more than one type of response? In other words, is it not possible that all the diverse effects to any one type of plant hormone are simply the manifestation of cell differentiation within the responsive plant and that only one type of primary response is induced in any responsive cell? If this possibility is realistic, the idea of a common primary mechanism of action deserves serious consideration as the simplest model.

Structure-activity relationships. Another piece of evidence against the idea of a number of unrelated receptors for each plant hormone comes from studies of structure-activity relationships, where there is little evidence for any response-specific relationships. Although some very marked differences are found in the relative effectiveness of different hormone analogues in certain bioassays, these differences seem to be more species specific than response specific. Given that the relative effectiveness of analogues would also be expected to be influenced by differing rates of uptake, transport and metabolism, structure-activity relationships do not argue strongly in favour of a number of unrelated receptors with different sites of action. If there are more than one receptor for each hormone it seems much more likely that they are related in some way, possibly having large homologous regions involving the active sites. This is a prediction which should be relatively easy to test - tryptic digests of the receptors should produce some similar fragments.

Interestingly, those working on animal hormones are engaged in a similar debate as to whether there is one flexible response to a single primary hormone action (Rodbell, 1985) or whether there are multiple (but evolutionarily related) receptors and mechanisms of action (Michell & Houslay, 1986).

What else could binding proteins be if not receptors?

It has just been argued that there are reasonable grounds for expecting a very similar receptor, of universal occurrence, in all cells capable of showing an equivalent response to a particular hormone. It has also been proposed that available evidence does not preclude the possibility that there is only one primary mechanism of action for each hormone. If this were true, one would have to find some possible explanation for the fact that a number of different binding proteins have been found for some hormones. The simplest, and possibly less popular, explanation would be that some of the binding proteins found are not physiologically relevant.

It would seem that those seeking plant hormone receptors have had a much worse experience with "artifactual" binding than those working with animal hormones. "Saturable" or "specific" binding of plant hormones has been found to talc, plastic test-tubes and BSA (bovine serum albumin). Fortunately, this binding is easily dismissed simply because the binding site is obviously not of plant origin. However, it is the origin of the material, rather than the properties of the binding, which readily distinguishes this artifactual binding from some types of binding reported to plant derived material. The fact that the commonest major plant protein (RUBISCO) showed saturable binding of auxin encouraged scepticism largely because the real role of that protein was known. One wonders whether similar scepticism would have been so evident if the binding protein had been a minor constituent, of unknown role, and found in some more likely organ? In other words we know already that artifactual binding can occur and it may be more common than we care to believe. It is true that the chances of detecting such artifactual binding are greatly reduced if the choice of binding assay is made carefully (Venis, 1984) but obviously it must be recognised that plants do contain proteins which bind hormones yet do not play any known role in hormone action. The possibility that hormone metabolising enzymes or transport proteins could be confused with receptors has long been recognised but it is possible that other proteins could fortuitously bind plant hormones. The diversity of plant biochemistry makes this a not unrealistic possibility. This diversity is such that those seeking receptors for a plant hormone in, for instance, the roots of 5 different species will encounter many more proteins than an animal cell biologist would when looking for the receptor for an animal hormone in a particular organ from five mammals. This great biochemical diversity in plants will include a range of proteins capable of transforming secondary plant products, some of which may be structurally similar to some plant hormones. Combined with this diversity of enzyme activity, some organs will also possess a range of storage proteins. It is just possible that some binding proteins described in the literature are just such proteins. In support of this pessimistic view, is the very limited taxonomic distribution of some plant hormone binding proteins found to date. This is precisely what would be expected if the binding proteins were really proteins involved in secondary plant metabolism or storage proteins because both types of compound are often restricted to a small group of plants.

The benefits of seeking receptors in a purposeful way

One possible way of avoiding some of the possible problems outlined above is to place "physiological limits" on the search for any receptor. Instead of seeking any receptor for auxin, for instance, one would set out to find the receptor for auxin which is involved in the rapid growth reponse of cells to auxin. The physiological limits would include all the usual criteria (known structure activity relationships for this response, the estimated K_D, etc.) and the criteria that all cells capable of showing this response should have a similar receptor. Given that all fundamental hormone responses can be shown in a wide range of species, the receptor would have to be found in all these species. If only one of the chosen species contains a binding protein, it would be necessary to conclude that the binding protein was not the one sought for the response being studied. Whether the binding protein was of interest to some other hormone response or simply physiologically irrelevant would require more experiments based on different criteria and predictions.

NPA binding - living proof of such logic? It is instructive to consider the example of NPA binding, even though it is not a plant hormone itself. The interesting feature of NPA binding is that it has been found in a wider range of plants than any hormone binding protein. Although for some reason, maize coleoptiles are usually the organ of choice for those seeking to characterise the NPA binding protein, similar binding has been found quite easily in other species and organs (Table 3). This is as expected if the

Table 3. The known distribution of NPA binding to particulate fraction

Maize coleoptiles
Zuccini hypocotyls
Tobacco callus cultures
Tobacco suspension cultures
Pea internode
Sugar cane suspension cultures
Sunflower hypocotyls
Marrow hypocotyls

NPA binding protein is in fact the site at which the compound produces its physiological effect, because the effect of NPA has been found in a wide range of species. In my limited experience with NPA binding, there is no need to optimise the extraction or binding assay conditions in order to find specific NPA binding in extracts from various species. The conditions used for work in maize coleoptiles are adequate. This also builds confidence in the NPA site. Some recent preliminary experiments have further supported the logic that important, physiologically meaningful sites will be found readily in all responsive organs and may be absent from organs where the ligand has no known effect. I selected a number of plant organs from my garden or the local vegetable shop - all chosen because NPA binding was unknown in them and because they lacked intense pigmentation which might interfere with the liquid scintillation counting. Using the methods developed by others to measure NPA binding in maize coleoptiles, the results shown in Table 4 were obtained. It was found that organs likely to conduct auxin transport (celery stalks and mungbean hypocotyls) contained large amounts of binding. Cauliflower curd also contained significant NPA binding but auxin transport in that tissue is, to my knowledge, unknown. No significant binding was detected in the inner tissues of cucumber fruit. The small amount of binding in mushrooms is unexpected and intriguing. Obviously some studies of the effect of NPA on IAA uptake would be of interest in these tissues, such studies being needed to probe further the physiological role of NPA. However,

<u>Table 4. NPA binding to washed, high speed pellet from homogenates of various organs. The methods of Sussman & Goldsmith (1981) were used and 10g of each organ were homogenised. 95% confidence limits shown.</u>

	Non-specific binding (Bq)	Specific binding (Bq)
Cucumber fruit	1.6	0.1 ± 0.3
Celery stalks	4.0	10.8 ± 2.8
Mung bean hypocotyls	11.4	18.6 ± 0.8
Cauliflower curd	16.8	25.2 ± 0.8
Agaricus bisporus stalks & caps	14.3	2.2 ± 0.7

it would seem that when a convincing receptor is found, it really does behave in a fairly predictable manner and can be found readily in a wide range of species. So far this cannot yet be said of any plant hormone binding protein.

Summary

Those seeking plant hormone receptors have an important and difficult task. It would seem that proving the physiological relevance of a binding protein is much harder than characterising it but more attention must be given to the former in future. It is to be hoped that the novelty of finding any binding protein has gone and that future studies are linked much more obviously to known phsyiological effects of the hormones.

Acknowledgements. I would like to thank Hans Kende for stimulating my interest in plant hormone receptors many years ago. The experiments reported were skilfully conducted by Mike Hopgood.

References

Reviews

1. M. Venis (1985) Hormone binding sites in plants. Longman. London.
2. K.R. Libbenga, A.C. Mann, P.C.G.v.d. Linde & A. Mennes (1986) Auxin receptors. In "Hormones, receptors and cellular interactions in plants". Ed. C.M. Chadwick & D.R. Garrod. CUP. Cambridge. pp. 1-68.
3. M.A. Hall (1986) Ethylene receptors. See Ref 2. pp. 69-87.
4. J.L. Stoddart (1986) Gibberellin receptors. See Ref 2. pp. 91-111.

Other references

5. R.D. Firn (1986) Physiol. Plant. 67, 267-272.
6. R. Michell & M. Houslay (1986) Trends Biochem. Sci. 11, 239-241.
7. M. Rodbell (1985) Trends Biochem. Sci. 10, 461-464.
8. A. Trewavas (1983) Trends Biochem. Sci. 7, 354-357.

HORMONE RECOGNITION IN PLANTS

Gerard F. Katekar*, David A. Winkler# and Art E. Geissler*
* CSIRO Division of Plant Industry, Canberra, Australia
# CSIRO Division of Applied Organic Chemistry, Melbourne, Australia

INTRODUCTION:

It is axiomatic that for a substance to produce an effect in a biological organism, there must be some interaction between its molecules and certain counterparts in the organism. In the case of a hormone, the substance is made by the organism itself, and it is there to control specific functions. To have any value, therefore, these counterparts must be able to interact with the hormone only, and not with other substances, including other hormones, which may be present. In other words, the counterparts - receptors - have a recognition characteristic. This would hold regardless of how the effect is achieved, or the nature of the effect.

The known plant hormones have many similarities to animal hormones, in that they are low molecular weight substances, can produce responses in tissues remote from their site of synthesis, and are active at very low concentrations. There would seem, therefore, to be no *a priori* reason for the recognition characteristic to be different between plants and animals, even if the responses evoked by hormone-receptor interaction are different. On this basis, using techniques and concepts developed in animal physiology for determining the recognition characteristic can be justified.

STRATEGY:

Detailed knowledge of receptors, especially hormone receptors, is generally unavailable. One therefore uses differing molecules and tests their ability to bind to the receptor macromolecule, if it can be done, and/or examines the physiological response when it is apparent that such an interaction has occurred. The key feature is the shape of the molecule when it interacts with the receptor. There are various physico-chemical techniques which can be used to determine molecular shape, such as NMR and X-ray crystallography. These techniques are important as far as they go, but they can only tell the shape of the molecule, and perhaps only one shape, when there are many possibilities, and in the absence of the receptor. To go further, the approach we have taken is to use the

NATO ASI Series, Vol. H10
Plant Hormone Receptors. Edited by D. Klämbt

pharmacophore hypothesis, developed for drug design, together with its subsequent refinements. The basic hypothesis is that there may be very few, perhaps only two, functional groups in a molecule which are the primary determinants of recognition. Some groups are considered more important than others: hetero atoms, ion pairs, pi-centres, planar groups, etc. To interact with the receptor, they have to be in a certain configuration with respect to each other. On the negative side, too many functional groups may interfere with the recognition process. A third element is the spatial element, in that no part of the interacting molecule can occupy the space used by the receptor: there is receptor essential volume. Accessory binding areas are also a possibility. These are sites, adjacent to the recognition sites, which are part of the structure of the receptor macromolecule to which candidate molecules are capable of binding when they interact with the recognition site proper. In a strict sense they are irrelevant because they are not part of the recognition site and are not involved in binding the natural hormone. On the other hand, they are important, firstly, because they may well become involved when synthetic molecules are used to define the recognition site, secondly, because they may give rise to antagonist activity, and thirdly, by analogy with animals, they may differ between species, and between different regions of the organism. The first aspect must be adverted to when assessing candidate molecules, while the second and third aspects have been availed upon as very useful means to examine particular types and sub-types of receptors.

The integration of all these aspects, and the design of molecules to test hypotheses developed, can best be described as chemical intuition, and the techniques used to assist in this regard range from stick- and space-filling models to multi-regressional analysis of the physico-chemical properties of a series of molecules, such properties being electronic effects, steric effects, lipophilicity, pK, etc. More recently, because of increases in theoretical knowledge, the availability of computers, and the graphics capabilities of computers, the chemist has been given a tool which more easily enables the synthesis of ideas and their communication, together with the ability to determine whether what one has conceived may correlate with physical reality. The group of programmes used here is MORPHEUS as used by Lloyd and Andrews 1986. For the graphics programmes, non-bonded interactions can be calculated using bond angles, bond lengths, and van der Waals radii. Using empirical force fields rather than *ab initio* methods has been found effective in drug design, and has the advantage of simplicity, speed, and ease of operation. CRYSX is the central programme and uses this method. It produces a stick drawing of the screen as shown in Fig. 1, which structure can be manipulated. Guide points are also included in the drawing. These are explained later.

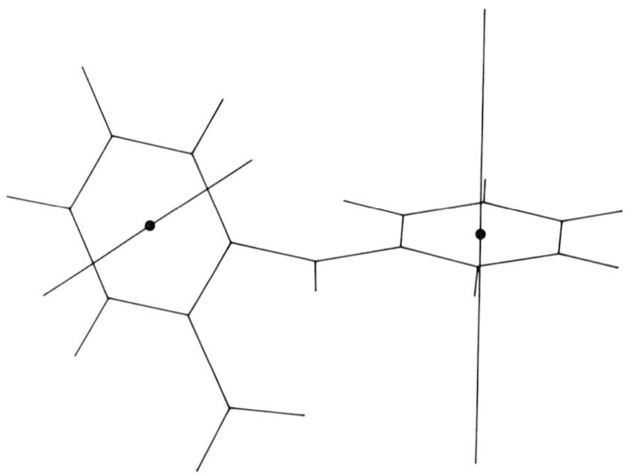

Figure 1. CRYSX screen image of benzoyl benzoic acid 6-1 with guide points.

Models can be constructed manually or from a fragment library. It can also use crystallographic data as input. This programme also outputs files to be used by the other programmes.

EXAMPLES:

1. Auxins. Indoleacetic acid is, of course, the pre-eminent plant hormone, and the stereochemistry of its interaction with its presumed receptors has long been a puzzle. The data available date back many decades. Such data are quite variable as to quality and as to the systems used to assess activity. An auxin receptor, of course, was not available. The data used here are taken from Katekar 1979 and references there cited. Using such *in vivo* data as measures of agonist activity is open to criticism, because such results incorporate non-receptor factors as uptake, sequestration, metabolism, etc. On the other hand, shape is obviously

important, and this is a characteristic of a receptor interaction. It will also be of interest to compare such conclusions as can be drawn with those from candidate auxin binding sites as they become available. They may also be an indicator of whether such a candidate site is in fact an auxin receptor.

Farrimond et al. 1978 were the first to apply computer chemistry to auxin structure-activity. They demonstrated that the charge separation theory of Thimann could not be correct, because the charge on the nitrogen, as calculated by them, is not positive, as required by the theory, but negative. In a later paper, (Farrimond et al. 1981), they also used computer calculations and could not find a correlation between charge and activity.

The approach now taken is very similar to that used by Lloyd and Andrews cited above, in that a common pharmacophore is selected and the stereochemistry is then based on this model. The pharmacophore used has been postulated previously (Katekar 1979) and is shown in Fig. 2.

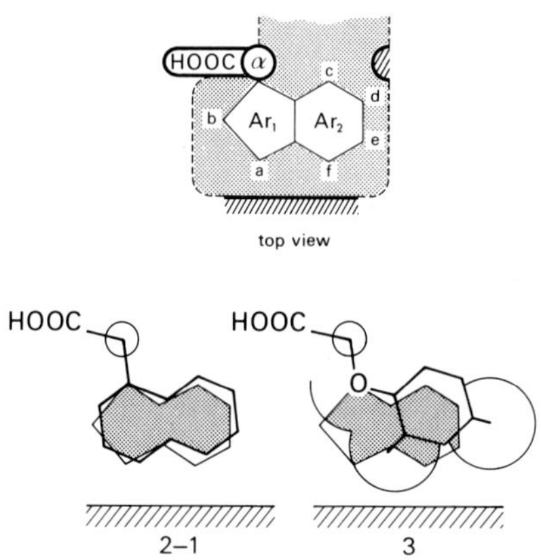

Figure 2. Auxin site and overlay of selected molecules.

The indole ring is assumed to engage in a pi-pi interaction with an aromatic ring of the receptor protein. The areas surrounding this have been considered as accessory binding areas, while the carboxyl group would interact with either a one- or two- point attachment (e.g. each carboxy oxygen) with other atoms on the receptor macromolecule. Assuming for the moment the planar extended conformation depicted, agonist molecules would fit on the site as shown. The possible conformations considered by the more recent theories (Kaethner, 1977; Farrimond _et al_. 1978; Rackhaminova _et al_. 1978; Katekar, 1979) can be classified as forms I IV (generally active) and their mirror-images (generally inactive) for asymmetric molecules. That is, they have been considered at 90° intervals. It is possible that none of these conformations are correct, and two extra conformations need to be considered, viz. the 60° and 300° conformations as shown in Fig. 3.

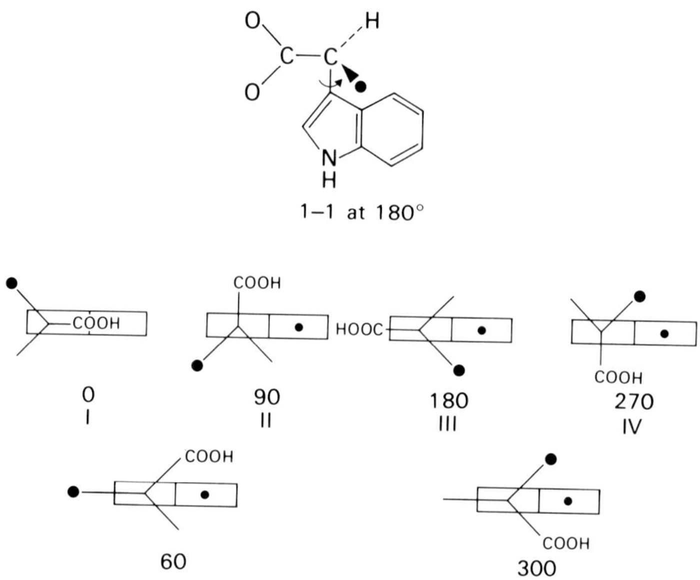

Figure 3. Possible auxin conformations

● = H; IAA

● = CH_3; S-indolepropionic acid

Figure 4. (Next page) Conformational energy maps of selected molecules. Bonds rotated correspond to those shown for 1-2.

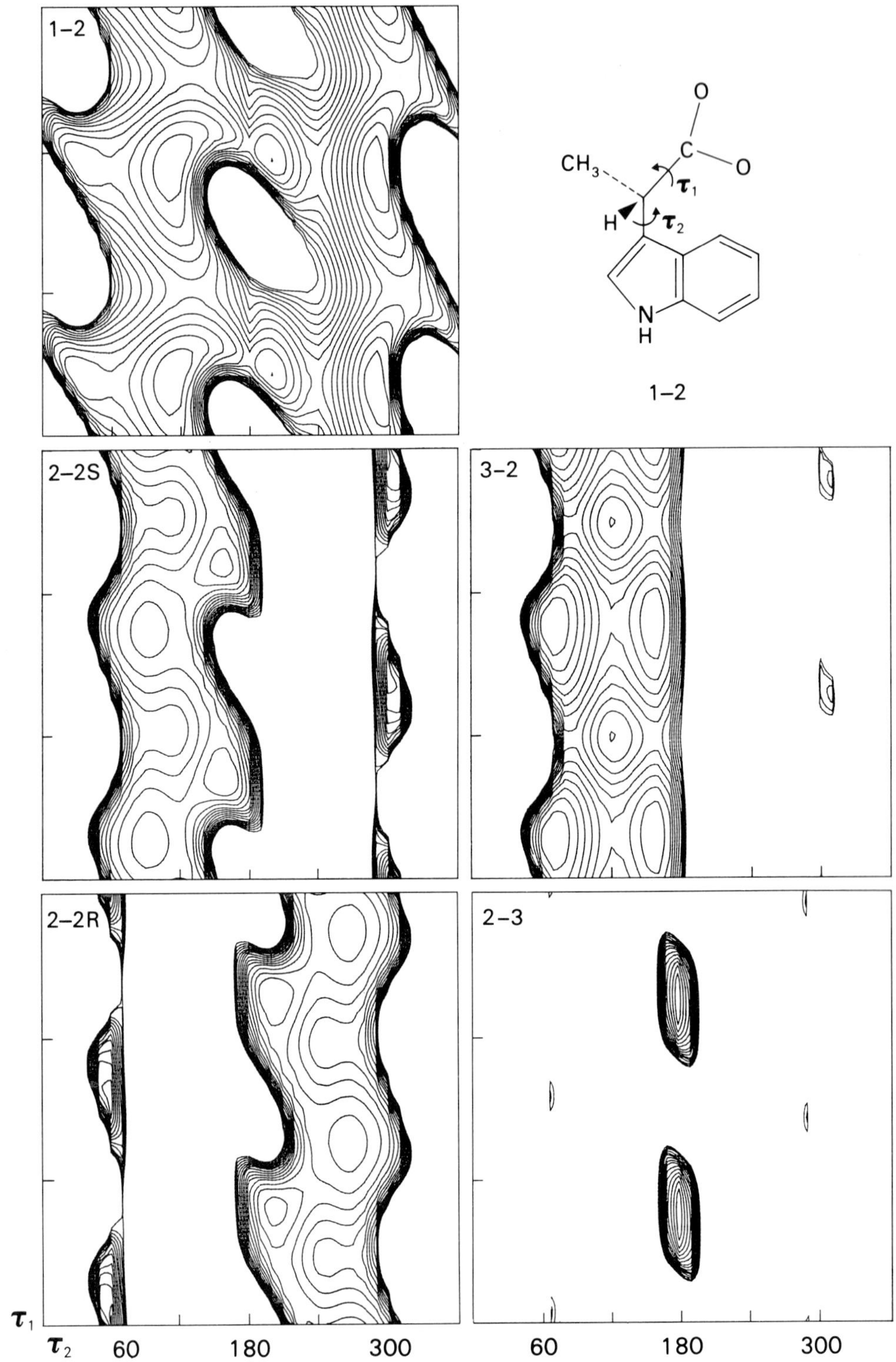
1–2
CH3
H
C
O
O
N
H
τ1
τ2
1–2
2–2S
3–2
2–2R
2–3
τ1
τ2
60
180
300
60
180
300

The key compounds to consider are shown in Table 1.

Table 1

Conformational energies (K.cal.) above global energy minimum for differing configurations. Carboxyl is at optimum configuration.

Compd.		Activity	Configuration 60°	180°	300°
1-3	S	+++	0	6	3
	R	+++	3	6	0
2-2	S	+++	0	8	5
	R	++	5	8	0
2-3		-	12	0	12
3-2	R	+++	1	14	14
	S	-	14	14	1
5	S	++	0	*	*
	R	+	*	*	0

*: Compounds cannot exist in this conformation because of molecular rigidity.

Conformational energy maps of some of the compounds are shown in Fig. 4. These are computer-calculated potential energies obtained by rotating the carboxyl(τ_1, y-axis) and the methyl group (τ_2, x-axis) through all possible conformations. It is considered that a conformation is thermally available if it has an energy of less than 10 K.cal above its minimum energy conformation. In the maps shown, the contours are 1 K.cal apart and they range from 0-15 K.cal. Where the contour lines appear, therefore, is a generous estimate of the conformations available. Where there are no contours, those conformations are not available. Fig. 4a shows the energy surface for S-indole propionic acid, 1-3S. Provided the carboxyl is optimised, it can adopt any conformation from 60°-300°. Its enantiomer would have a map which is approximately the mirror-image of the one shown, and could also adopt the same conformations. It is not surprising that both forms are active with such a large degree of conformational mobility. The S- and R- naphthalene propionic acids 2-2 are also both active, but their available conformations are far more restricted. It can be seen that they only have the 60°, 180° and 300° conformations in common. The reason for this is the increased interference between the methyl group and the 8-hydrogen as compared with the analogous interference in the IAA compounds. The active conformation must therefore be one of these three. Conformational energies of other compounds in these orientations are shown in Table 1.

It can be seen that for R-2,4,5-trichlorophenoxy propionic acid 3, which is active, the 300 conformation is not permissible, so that this conformation can also be eliminated. This is confirmed by the inactivity of its mirror image, which can adopt this conformation. The inactive 8-chloro naphthalene acetic acid, 2-3 can only adopt the 180° conformation. This eliminates the 180° conformation. It is concluded that compounds which can adopt the 60° conformation will have activity. The activity of S-acenaphthene carboxylic acid 5 can be accommodated by placing it 'the other way round' on the Ar1 Ar2 area so that its carboxyl can engage the carboxyl acceptor. The low but significant activity reported for the R-isomer would remain anomalous. The activity of the benzoic acids, which must be planar, and can have high activity, is also difficult to explain. It is concluded that while the stereochemical analysis through calculation of non-bonded interactions has provided more detail as to possibilities of what may be the interacting conformations of auxin molecules, a convincing explanation as to the precise conformational requirement remains to be achieved.

2. Phytotropins and the NPA Receptor. In the case of the NPA receptor, binding data are available. A simple model has been proposed for compounds to have phytotropin activity (Katekar and Geissler 1985). This is shown in Fig. 5.

The elements required for activity are a carboxylic acid interacting with a carboxyl acceptor, which is attached to an aromatic ring (engaging Ar1), this in turn being connected to a group of atoms which were postulated to engage a third region on the site - Ar2. Ar1 and Ar2 were postulated as being electrophilic in nature, being able to bind to aromatic rings. A 'sensitive area' was proposed - the shaded portion of Ar2 - which, if a candidate molecule could overlay this area, would have enhanced activity. The dotted outline represents the area over which the various phytotropins could lie after engaging Ar1 and the carboxyl acceptor, assuming that the part which was available to impinge on Ar2 would lie flat. Thus, the Ar2 area was originally proposed as being flat, but as can be seen from the structures, conformational flexibility was possible, so these molecules were also analysed by the methods outlined above. The strategy, as before, was to take the most sterically restrained model which had high activity, because the conformations it could adopt must be permissible for high activity.

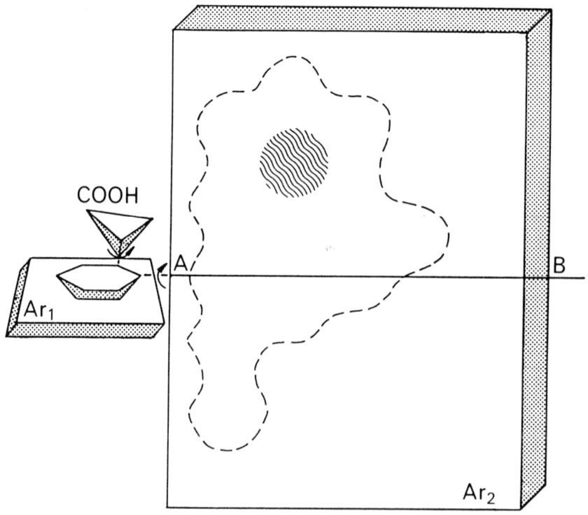

Figure 5. Preliminary Model of the Phytotropin Recognition Site.

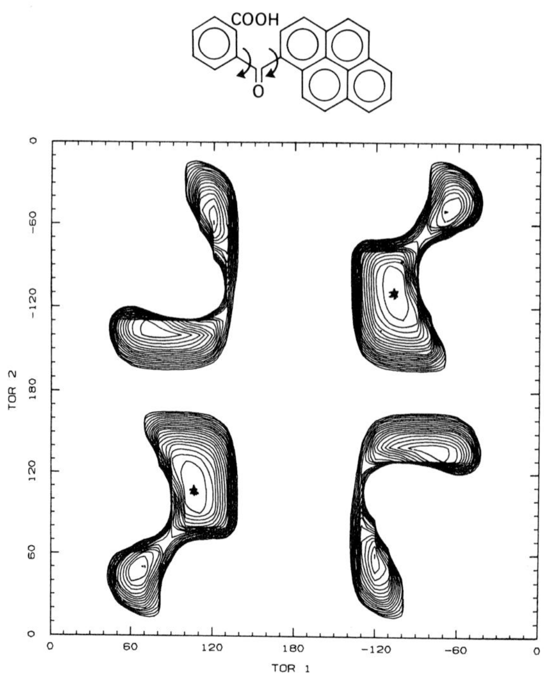

Figure 6. Energy surface of PBA * = Global energy minimum
Contours are at intervals of 2.5 K.cal.

The pyrene derivative PBA is highly active and is also quite conformationally restricted. It was therefore used as a template in determining the most likely active topography which phytotropins must adopt at the receptor. Fig. 6 shows the conformational energy surface arising from the rotation of the two torsion angles illustrated.

The molecule is capable of adopting essentially only one conformation. The other conformation which appears on the energy surface is the mirror image of the first. The global energy minimum for this molecule was found at τ_1=110 and τ_2=110 and this conformation is therefore close to, or coincident with the binding conformation. This conformation is illustrated in Fig. 7. The molecule is sufficiently hindered that moderate departures

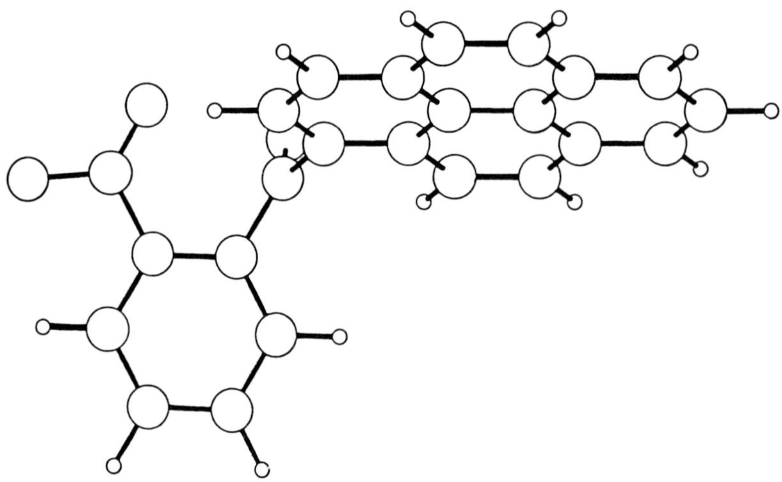

Figure 7. Low energy (binding) conformation of PBA.

from this conformation entail significant energy penalties. The other molecules in the set were then superimposed on this conformation of PBA. This was done by defining aryl receptor points - guide points - perpendicular to each aromatic ring (above and below the plane) at a distance of 3.4A from the ring as shown in Fig 1. The oxygen atoms in the benzoic acid moiety were also used as superimposition points. The superimposition of the four receptor binding points on the benzoic acid moiety can be done unambiguously. The energy of the conformation corresponding to the best fit for each molecule was then calculated and compared with the global energy minimum for that molecule. The energies of the best fit conformations are given in Table 2. How molecules can overlay the PBA conformation is shown in Fig 8.

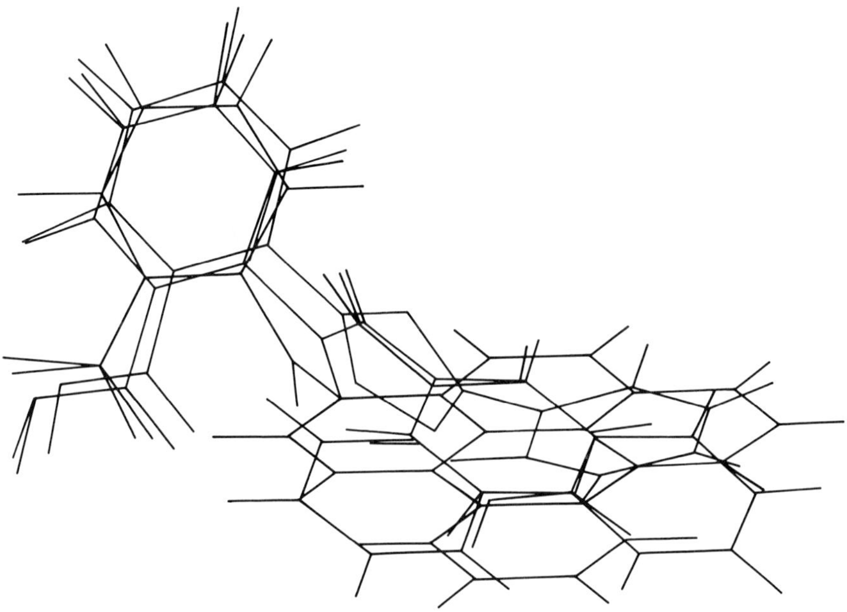

Figure 8. Overlay of active phytotropins on the PBA molecule.

There was found to be a good overlap of the benzoic acid moiety in each case with differences in the carboxyl oxygen positions being sufficiently small that hydrogen bonding interactions with a receptor should still be possible. The overlap of the aromatic moiety at the Ar2 binding position was also good. The aryl phthalamic acids all produced good overlap with the amide linkage locked trans, although the Ar2 binding occurs in a slightly different plane from that of the aryl benzoic acids and the heterocyclic analogues represented by 8 and 9. All molecules in the test series except for 2,6-dichloro phenylphthalamic acid 7-4 have low energies when adopting conformations corresponding to that shown in Fig. 7. The 2,6-dichloro compound cannot adopt this conformation due to steric hindrance from the two ortho chlorine atoms. This is consistent with its negligible activity.

Table 2

Relative energies of phytotropins in binding conformation

Compd.	Binding pKd	Energy relative to Global Minimum KJ/MOLE (= Kcal.x40)
6-1	3.89	3
6-2	5.01	3
6-3	5.11	15
PBA	8.47	0
7-1	4.42	22
7-2	7.60	16
7-3	7.28	13
7-4	<2	>500
8	6.67	21
9	7.47	26

The topical model based on PBA therefore appears better able to account for the *in vitro* activities of the phytotropins considered.

CONCLUSIONS:

The techniques and concepts developed in animal physiology for determining binding site characteristics would seem applicable to determining the hormone recognition characteristic in plants, at least with respect to the auxin and phytotropin areas. Computational analysis enables stereochemical properties of compounds to be calculated in a way that is useful, and can give insights into the types of molecules which interact, and perhaps the nature of what those interactions may be. Once data can be obtained from receptor binding in the auxin area, we may be able to look at agonist/antagonist activities, and by analogy with what has happened in animal pharmacology, use of such molecules may enable us to gain a greater understanding of the underlying mechanisms which auxin controls.

REFERENCES:

Farrimond, J.A., M.C. Elliott and D.W. Clack. Charge separation as a component of the structural requirements of hormone activity. Nature, 274 401-2 (1978).

Farrimond, J.A., M.C. Elliott and D.W. Clack. Auxin structure-activity relationships. Aryloxyacetic acids. Phytochemistry 20 1185-90 (1981).

Kaethner, J.M. Conformational change theory for auxin structure-activity relationships. Nature 267 19-23 (1977).

Katekar, G.F. Auxins: on the nature of the receptor site and molecular requirements for auxin activity. Phytochemistry 18 223-33 (1979).

Katekar, G.F. and A.E. Geissler. Recognition of phytotropins by the receptor for 1-N-naphthylphthalamic acid. Phytochemistry. In press.

Lloyd, E.J. and P.R. Andrews. A common structural model for central nervous system drugs and their receptors. J. Med. Chem. 29 453-62 (1986).

Rakhaminova, A.B., E.E. Khavkin, and L.S. Yaguzhinskii. Construction of a model of the auxin receptor. Biochimia 43 639-53 (1978).

CHEMICAL STRUCTURES

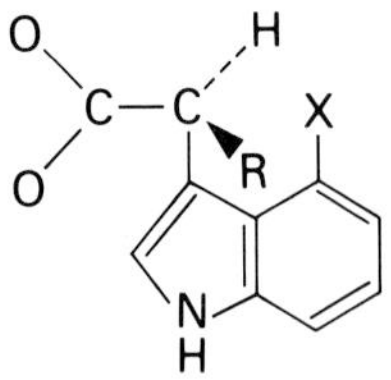

1

1–1 R=H; X=H
–2 R=H; X=Cl
–3 R=CH_3; X=H

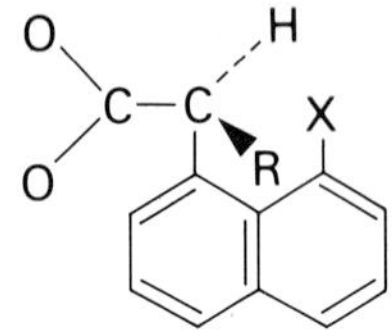

2

2–1 R=H; X=H
–2 R=CH_3; X = H
–3 R=H; X=Cl

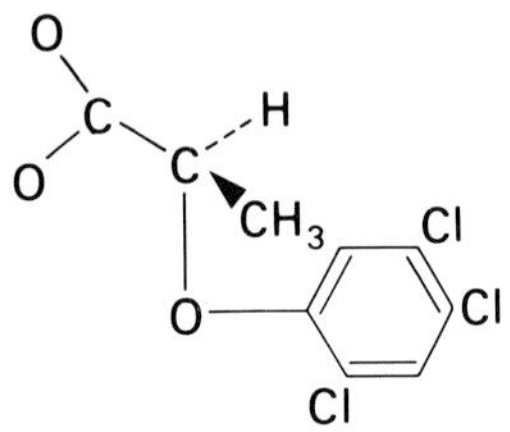

3

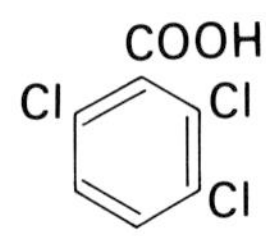

4

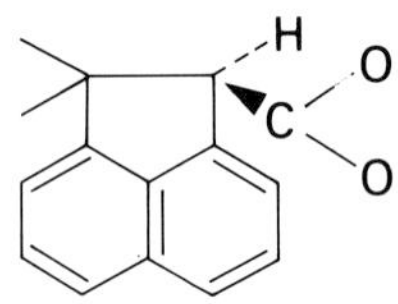

5

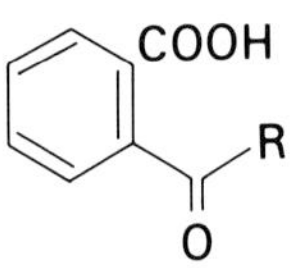

6

6–1 R = Phenyl
–2 R = 4 Chlorophenyl
–3 R = 1–Naphthyl
–4 R = 1–Pyrenoyl:PBA

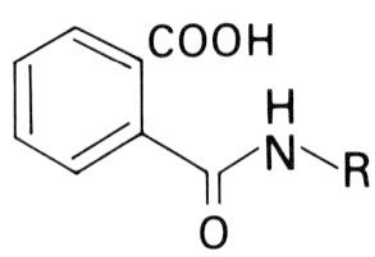

7

7–1 R = Phenyl
–2 R = 1–Naphthyl
–3 R = 2–Naphthyl
–4 R = 2,6 dichlorophenyl

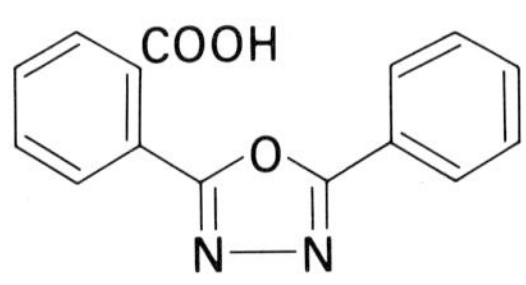

8

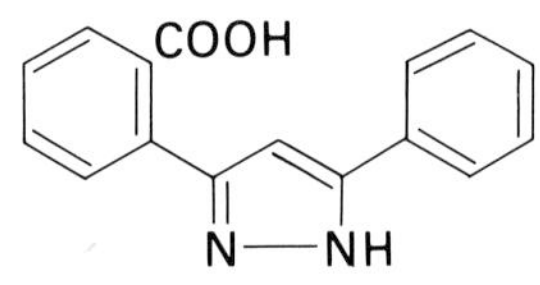

9

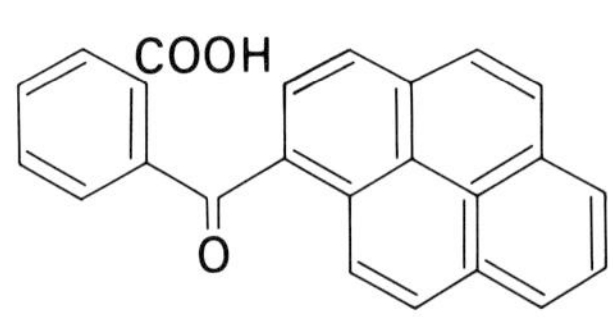

PBA

CAN AUXIN RECEPTORS BE PURIFIED BY AFFINITY CHROMATOGRAPHY?

Michael A. Venis
Institute of Horticultural Research,
East Malling, Maidstone, Kent, ME19 6BJ, UK.

INTRODUCTION

Several attempts have been made to purify auxin receptors by affinity chromatography, beginning with the use of 2,4-D-lysine coupled to activated Sepharose to isolate from maize and pea extracts proteins that promoted DNA-dependent RNA synthesis (Venis, 1971). Similar fractions were subsequently isolated from soybean on the same matrix (Rizzo _et al_., 1977). However, in neither case was activity auxin-dependent, nor could auxin binding be demonstrated. The analogous IAA-lysine adsorbent was reported to isolate homogeneous auxin-binding protein from coconut endosperm (Roy and Biswas, 1977), but many questions have been raised about this system (Venis, 1985).

Auxin-binding sites in maize membranes, first studied by Hertel _et al_., (1972), have been extensively investigated and appear to fulfil many of the criteria expected of _bona fide_ receptors (Venis, 1985). With the development of a convenient non-detergent method of solubilising the binding activity (Venis, 1977), it is now possible to evaluate critically the potential of affinity adsorbents for the isolation of putative auxin receptor proteins. Löbler and Klämbt (1985 a,b) obtained partial purification of the maize proteins on a column of 3,5-diiodosalicylic acid-Sepharose, and by an ingenious though circuitous sequence involving immunoaffinity columns they succeeded in deriving both a purified binding protein contaminated only by IgG fragments, and a monospecific antiserum fraction.

It would be advantageous to have a more straightforward and reliable procedure for the purification to homogeneity of an auxin receptor protein to permit biochemical characterisation and allow direct preparation of polyclonal antibodies. The possible role of auxin affinity chromatography in such a procedure has

NATO ASI Series, Vol. H10
Plant Hormone Receptors. Edited by D. Klämbt

been evaluated using a range of analogues, and compared with 'conventional' and FPLC (Fast Protein Liquid Chromatography) methods.

MATERIALS AND METHODS

Solubilisation of Membrane Binding Proteins

A membrane fraction was prepared from *Zea mays* (cv. Beaupre) coleoptiles and enclosed leaves by differential centrifugation from 4000g x 20 min - 80000g x 30 min (Batt *et al.*, 1976). The resuspended membrane pellet was injected into acetone at -15°C, yielding a preparation from which the auxin-binding activity could be solubilised quantitatively with buffer (Venis, 1977). The buffer (1 ml per 6g fresh wt.) was either citrate binding buffer (0.25M sucrose - 10mM sodium citrate - acetate pH 5.5 - 5mM $MgSO_4$) containing 0.25mM PMSF (phenylmethanesulphonyl-fluoride) or, more usually MES binding buffer, of the same composition except that citrate-acetate was replaced by 10mM morpholino-ethanesulfonic acid-NaOH.

Affinity Column Preparation

Epoxy-Sepharose. The ligands shown in Fig.1 were coupled to Epoxy-activated Sepharose 6B (Pharmacia) as follows. Freeze-dried gel (1.5g) was swollen and washed in water, then added to 15ml of 2M Na_2CO_3 followed by the ligand (450µmol) dissolved in 2ml of 1N NaOH, and the mixture adjusted to pH 12. After incubation for 16-18h at 30-40°C, the gel was filtered off and washed on a Buchner funnel with several cycles of 0.2M Na_2CO_3 and 0.1M NaOAc-0.5M NaCl, pH4, followed by water and 0.02% w/v sodium azide. The amount coupled was determined spectrophotometrically from the washings. Lower substitution gels were prepared by reducing the ligand:gel ratio ten-fold. 5-hydroxy-IAA(5-HIAA) was coupled using degassed solutions under nitrogen. Remaining active groups on the gels were blocked with 1M ethanolamine-HCl, pH11 at 30°C for 4h, followed by the same washing cycle. For 'unblocked' gels this step was omitted. A 'spacer' gel was prepared using ethanolamine only, without any auxin ligand.

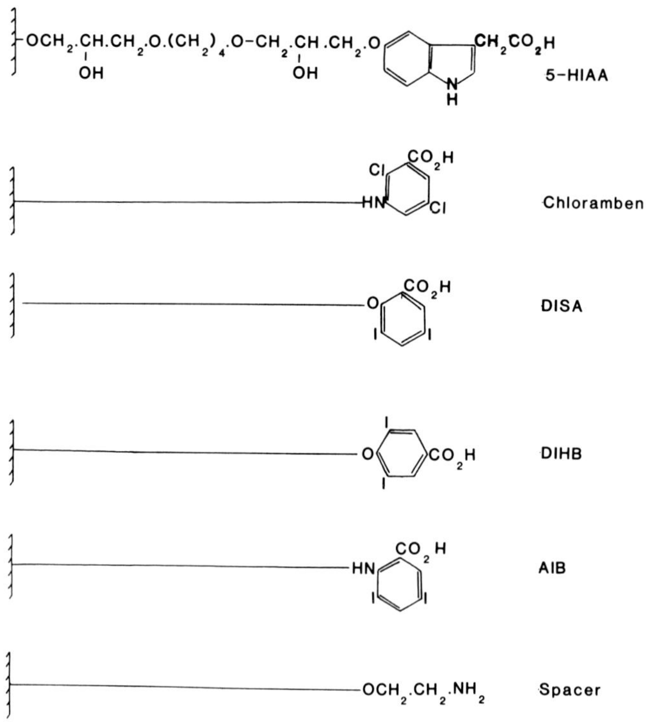

Fig. 1 Column matrices based on Epoxy-Sepharose

CNBr-Sepharose.2,4-dichlorophenoxyacetyl-ϵ-L-lysine (2,4-D-lysine) and 6-amino-2-(naphth-2-oxy)caproic acid (ANOC) were coupled by adding 40μmol in 800μl of 0.4N NaOH to 1.5g of swollen, washed CNBr-activated Sepharose 4B(Pharmacia) in 10 ml of 0.1M $NaHCO_3$, final pH 9.2, for 2h at 22°C, followed by 16h at 4°C to permit hydrolysis of residual active groups. Washing and estimation of coupling were as described above. Aminohexyl (AH)-Sepharose was prepared in an analogous manner using diaminohexane, or was purchased from Pharmacia. AH-Sepharose was further reacted to yield the p-aminobenzamido-derivative (Cuatrecasas, 1970), which was then diazotised and coupled to 5-HIAA or to Chloramben (2,5-dichloro-3-aminobenzoic acid) in 0.2M sodium borate, pH10 for 16h at 4°C. The various CNBr-Sepharose matrices are shown in Fig.2.

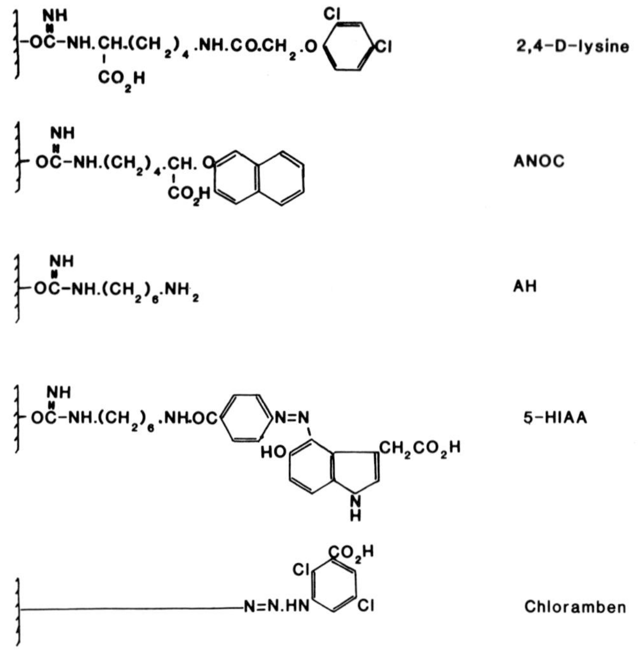

Fig. 2 Column matrices based on CNBr-Sepharose

Chromatography

Affinity Columns. Adsorbents were packed in 5cm x 1cm columns and equilibrated in either MES or citrate binding buffer. Solubilised membrane preparations (10-15g tissue equivalents) were applied in the appropriate buffer (sometimes after buffer exchange on a short Sephadex G25 column) and absorbance at 280nm monitored (Pharmacia UV-1). The non-retained peak was collected, followed by the peaks eluted step-wise with increasing concentrations of NaCl in column buffer. The various fractions were assayed for auxin binding activity (see below).

FPLC. Solubilised preparations (up to 140g equivalents) were chromatographed on 5cm x 1.5cm DEAE-Bio Gel columns in MES binding buffer. The fraction eluted step-wise with buffer containing 0.1M NaCl contained the binding activity and was precipitated with ammonium sulphate (75% saturation). The pellet was redissolved in 1ml of citrate binding buffer and subjected to FPLC gel filtration on Superose 12 (Pharmacia) in 500μl aliquots.

The column buffer contained 20mM piperazine-HCl pH 5.5-0.25M sucrose-50mM NaCl-0.25mM PMSF and 500μl fractions were collected at 1ml min^{-1}. Binding activity appeared in three fractions centred at 40-45 kDa. These were pooled, diluted with 0.5 volumes of the same buffer minus NaCl and applied to a Mono Q (Pharmacia) FPLC anion exchange column. The column was eluted at 1.5ml min^{-1} in a linear 20ml gradient from 0-0.35M NaCl in piperazine buffer, followed by 5ml of 1M NaCl in buffer. Fractions (1ml) were collected and assayed for auxin binding. Alternatively, Mono Q columns were run in buffers containing 20mM sodium phosphate pH 5.5 in place of piperazine.

Binding Assays

These were performed with naphthaleneacetic acid (NAA)-^{14}C (Amersham, 61mCi/m mol) by one of three methods (described in Venis, 1984), depending on the speed required: equilibrium dialysis (slowest), ammonium sulphate precipitation, or centrifugal ultrafiltration (fastest).

Electrophoresis

After dialysis against 0.25mM PMSF and lyophilization, fractions were separated on 12% SDS gels (Laemmli, 1970) which were either stained directly or else blotted onto a 0.22μ nitrocellulose membrane (100-120ma, 16h at 4°C). Blots were stained for total protein by iodine-starch (Kumar *et al*., 1985) or for glycoproteins by the concanavalin A-peroxidase method of Faye and Chrispeels(1985).

Chemicals

3,5-diiodosalicylic acid (DISA) and 5-HIAA were obtained from Sigma, 3,5-diiodo-4-hydroxybenzoic acid (DIHB) and 2-amino-3,5-diiodobenzoic acid (AIB) from ICN. Chloramben was a gift from Amchem and ANOC was synthesised and provided by Dr. E.W. Thomas (Salford). 2,4-D-lysine was synthesised by the method of Hutzinger and Kosuge (1968).

RESULTS

Affinity Column Performance

Both 5-HIAA matrices performed erratically, probably because of poor stability characteristics; gradual darkening of the columns was evident, even though they were exposed only to relatively weak cold room illumination and were shielded by foil between runs.

Low substitution (1-2μ mol/ml) ANOC and 2,4-D-lysine columns failed to retain any binding activity using either citrate or MES buffers. With higher substitution gels (6-7μmol/ml) activity was retained, but recovery was poor (Table 1). When the active 0.2M NaCl eluate was combined with the inactive non-retained (column

Table 1. Elution behaviour of ANOC column in citrate binding buffer.
NAA-^{14}C binding activity applied = 21900 dpm

	Eluate	Binding activity, dpm/fraction Coupling ratio, μmol/ml 1.5	 7
1.	Column buffer	19200	0
2.	0.2M NaCl	-	2920
3.	1M NaCl	0	0
4.	1M NaSCN	-	0
5.	1 + 2 + 3	-	6710

buffer eluate) and 1M NaCl fractions, precipitated with ammonium sulphate and assayed, activity was more than doubled, though recovery was still only 30% of that applied (Table 1). Similar findings were obtained with the 2,4-D-lysine column. A more strongly retained protein fraction could be eluted with 1M NaSCN, though without any binding activity (Table 1.).

When an ethanolamine-blocked DISA column, analogous to that used by Löbler and Klämbt (1985 a) was run in MES buffer,

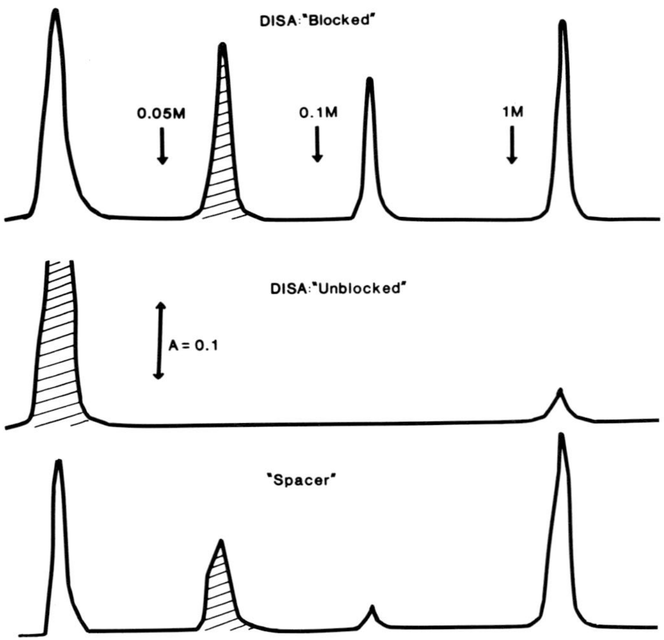

Fig. 3 Elution profiles (A280) of solubilised maize membrane proteins on 'blocked' and 'unblocked' DISA columns, and on an ethanolamine-blocked 'spacer' column. Columns run in MES binding buffer. Arrows indicate step-wise increases in NaCl molarities. Peaks with binding activity are shaded.

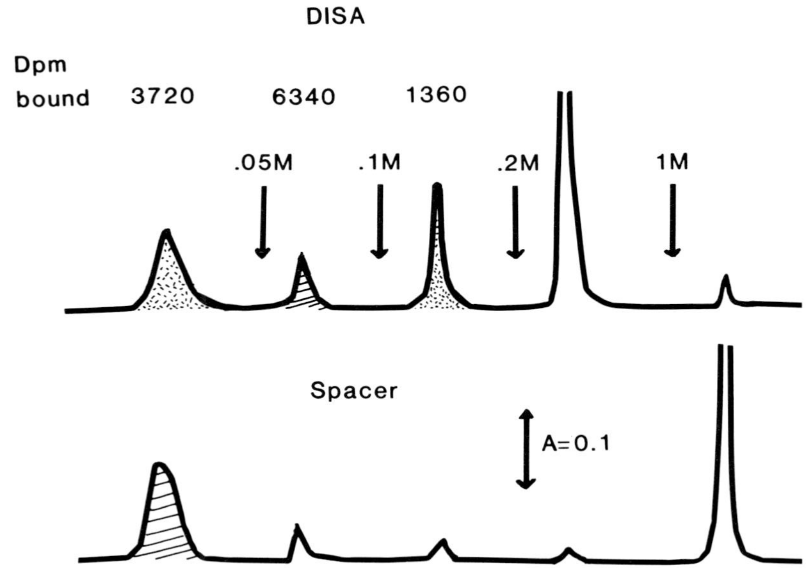

Fig. 4 Blocked DISA and ethanolamine columns run in citrate binding buffer. Arrows and shading as in Fig.3.

activity was eluted with good recovery at 0.05M NaCl (Fig. 3). However, the same column without ethanolamine blocking failed to retain the binding proteins, all activity appearing in the initial fraction eluted with column buffer. On the other hand, a simple 'spacer' column, blocked with ethanolamine but with no auxin analogue attached, behaved in a qualitatively similar manner to the equivalent column substituted with DISA, in that activity was retained and eluted at 0.05M NaCl (Fig.3). In citrate binding buffer (as used by Löbler and Klämbt, 1985 a), activity was not retained on the spacer column, while with the blocked DISA column, activity was found predominantly in the 0.05M NaCl eluate, but with significant binding activity also in the 0.1M NaCl fraction and more particularly in the starting buffer fraction (Fig.4). Blocked and unblocked Chloramben columns run in MES performed in a similar manner to the DISA columns of Fig.3. With the diazo-linked Chloramben matrix (Fig. 2), which was unblocked, all activity emerged with the starting buffer.

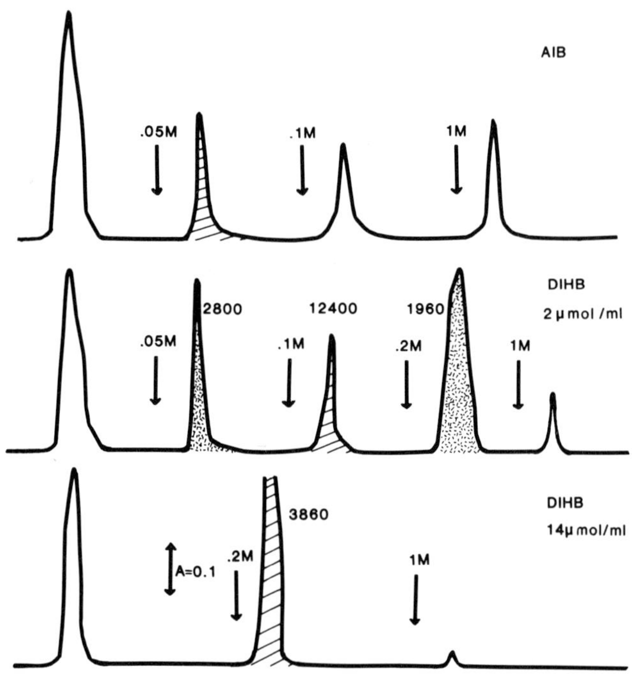

Fig. 5 Elution profiles on AIB and two DIHB columns of differing substitution (all ethanolamine-blocked) run in MES. Arrows and shading as in Fig.3. Numbers by peaks represent total NAA- binding activity as dpm/ fraction.

The AIB matrix (blocked) also behaved comparably to the equivalent DISA column (Fig.5). However, when DIHB, the 4-hydroxy analogue of DISA was used at similar substitution (14)mol/ml), recovery of activity was very poor. Direct elution with 0.2M salt was better than incremental changes, but even then recovery was still under 20%. With a lower substitution matrix recovery was greatly improved, though activity was distributed over several fractions (Fig.5).

FPLC

Chromatography on DEAE Bio-Gel is a useful intitial step, giving around 10-fold purification (Venis, 1977). Originally, columns were run in diluted standard binding buffer (citrate-acetate), but subsequent experience indicated that retention of the binding proteins was sometimes incomplete. This problem was circumvented by changing the buffer to MES, which provides reliable retention and enables elution of binding activity at 0.1M NaCl. It is possible to apply this fraction (after salting out and buffer exchange) directly to a Mono Q FPLC anion exchange column, but to obtain consistent, non-aggregating elution profiles it was found necesary to introduce a preceding FPLC gel filtration step in Superose 12 (Fig.6), even though this provides only modest further purification (Fig.7). The binding proteins elute from Superose 12 close to the ovalbumin 45 kilodalton marker, in accordance with previous gel filtration estimates of native mol.wt. (Venis 1977,1980). When the active eluate is rechromatographed on Mono Q, extensive further purification is obtained (Fig.6). Gel electrophoresis (Fig.7) reveals that band enrichment is predominantly in a single polypeptide at 22 kilo-daltons, slightly larger than the 20 kilodaltons reported by Löbler and Klämbt (1985 a) and suggesting a dimeric structure for the native receptor. In agreement with Löbler et al.(1986), this polypeptide is highly glycosylated and in the most purified fractions (Fig.7, track 6) it represents the sole Con A-binding glycoprotein . It will be seen from Fig.7 that in fractions from Superose and from Mono Q run in piperazine (though not in phosphate) the 22 kilodalton band is associated with a slightly smaller polypeptide at 21 kilodaltons. Whether or not this represents a subunit of another binding protein is uncertain, but it is of interest that it too appears to be highly glycosylated.

DISCUSSION

From Table 1 and Figs.3-5, it is clear that none of the columns evaluated provided satisfactory affinity chromatography. Performance is influenced, not surprisingly, by the degree of ligand substitution and by the nature of the column buffer.High substitution ANOC, 2,4-D.lysine and DIHB columns gave poor recoveries. There is some indication from Table 2 that components required for full activity may have been separated (though this has not been found in other forms of chromatographic resolution) and it may be that (denatured) activity resides in the fraction eluted with NaSCN. Nevertheless, it is evident that none of these columns is satisfactory, while at lower coupling ratios they provide little (DIHB, Fig.5) or no (ANOC, 2,4-D-lysine) purification.

The major influence of the ethanolamine blocking groups (Figs.3,4) is clear. In MES buffer it is evident that these dominate the performance of the DISA column (Fig.3). Indeed, the ethanolamine spacer column itself provides at least as effective purification as any of the other matrices tested. This is not surprising, as it is already known that useful purification can be obtained on AH-Sepharose (Venis, 1980). In agreement with Löbler and Klämbt (1985 a), the receptor proteins did not bind to the spacer column when run in citrate-acetate, but were retained on the DISA matrix. However, in our hands about 30% of the activity was not retained and appeared in the start buffer eluate, even at one sixth the loading used for DEAE Bio-Gel. Löbler and Klämbt (1985 a) did not indicate either their column loading or coupling ratio, but it seems likely that column performance will be critically governed by the balance of ethanolamine and DISA substitution. If this is just right, it may be possible to achieve the differential performance of DISA ethanolamine and 'pure ethanolamine' columns that Löbler and Klämbt (1985 a) exploited successfully, but it may be difficult to reproduce this balance consistently, and in any event the immunological route that must then be pursued to pure receptor

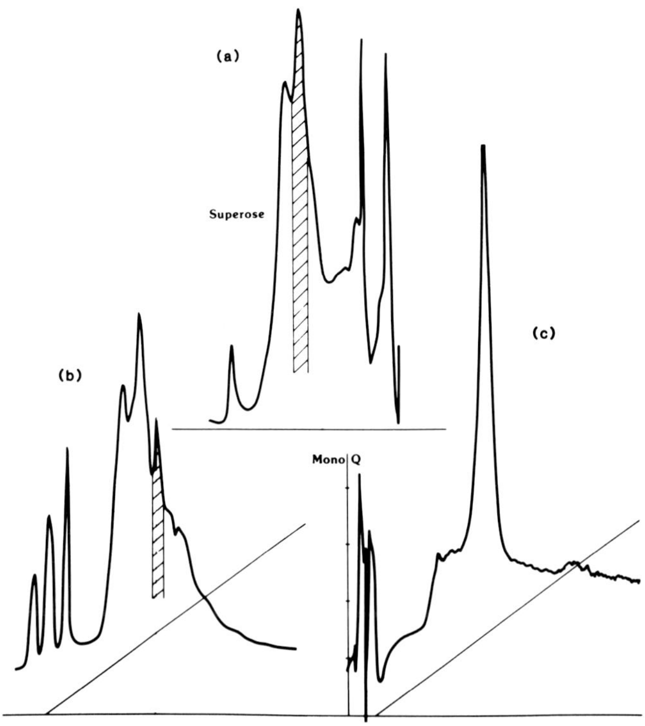

Fig. 6 Receptor purification by FPLC. The active eluate from Superose 12 (shaded) was applied to Mono Q in piperazine (b) and eluted in a 0-0.35M NaCl gradient. Binding activity eluted at 0.12-0.14M NaCl (shaded) and rechromatographed (c) as a single absorbance peak.

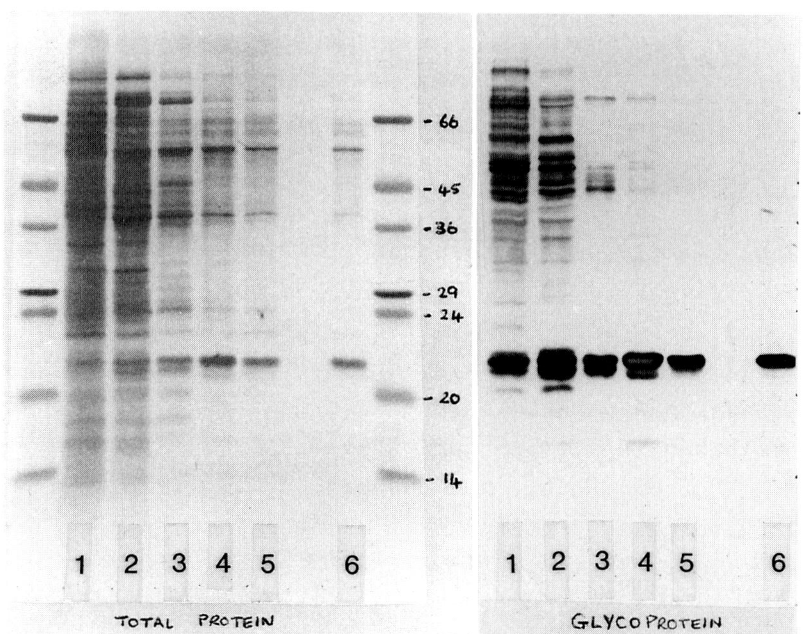

Fig. 7 Electrophoretic monitoring of receptor purification. Identical gels were blotted and stained either for total protein or for glycoprotein (see Methods). Fractions from: 1. DEAE-BioGel. 2. Superose 12. 3,4. Mono Q in piperazine, adjacent active fractions. 5. Mono Q in phosphate. 6. Rechromatography of 5 under same conditions.

remains a laborious one. On the other hand, with minor improvements to the non-affinity methods outlined here, it is anticipated that a reliable direct route to homogeneous receptor will be attainable.

Very recently Shimomura *et al.*, (1986) obtained 29-fold purification of the maize receptor by ligand elution of a column consisting of NAA coupled *via* the carboxyl group to AH-Sepharose. In combination with ion exchange and gel filtration the resulting preparation showed on SDS gels a major 21 kilodalton and a minor 20 kilodalton band, in general agreement with our own findings (Fig.7). Clearly this is a valuable protocol, though the absolute requirement of an acidic function for auxin activity and the absence of such a function on the NAA column makes it likely that hydrophobic rather than ligand-specific interactions governed the chromatography, especially as elution was carried out at high salt concentration.

Acknowledgements: I am indebted to Mike Bolton for valuable technical assistance and to Dr. E.W. Thomas for the sample of ANOC.

REFERENCES

Batt, S., Wilkins, M.B. and Venis, M.A., (1976). Auxin binding to corn coleoptile membranes: kinetics and specificity. *Planta 130*, 7-13.

Cuatrecasas, P., (1970). Protein purification by affinity chromatography. *J. Biol. Chem.* 3059-3065.

Faye, L., and Chrispeels, M.J., (1985). Characterisation of N-linked oligosaccharides by affinoblotting with concanavalin A-peroxidase and treatment of the blots with glycosidases. *Anal. Biochem.* *149*, 218-224.

Hertel, R., Thomson, K-St., and Russo, V.E.A., (1972). *In vitro* auxin binding to particulate cell fractions from corn coleoptiles. *Planta 107*, 325-340.

Hutzinger, O. and Kosuge, T., (1968). Microbial synthesis and degradation of indole-3-acetic acid.III. The isolation and characterisation of indole-3-acetyl-ε-L-lysine. *Biochemistry 7*, 601-605.

Kumar, B.V., Lakshmi, M.V. and Atkinson, J.P., (1985). Fast and efficient method for detection and estimation of proteins. Biochem. Biophys. Res. Comm. 131, 883-891.

Laemmli, U.K. (1970). Cleavage of structural proteins during the assembly of the head of bacteriophage T4. Nature (London) 227, 680-685.

Löbler, M., and Klämbt, D., (1985). Auxin-binding protein from coleoptile membranes of corn. I. Purification by immunological methods. J. Biol. Chem. 260, 9848-9853.

Löbler, M., and Klämbt, D., (1985). Auxin-binding protein from coleoptile membranes of corn. II. Localisation of a putative auxin receptor. J. Biol. Chem. 260, 9854-9859.

Löbler, M., Klämbt, D., and Simon, K., (1986). Auxin-binding in target tissue. J.Cell. Biochem. 10B (Suppl.), 11.

Rizzo, P.J., Pederson, K., and Cherry, J.H., (1977). Stimulation of transcription by a soluble factor isolated from soybean hypocotyl by 2,4-D affinity chromatography. Plant Sci. Lett. 8, 205-211.

Roy, P, and Biswas, B.B., (1977). A receptor protein for indoleacetic acid from plant chromatin and its role in transcription. Biochem. Biophys. Res. Commun. 74, 1597-1606.

Shimomura, S., Sotobayashi, T., Futai, M., and Fukui, T., (1986). Purification and properties of an auxin-binding protein from maize shoot membranes. J. Biochem. 99, 1513-1524.

Venis, M.A., (1971). Stimulation of RNA transcription from pea and corn DNA by protein retained on Sepharose coupled to 2,4-dichlorophenoxyacetic acid. Proc. Natl. Acad. Sci. USA 68, 1824-1827.

Venis, M.A., (1977). Solubilisation and partial purification of auxin-binding sites of corn membranes. Nature (London) 66, 268-269.

Venis, M.A., (1980). Purification and properties of membrane-bound auxin receptors in corn. In "Plant Growth Substances 1979" (F. Skoog, ed.), pp.61-70. Springer-Verlag, Berlin - Heidelberg - New York.

Venis, M.A., (1984). Hormone-binding studies and the misuse of precipitation assays. Planta 162, 502-505.

Venis, M.A., (1985). Hormone-binding Sites in Plants. Longman, Harlow, 191 pp.

THE AUXIN RECEPTOR IN CORN COLEOPTILES

Marian Löbler[+], Dieter Klämbt[++]
[+]University of California, Division of
Molecular Plant Biology, Berkeley CA 94720, USA
[++]Botanisches Institut der Universität Bonn
Meckenheimer Allee 170, D-5300 Bonn 1, FRG

INTRODUCTION

A common approach to isolate hormone receptors is the preparation of a protein fraction with high affinity to a particular hormone. The search for an auxin receptor can be traced back to 1972 when Hertel and his coworkers prepared a membrane fraction with auxin binding activity. In subsequent publications up to three kinetically distinguishable auxin binding sites have been described and thoroughly been characterized (Batt & Venis 1976, Batt et al 1976, Cross & Briggs 1978,1979, Dohrmann et al 1978, Löbler & Klämbt 1985a, Murphy 1980, Ray et al 1977a/b, Shimomura et al 1986, Tappeser et al 1981, Venis 1977 a/b, 1980). However, the only criterion for an auxin receptor has been its high affinity for auxin, but the data for a signal transduction after auxin binding, leading to a physiological response, was lacking. Indirect evidence for a receptor function of the site I auxin binding was given by the modulation of the auxin binding capacity by light (Walton & Ray 1981). An inhibition of auxin induced elongation growth by anti receptor antibodies clearly showed the involvement of the auxin receptor in signal transduction (Löbler & Klämbt 1985b).

PREPARATION

The auxin receptor was purified using a combination of affinity and immunoaffinity matrices (Löbler & Klämbt 1985a). The flow diagram in figure 1 summarizes the purification procedure. The salt eluate of 2-OH-3,5-diiodobenzoic acid-Sepharose (DIBA-Sepharose, Fig 2 lane 2) was sieved on ultrogel (Fig 1) and the fractions

NATO ASI Series, Vol. H10
Plant Hormone Receptors. Edited by D. Klämbt

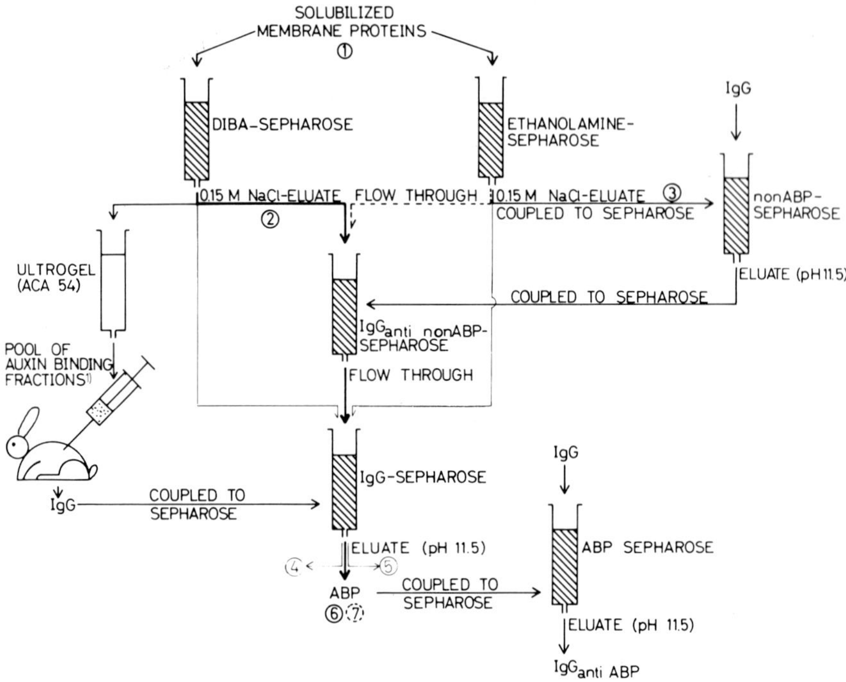

FIGURE 1 Purification diagram for the auxin binding protein (ABP) and monospecific antibodies (IgG anti ABP). Routes for the preparation of immunoaffinity matrices (——), main route (——) and side route (- - -) for ABP preparation, initial routes for the purification of the salt eluates of either DIBA-Sepharose or ethanolamine-Sepharose. The circled numbers refer to the lanes of the electrophoresis shown in figure 2. [1)] Elution at a molecular weight position of 40,000 dalton. From Löbler & Klämbt 1986 with permission of the Czechoslovak Academy of Sciences.

with auxin binding activity were pooled and used as antigen (Fig 1). These fractions correspond to a molecular weight of 40 kDa whereas in sodiumdodecyl sulfate-polyacrylamide-gel electrophoresis (SDS-PAGE) the most prominent protein band corresponds to 20 kDa (Löbler & Klämbt 1985a). In parallel to the DIBA-Sepharose an affinity matrix without the auxin analogue was used (ethanolamine-Sepharose). The proteins eluted from ethanolamine-Sepharose did not show auxin binding activity (nonABP, Fig 2 lane 3). The salt eluates of DIBA-Sepharose and ethanolamine-Sepharose (Fig 2 lanes 2 and 3 resp.) were further processed on IgG-Sepharose (Fig 1, Fig 2 lanes 4 and 5 resp.). The only difference is a 20 kDa protein which must posess the auxin binding activity measured (Fig 2 lane 4) All other proteins present in this eluate (Fig 2 lane 5) were retained on IgG

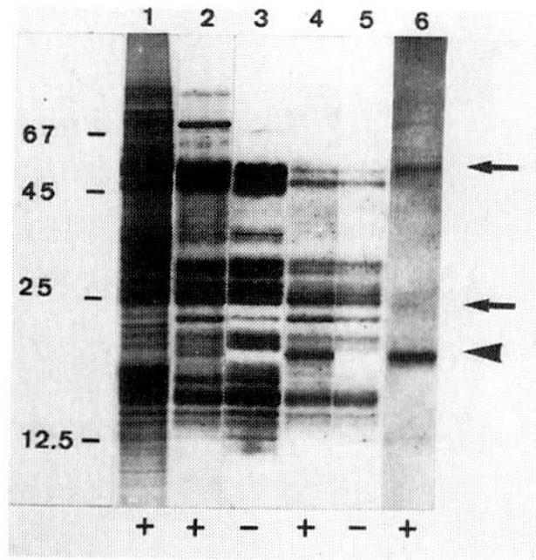

FIGURE 2 SDS-PAGE of the protein fractions obtained during ABP preparation (comp. Fig 1). Solubilized membrane proteins (1), 0.15 M NaCl eluate of DIBA-Sepharose (2) of ethanolamine-Sepharose (3), eluate of IgG-Sepharose which was incubated with the salt eluate of DIBA-Sepharose (4) or ethanolamine-Sepharose (5), ABP from the preparation: salt eluate of DIBA-Sepharose, flow through of IgG anti nonABP-Sepharose, eluate of IgG-Sepharose (6). Molecular weight markers are given in kDa. The arrows indicate the positions of IgG, the arrowhead the position of ABP. The fractions in which auxin binding was found (not found) are marked by + (-). From Löbler & Klämbt 1986 with the permission of the Czechoslovak Academy of Sciences.

anti nonABP-Sepharose and the high molecular weight proteins (Fig 2 lane 2) were lost in a final chromatography on IgG-Sepharose (Fig 2 lane 6). The contamination of the auxin receptor is due to IgG molecules eluted from IgG-Sepharose.
To obtain monospecific antibodies against the auxin receptor the ABP was used to isolate IgG anti ABP by affinity chromatography (Fig 1). Another approach was to immunize rabbits with the denatured and stained ABP cut from polyacrylamide gels (Löbler et al 1986). The obtained IgG fraction is named anti SDS-ABP in order to differentiate between the two preparations.

MOLECULAR PROPERTIES

The molecular weight estimations for the native auxin receptor range from 40 to 80 kDa (Cross & Briggs 1978, Löbler & Klämbt 1985a, Venis 1977b, 1980), whereas for the denatured protein molecular weights of 20 kDa (Löbler & Klämbt 1985a) and 50-54 kDa (Venis 1977a) were reported. Only recently Shimomura et al (1986) and Venis (1986) confirmed our data on the denatured receptor by de-

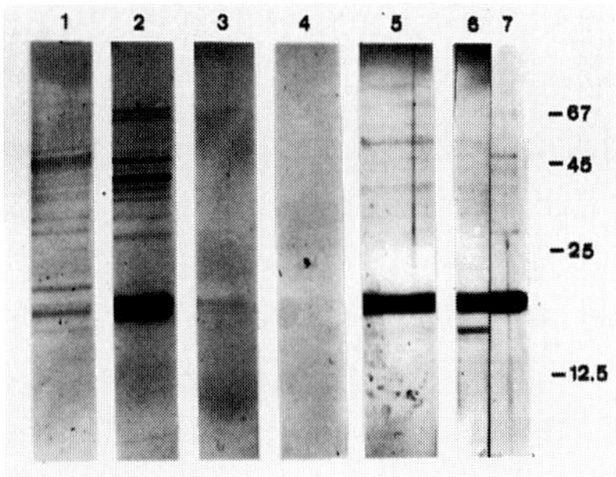

FIGURE 3 Western blot of receptor protein fractions after SDS-PAGE stained with amido black (1), concanavalin A and horseradish peroxidase (2,7), concanavalin A in the presence of α-methyl mannopyranoside and horseradish peroxidase (3), horseradish peroxidase only (4), anti ABP and anti rabbit IgG-peroxidase conjugate (5,6). The receptor protein fraction was treated with endoglucosidase H prior to electrophoresis (6,7). Molecular weight markers in kDa. From Löbler et al 1986 with permission of Alan R. Liss Publ.

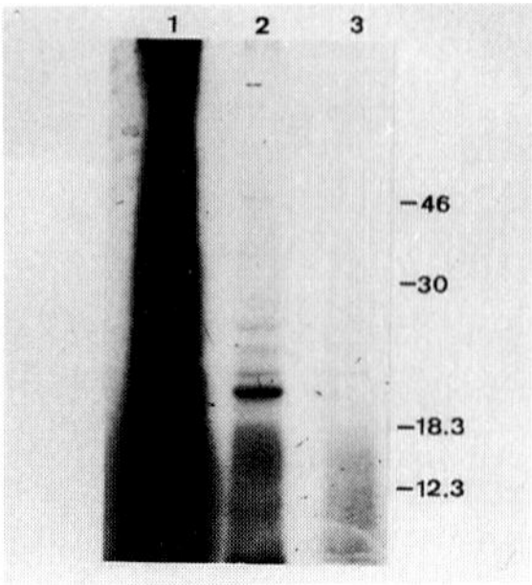

FIGURE 4 Fluorography of translation products from a wheat germ in vitro protein synthesis after SDS-PAGE. Trichloracetic acid precipitable proteins (1), immunoprecipitated proteins (2), competition of immunoprecipitation with unlabeled auxin receptor protein (3). Molecular weight markers in kDa. From Löbler et al 1986 with permission of Alan R. Liss Publ.

monstrating a 21-22 kDa auxin binding protein on SDS-PAGE. The 80 kDa molecular weight of the native receptor was discussed to be due to high pH and high salt present during gel filtration (Venis 1980). The native auxin receptor is a dimer of 20-21 kDa monomers with one auxin binding site per dimer (Löbler &Klämbt 1985a, Shimoura et al 1986). The receptor protein is glycosylated as is shown in figure 3 lane 2. After incomplete endoglucosidase H digestion no concanavalin A binding is observed at the 17 kDa protein (Fig 3 lane 7) which still is detected by anti ABP (Fig 3 lane 6). From the molecular weight difference of 3 kDa it can be deduced that one glycosylation site is present (Löbler et al 1986). It is likely that the glycosylation core has the common structure Asn-GlcNAc-GlcNAc-Man because endoglucosidase H cleaves between the two GlcNAc residues (Tarentino & Maley 1974). The auxin receptor is synthesized as a polypeptide with a signal sequence of 25-35 amino acids. this was shown by in vitro translation of poly $(A)^+$

RNA in a wheat germ system (Löbler et al 1986) which is not capable of polypeptide processing (Burr & Burr 1983). The immunoprecipitated polypeptide is 20 kDa in size and unglycosylated, whereas the deglycosylated mature protein is 17 kDa in size (Fig 3 and 4). This difference in size can only be due to a signalpeptide that is cleaved off in vivo cotranslationally.

BIOCHEMICAL PROPERTIES

The native form of the auxin receptor is a dimer of 20 kDa subunits as revealed by gel filtration (comp. Fig 1, Shimomura et al 1986). The oligomeric nature of the auxin binding site was suggested earlier (Venis 1977a, Tappeser et al 1981). The affinity of the auxin receptor to 1-naphthylacetic acid (NAA) was determined by linear regression of the Scatchard plot data which yielded a dissociation constant of 5.7×10^{-8} M (Löbler & Klämbt 1985a). This is in reasonable agreement with the calculations of others (Cross & Briggs 1978, 4.6×10^{-8} M, Dohrmann et al 1978, $4\text{-}8 \times 10^{-7}$ M, Batt et al 1976 1.8×10^{-7}M, Tappeser et al 1981 1.2×10^{-7} M, Venis 1977b 1.7×10^{-7} M, Shimomura et al 1986 5.9×10^{-8} M) taking into account that less purified preparations were used. The sharp pH optimum at 5.5 (Cross & Briggs 1978, Batt & Venis 1976, Ray et al 1977a) could be confirmed (Löbler & Klämbt 1985a) whereas Shimomura et al (1986) report a optimum at pH 5.0. The optimal temperature for the binding assay is at 0 C (Löbler & Klämbt 1985a, Ray et al 1977a, Shimomura et al 1986).The data from competition experiments with auxins and auxin analogues reveal that all cited reports concern the same auxin binding site from corn coleoptiles (Löbler & Klämbt 1985a their table 2, Shimomura et al 1986 their table iV).

PHYSIOLOGICAL SIGNIFICANCE

The concentration of membrane associated auxin binding sites in maize coleoptile tissue was calculated by 50 pmol/g fresh weight (Dohrmann et al 1978, Batt et al 1976, Cross & Briggs 1978, Löbler & Klämbt 1985a, Ray et al 1977a). Assuming an evenly distributed

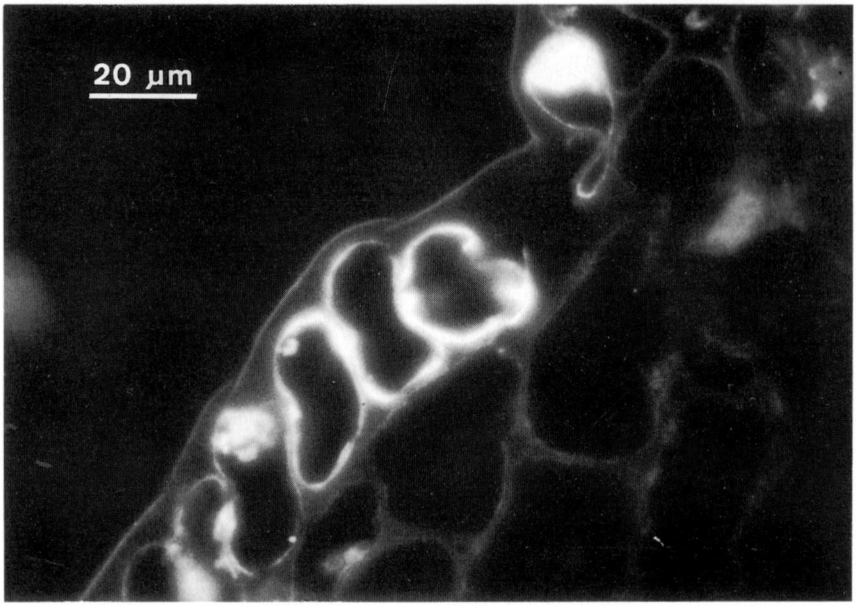

Figure 5 Cross section of an aldehyde-fixed coleoptile 4 mm below the tip. The section was trypsin treated and successively incubated with IgG anti ABP and FITC anti rabbit IgG. From Löbler & Klämbt 1985b with permission of The American Society of Biological Chemists, Inc.

auxin receptor on the plasmalemma of all coleoptile cells a density of 1.4 receptor molecules/um^2 was calculated (Rubery 1981). It was shown, however, that the auxin receptor receptor mainly resides in the outer epidermal cells of the coeoptile (Fig 5, Löbler & Klämbt 1985b). This is in agreement with physiological experiments showing that these cells are the auxin responsive ones (Thimann & Schneider 1938, Pope 1982, Löbler & Klämbt 1985b). If only the plasmalemma of the outer epidermal cells is taken into consideration a receptor density of 2000/um^2 was calculated (Löbler & Klämbt 1985b) which is close to the Abscisic acid receptor density in Vicia faba guard cells (Hornberg & Weiler 1984). The localization of the auxin receptor at the outer surface of the plasmalemma was concluded from inhibition experiments with monospecific IgG anti ABP (Fig 6, Löbler & Klämbt 1985b). These inhibition studies further revealed that the described auxin receptor is involved in auxin induced elongation growth and thus is a true auxin receptor.

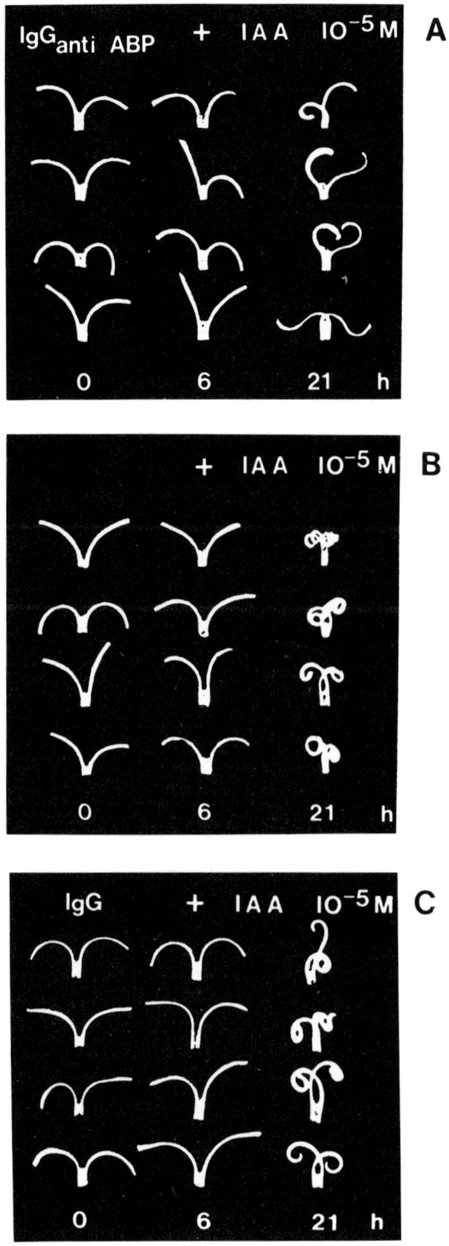

Figure 6 Coleoptile split test. V -shaped coleoptile pieces were preincubated for 6 h in 10^{-8} M IgG anti ABP (A) or buffer (12 mM Na phosphate, 8 mM Tris-HCl, pH 7.4) (B) or 10^{-7} M IgG (C), and then IAA (10^{-5} M final concentration) was added. Xerocopies were taken at indicated times. The same results were obtained with NAA instead of IAA. From Löbler & Klämbt 1985b with permission of The American Society of Biological Chemists, Inc.

CELLULAR LOCALIZATION

Most variable and controversial are the data concerning the cellular localization of the auxin receptor. ER-, Golgi-, and plasmamembrane as well as the tonoplast were discussed (Batt & Venis 1976, Cross & Briggs 1978, 1979, Dohrmann et al 1978, Ray 1977). The plasmalemma localization reported by Löbler & Klämbt (1985b) only takes into account the functional receptor whereas others correlated auxin binding activity with marker enzymes in membrane fractions from sucrose density gradients (Batt & Venis 1976, Dohrmann et al 1978, Ray 1977). Although the precursor of the mature auxin receptor is synthesized at the ER (cleavage of the signal peptide and glycosylation) and transported via the Golgi cisternae to the plasmalemma it is unlikely that a high concentration of immature auxin receptor is stored at the ER. In order to solve this controversy only immunohistochemical investigations at the electron microscope level will be helpful.

LITERATURE

Batt, S. & Venis M.A. (1976) Planta 130, 15-21

Batt, S., Wilkins, M.B. & Venis, M.A. (1976) Planta 130, 7-13

Cross, J.W. & Briggs, W.R. (1978) Plant Physiol. 62, 152-157

Cross, J.W. & Briggs, W.R. (1979) Planta 146, 263-270

Dohrmann, U., Hertel, R. & Kowalik, H. (1978) Planta 140, 97-106

Hertel, R. Thomson, K.-S. & Russo, V.E.A. (1972) Planta 107, 325-340

Hornberg, C. & Weiler, E.W. (1984) Nature 310, 321-324

Löbler, M. & Klämbt, D. (1985a) J. Biol. Chem. 260, 9848-9853

Löbler, M. & Klämbt, D. (1985b) J. Biol. Chem. 260, 9854-9859

Löbler, M. & Klämbt, D. (1986) Biol. Plantarum in press

Löbler, M., Simon, K., Hesse, T. & Klämbt, D. 1986 in Molecular Biology of Plant Growth Control, J.E. Fox & M. Jacobs (eds) New York in press

Murphy, G.J.P. (1980) Planta 149, 417-426

Ray, P.M. (1977) Plant Physiol 59, 594-599

Ray, P.M., Dohrmann, U. & Hertel, R. (1977a) Plant Physiol 59, 357-364

Ray, P.M., Dohrmann, U. & Hertel, R. (1977b) Plant Physiol 60, 585-591

Shimomura,S., Sotobayashi, T., Futai, M. & Fukui, T. (1986) J. Biochem. 99, 1513-1524

Tarentino, A.L. & Maley, F. (1974) J. Biol. Chem. 249, 811-817

Tappeser, B., Wellnitz, D. & Klämbt, D. (1981) Z. Pflanzenphysiol. 101, 295-302

Venis, M.A. 1977a in Plant Growth Regulation, P.E. Pilet (ed), Berlin, Heidelberg, New York, 27-36
Venis, M.A. (1977b) Nature 266, 268-269
Venis, M.A. 1980 in Plant Growth Substances 1979, F. Skoog (ed) Berlin, Heidelberg, New York, 61-70
Venis, M.A. (1986) in Molecular Biology of Plant Growth Control, J.E. Fox & M. Jacobs (eds), New York in press
Walton J.D. & Ray, P.M. (1981) Plant Physiol. 68, 1334-1338

CYTOSOLIC AND MEMBRANE-BOUND HIGH-AFFINITY AUXIN-BINDING PROTEINS IN TOBACCO

A.M. Mennes, C. Nakamura, P.C.G. van der Linde, E.J. van der Zaal, H-J. van Telgen, A. Quint and K.R. Libbenga
Department of Plant Molecular Biology
Botanical Laboratory
Nonnensteeg 3 2311 VJ LEIDEN
The Netherlands

INTRODUCTION

The first question that arises in the study of the molecular action mechanism of hormone signals, like auxins, will be the initial interaction of the signal within the cell. Cells can detect signals which they receive by means of receptor proteins. These proteins specifically and reversibly bind the signals without changing them chemically. Upon binding the receptors are, through a conformational change, transformed into an activated state, thus initiating a molecular programme that ultimately leads to the (signal-)characteristic response. Thus, receptor proteins act both as primary detectors and transducers of the hormonal signal.

For animal steroid hormones, which like the plant hormones can easily be taken up by target cells, we know that they are detected by both plasmamembrane-bound and cytoplasmic/nuclear receptors. For the cytoplasmic/nuclear receptor system the signal has been shown to be transduced into an activation of nuclear gene transcription (1).

Also for plant hormones there has been ample evidence that they can combine with membrane-bound and cytoplasmic/nucleoplasmic high-affinity binding proteins (2). For only a few of these proteins, however, there exists evidence that they might have a receptor function.

For many years we have studied auxin-binding proteins in tobacco

NATO ASI Series, Vol. H10
Plant Hormone Receptors. Edited by D. Klämbt

and we will describe the results of our research on two putative auxin-receptors. The soluble auxin-receptor, present in cytosolic fractions from tobacco cells or in high-salt extracts from nuclei isolated from these cells, is thought to have a receptor function in the activation of nuclear transcription. The plasmamembrane-bound auxin-receptor from tobacco is probably involved in auxin-induced root regeneration.

METHODS

The methods used in the study of the auxin-binding proteins in tobacco are described in the second part of these Proceedings (3).

RESULTS

Soluble receptor

As experimental system we selected a well-established batch-cultured cell line from tobacco which requires only 2,4-D as growth factor. When the cells are depleted from auxin by sub-culturing early stationary-phase cells in 2,4-D lacking medium, they grow to a certain density and reach a new stationary phase after 5 days (Fig. 1). Injection of a small (0.3 ml) volume of a concentrated 2,4-D solution into the culture medium (to reach a final concentration of 4.4 μM 2,4-D) restores cell division after a lag-phase of ca. 10 h. In control experiments, when H_2O is injected, cell division is not restored. Because this cell-suspension culture apparently misses the membrane-bound auxin receptor it seems a good auxin-target system to study the perception-transduction mechanism of the soluble receptor.

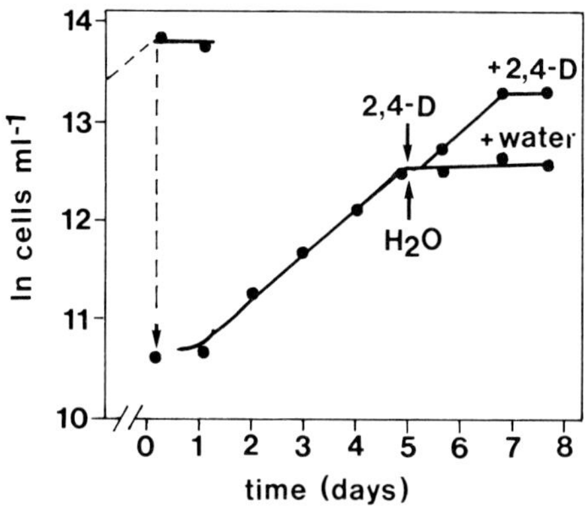

Figure 1.
Growth of tobacco cells after transfer to 2,4-D-free medium.

In order to study the influence of auxin on nuclear transcription in our experimental system, RNA was extracted 1,2 and 4h after addition of 2,4-D or H_2O, respectively. These RNAs were translated in vitro using a rabbit reticulocyte translation system and the $[^{35}S]$ -methionine-labelled polypeptides were analyzed by 2D-gel electrophoresis.We found that at least 3 polypeptides with a MW of 35 KD, 24,5 KD, and 27.8 KD and a pI of 7.1, 6.3, and 5.6, respectively, were increased within 1h in the 2,4-D-treated cells as compared to the H_2O-control. A c-DNA library was constructed from mRNAs isolated 4h after 2,4-D addition. After differential screening of 1600 colonies with ssc-DNA several c-DNA clones were isolated and further characterized. This resulted in the isolation of 7 non-crosshybridizing c-DNA clones from 2,4-D-induced mRNAs. Within 1h after 2,4-D addition to the cells and increase in the mRNAs was detected on Northern-blots (Fig. 2). This response-time on 2,4-D addition correlates nicely with the data from the in vitro translation mentioned above. Further screening and characterization of the auxin-induced mRNA-species is in progress. Preliminary experiments have shown that mRNAs selected by hybridization to some of the c-DNA clones gave rise to polypeptides with a MW corresponding to the MW of the auxin-induced polypeptides found in the earlier-mentioned translation experiment.

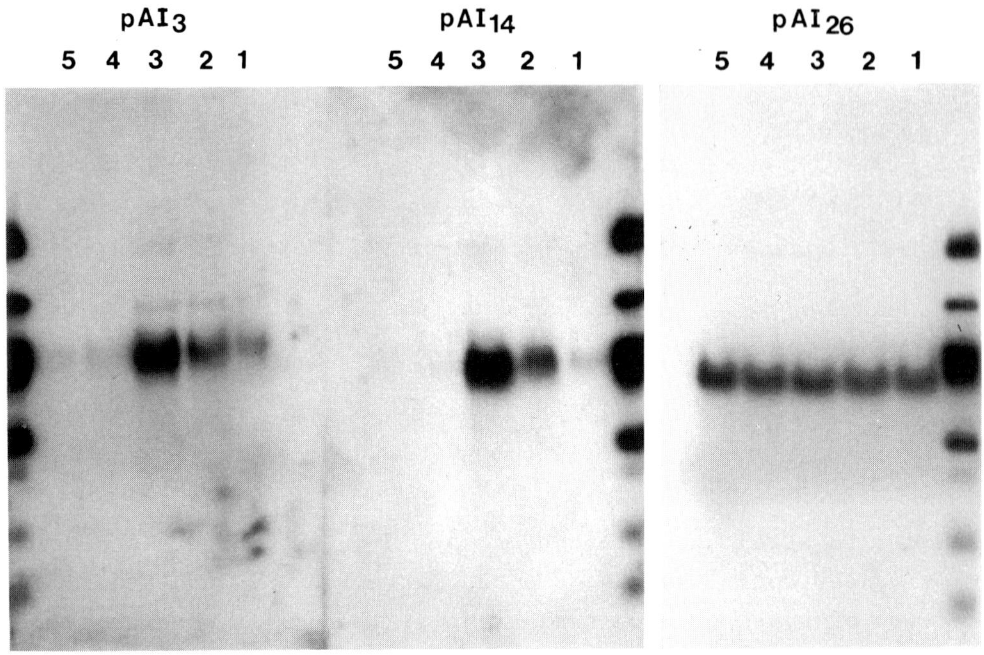

Figure 2.
Northern-blots probed with nick-translated c-DNA clones. Lanes 1,2,3 represent RNAs extracted from auxin-starved stationary-phase cells, respectively 1,2 and 4h after addition of 2,4-D, and lanes 4 and 5, respectively 1 and 4h after addition of H_2O. pAI3 and pAI14 are c-DNA clones from 2,4-D-induced mRNAs, while pAI_{26} represents a control clone.

Since the cells in our experimental system seem to transduce *in vivo* the auxin-signal into an effect on nuclear transcription the question arises how this signal is detected and transduced.

Over the past years we have found that a soluble auxin receptor can be detected in the cytoplasm and nucleus of cells from actively dividing tobacco tissues (4). The main characteristics of this binding protein are summarized in table 1.

TABLE 1

PROPERTIES OF THE SOLUBLE AUXIN RECEPTOR

$Ka(M^{-1})$ (IAA)	pH opt.	Temp. of binding assay	Max. conc. in pmol mg^{-1} protein	Location in cell	Present in
10^9 - 10^8	7.5	26°C	0.2	Cytoplasm Nucleus	Callus Cell suspension Shoot tips

There were considerable variations in the receptor levels that were often even below the detection level of the binding assay(DCC-method). However, we have shown that the number of detectable binding sites in the preparations could be substantially increased by the addition of MgATP and/or p-nitrophenylphosphate, a substrate for phosphatases (Fig. 3) (4). In the nuclear receptor preparations both protein kinase as well as phosphatase

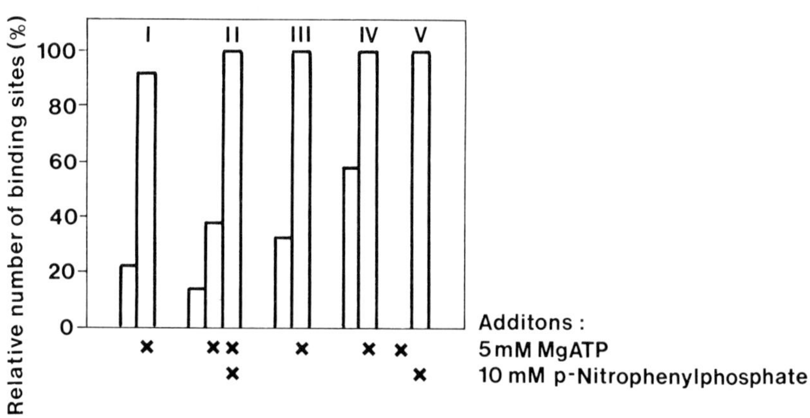

Figure 3.
Effect of the addition of MgATP and p-nitrophenylphosphate on the relative number of high-affinity IAA-binding sites.

activities were found to be present. This might suggest that the receptor is modulated by phosphorylation and dephosphorylation, transforming it into a high- or low-affinity form, re-

spectively (5).
In accordance with steroid receptors we might expect that addition of 2,4-D to 2,4-D-starved cells in our experimental system, would result in increased levels of receptor within the nuclei. Thus levels of receptor were determined in crude high-salt lysates from nuclei isolated at various times after addition of 2,4-D (see Methods). It was found that as early as

TABLE 2

SPECIFIC BINDING OF ^{3}H-IAA IN NUCLEAR RECEPTOR PREPARATIONS AT 5 MIN AFTER ADDITION OF EITHER H_2O OR 2,4-D TO STATIONARY-PHASE 2,4-D-STARVED CELLS

Treatment	pmoles ^{3}H-IAA bound per mg of protein
H_2O	0.052
2,4-D	0.118

5 min after 2,4-D addition the receptor levels in the nuclei were sustantially higher than in the nuclei from the H_2O-control (Table 2). This effect was also shown for auxin analogues; 2,4-D, 1-NAA and IAA were effective whereas 2-NAA and NPA had no significant effect (Table 3).

TABLE 3

EFFECT OF AUXIN ANALOGUES ON RECEPTOR LEVELS IN NUCLEI FROM STATIONARY-PHASE CELLS

Treatment (Final concentration of analogue is 4.4 μM)	Receptor levels relative to the level in 2,4-D treated cells at 30 min after addition of the analogue
2,4-D	100%
1-NAA	136%
2-NAA	0%
IAA	67%
NPA	6%

We assume that the relatively low effect of IAA might be due to the fact that it is easily metabolized.
Purification of the soluble receptor was studied using an affinity-column consisting of 5-hydroxy-IAA coupled to epoxy-activated Sepharose-6B (see Methods). Preparations obtained by this procedure exhibited high-affinity IAA-binding with a Ka of 10^8-10^9 M^{-1}. Due to the very low protein concentration (less than 10 μg/ml) of the preparation a filtration assay using polyethylenimine-treated glass fiber filters (6) was applied instead of the normally used DCC-method.
The auxin-binding protein fraction eluted from the affinity column and kept saturated with the hormone, was tested for its effect on overall _in vitro_ transcription, using nuclei isolated from the cell suspension in either the late log-phase or early stationary-phase. In most experiments a significant stimulation of total transcription was observed (Table 4). This result confirms our previous work in which auxin-dependent stimulation of RNA-polymerase II activity was shown with a receptor

TABLE 4
EFFECT OF PURIFIED RECEPTOR PREPARATIONS
ON OVERALL _IN VITRO_ TRANSCRIPTION

Experiment	^{3}H-UMP incorporation (Bq) nuclei + elution buffer	nuclei + eluted fraction	Stimulation (%)
A	28.8	37.6	31
B	24.9	28.4	14
C	24.0	40.8	70
D	22.7	18.3	-19
E	14.7	14.4	- 2
F	6.0	9.6	60
G	5.2	23.2	346
H	5.0	9.5	90
I	3.6	11.0	206

A-E = Late log-phase cells; F-I = early stationary cells

preparation partially purified in Sephadex-G200 gel filtration and where we found that the relative stimulation was correlated with the occupancy of the receptor (7).
The protein fraction eluted from the affinity column with IAA was concentrated and loaded on a SDS-polyacrylamide slab gel. After electrophoresis the gel was developed by silver staining. Of the few proteins detected, 2 with a MW of ca. 50 KD seem to be specifically eluted by IAA. When the crude protein fraction was incubated with ^{32}P-ATP prior to affinity chromatography (see Methods) a ^{32}P-labelled protein peak could be detected with an average MW of 51.7 KD (Fig. 4). These results indicate

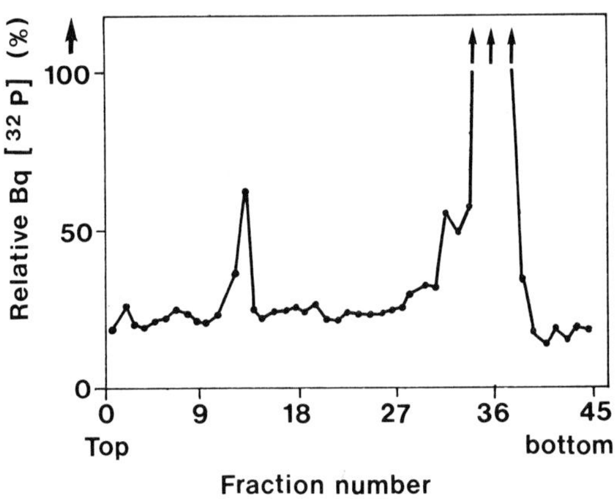

Figure 4.
Electropherograph of the ^{32}P-labelled protein fraction eluted from the affinity column with IAA

that from the affinity column a phosphorylated protein can be eluted with IAA and that in this IAA-eluate specific proteins are present with a MW around 50 KD. Since the MW of the native form of the soluble receptor was estimated at 150-200 KD, the 50-KD proteins might be subunits.

Membrane-bound receptor

A second auxin-binding protein found in tobacco cells is plasmamembrane-bound (MBR) (8). The main characteristics of this protein are summarized in table 5. The MBR could be detected

TABLE 5

PROPERTIES OF THE MEMBRANE-BOUND AUXIN RECEPTOR

$Ka(M^{-1})$	pH opt.	Temp.of binding assay	Max.conc. in pmol mg^{-1} protein	Location in cell	Present in
(IAA) 6×10^4 (1-NAA) 1×10^7	5.0	25°C	50	plasma-membrane	Stem-pith Callus Leaves Cell suspension

in all tissues, except for the 2,4-D-sustained cell-suspension culture used as experimental system for the soluble receptor. So apparently the MBR is not involved in auxin-induced cell division. Callus obtained from these cells lacks both, the MBR as well as the ability to regenerate roots on media with a high auxin concentration (10^{-4}M IAA). Upon subculturing the callus tissue on medium containing NAA and kinetin instead of 2,4-D, the MBR reappeared and the ability to regenerate roots was restored (Fig. 5). It was therefore postulated that the membrane-bound

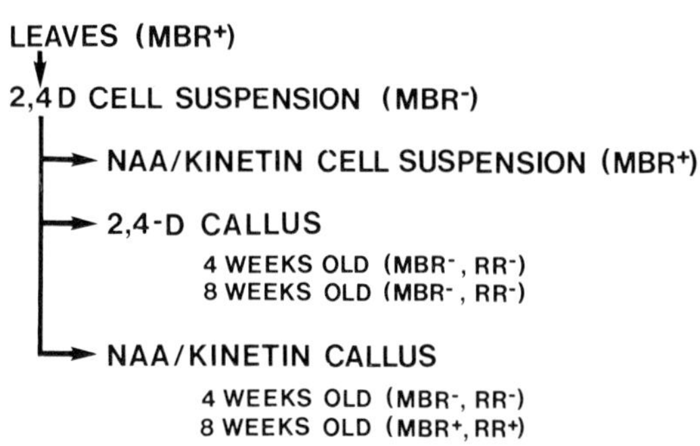

Figure 5.

Different culture conditions for cell suspensions and callus cultures derived from tobacco leaves. MBR^+/MBR^- = presence and absence of membrane-bound receptor; RR^+/RR^- = ability and inability to regenerate roots.

auxin-binding proteins are involved in auxin-induced root regeneration (9). This agrees well with the relatively high K_d of this binding site for the natural auxin IAA, as well as its location at the outerside of the plasmamembrane where it easily can detect the high concentration of auxin needed to saturate the binding sites and to induce root formation. We have studied the assumed correlation between the presence of MBR and root regeneration in some more detail (Table 6).
When MBR^--callus grown on 2,4-D was subcultured on a medium with only kinetin or with NAA + kinetin, MBR could already be detected after 2 weeks of culture. However, grown on a medium

TABLE 6

APPEARANCE OF MEMBRANE-BOUND RECEPTOR (MBR) IN CALLUS GROWN ON MEDIA WITH DIFFERENT COMBINATIONS OF HORMONES

Culture medium	Transfer medium	Transfer-period (2 weeks) 1	2	3	4	5
D -	N	-	-	-	±	±
	DK	-	+			
	NK	+				
	K	+				
NK +	D	-				
	N	±				

D = 2.26 μM 2,4-D; K = 1 μM kinetin; N = 10 μM NAA; DK = 10 μM 2,4-D + 1 μM kinetin; NK = 10 μM NAA + 1 μM kinetin; +/- = presence and absence of MBR; ± = Low activity of MBR; (for further explanation, see text).

with 2,4-D + kinetin, MBR could only be detected after the second transfer to this medium. Subcultured on a medium with only NAA, it lasted 8 weeks before some binding was detected. When MBR^+-callus, obtained by replacing the 2,4-D in the medium by NAA + kinetin (for 2 transfer periods), was transferred back to a 2,4-D-medium the callus was found to be MBR^- within 2 weeks of culture. However, transferred back to a NAA-medium for the same period, still some binding could be detected.

These results indicate that the appearence of MBR is regulated by both exogenous auxin and/or cytokinin and that in combination with kinetin NAA was less inhibitory than 2,4-D. In the presence of only kinetin, that did not interfere with the binding of auxin, MBR could be detected rapidly, whereas in the presence of only 2,4-D the tissues were always MBR^-.
To have an earlier marker for regeneration than the appearance of roots, we examined the presence or absence of a root-specific isoperoxidase (RSP) in the tissues (10). We found a strict correlation between the presence of RSP and MBR, but the opposite was not that strict. For example, tobacco leaves are MBR^+ but RSP^-.

CONCLUSIONS

In tobacco tissues two high-affinity auxin-binding proteins are present.

1. All evidence presented thus far suggest that the soluble auxin-binding protein is a receptor. As has been shown for steroid hormones, the affinity of the receptor is probably regulated by phosphorylation and dephosphorylation. The results from in vitro and in vivo experiments suggest that the protein is involved in auxin-regulated cell division and acts by activating nuclear transcription resulting in the accumulation of specific mRNA species in the cytoplasm. Since c-DNA clones are available it is now possible to analyse the effect of auxin-receptor complexes on 'overall' RNA-polymerase II activity and on specific gene transcription.
2. The membrane-bound binding protein probably is involved in auxin-induced root regeneration. The relatively low-affinity for IAA seems to be in accordance with the high auxin concentration needed for root induction. High concentrations of cytokinin inhibit root regeneration, but the cytokinin does not interfere with the auxin binding. Data obtained with the 2,4-D dependent cell line indicate that cytokinins may play a role in the activation and/or

synthesis of the membrane-bound auxin receptor. This cytokinin effect is inhibited by auxins, and especially 2,4-D seems to be a strong inhibitor.

REFERENCES

1. Anderson, J.N. (1984). The effect of steroid hormones on gene transcription. In: Biological regulation and development, Vol 3B. pp. 169-212, Goldberger, R.F., Yamamoto, K.R. eds. Plenum Publishing Comp. New York
2. Venis, M.A. (1985). Hormone binding sites in plants. Longman Inc. New York-London
3. Van Telgen, H.J., Mennes, A.M., Nakamura, C., Van der Linde, P.C.G., Van der Zaal, E.J., Quint, A., Libbenga, K.R. (1986) In: Proc. NASI "Plant Hormone Receptors". Klämbt, D. ed. Springer Verlag, Berlin, Heidelberg, New York
4. Van der Linde, P.C.G., Maan, A.C., Mennes, A.M., Libbenga, K.R., (1985). Auxin receptors in tobacco. In: Proc. 16th FEBS meeting at Moscow, Part C, pp. 397-403, Ovchinnikov, Y.A. ed.VNU Science Press
5. Grody, W.W., Schrader, W.T., O'Malley, B.W. (1982). Activation, transformation and subunit structure of steroid hormone receptors. Endocrine Reviews 3, 141-162
6. Bruns, R.F., Lawson-Wending, K., Pugsley, Th.A. (1983). A rapid filtration assay for soluble receptors using polyethylenimine-treated filters. Anal. Biochem. 132, 74-81
7. Van der Linde, P.C.G., Bouman, H., Mennes, A.M., Libbenga, K.R. (1984). A soluble auxin-binding protein from cultured tobacco tissue stimulates RNA synthesis _in vitro_. Planta 160, 102-106
8. Vreugdenhil, D., Burgers, A., Libbenga, K.R. (1979). A particle-bound auxin receptor from tobacco pith callus. Plant Sci. Lett. 16, 115-121
9. Maan,A.C., Van der Linde, P.C.G., Harkes, P.A.A., Libbenga, K.R. (1985). Correlation between the presence of membrane-bound auxin binding and root regeneration in cultured tobacco cells. Planta 164, 376-378
10. Thorpe, T.A., Tran-Thanh-van, M., Gaspar, T. (1978). Isoperoxidases in epidermal layers of tobacco and changes during organ formation _in vitro_. Physiol. Plant 44, 388-394

SOLUBLE AUXIN-BINDING: IS THERE A CORRELATION BETWEEN GROWTH-STAGE DEPENDENT HIGH-AFFINITY AUXIN-BINDING AND AUXIN COMPETENCE?

H.-J.Jacobsen, K.Hajek, R.Mayerbacher and B.Herber
Institut für Genetik, Universität Bonn
D-53oo Bonn - 1 West Germany

It is a general observation in plant tissue culture experiments that explants, depending on type, age, physiological state or genotype show specific and different reactions to the application of phytohormones in the medium: Tissues seem to possess different competence to react to the phytohormones applied, in most cases auxins and/or cytokinins. In large seeded legumes, for instance, high concentrations of NAA, 2,4-D or Picloram induce the formation of somatic embryos from immature zygotic embryos (Lazzari et al., 1985, Barwale et al. 1985, Ranch et al. 1986, Kysely et al. 1986), while the same concentrations applied to leaf - or epicotyl explants will lead to poor or nearly no callus formation. From numerous physiological experiments on the action of phytohormones the concept has been evolved that there might be changes in the "growth substance sensitivity" (see Firn 1986 and the literature cited in this paper). It was argued by Trewavas (1981 a,b, 1982, 1983) that the number of receptors may be altered thus causing changes in the sensitivity. Basing the discussion on our own data on soluble auxin binding proteins in pea and soybean cell suspensions as well as on the observations mentioned above, we will discuss the possibility that, besides changes in the number of receptors or modifications of the affinity of the receptor, also differences in the expression of different receptor sites may contribute to the observed phenomena of differential phytohormone sensitivity.

NATO ASI Series, Vol. H10
Plant Hormone Receptors. Edited by D. Klämbt

Soluble auxin binding proteins in pea have been described by our group on several occasions in the past five years (Jacobsen 1981, 1982, 1984, Jacobsen and Hajek 1985, 1986). The cytoplasmic binding of auxins in etiolated pea epicotyls has been shown to be auxin-specific, proteinase sensitive and very labile (Jacobsen 1982). The binding is rapidly saturable and can be assayed by ammonium-sulfate precipitation (Wardrop and Polya 1977), the use of polyethylene-imine-treated glassfiber filters (Bruns et al. 1983), and, in partially purified cytosol preparations, also by equilibrium dialysis (Jacobsen and Hajek 1985), all assays being performed between o-4 °C. Partial purification by a preparative chromatofocusing column (Jacobsen 1984, Jacobsen and Hajek 1985) revealed a second soluble auxin binding site ($sABP_2$), in pea epicotyls, which is expressed in seedlings older than 9-1o days. The binding site detected in young and old seedlings ($sABP_1$) appears in a protein fraction characterized by a pI-range of pH 5.4-5.7, while $sABP_2$ is found in another fraction between pH 6.o-6.5. In the tissues, $sABP_2$ predominantly is present in nodal tissues, while $sABP_1$ can be found both in nodes and internodes (Hajek, unpublished results). The optimal pH for binding, assayed at present only by ammonium-sulfate precipitation, in both cases is between pH 6.5 and pH 8.o. So $sABP_1$ and $sABP_2$ can be distinguished from each other by their

-time course of appearance

-pI

-tissue localization.

Furthermore, in all experiments performed so far, the affinity of $sABP_1$ for auxins was higher than that of $sABP_2$ (K_d-$sABP_1$: 1-2 x $1o^{-8}$M; K_d-$sABP_2$: 4-6 x $1o^{-8}$M).

Recently we have detected a third binding site in a rather neutral cytosol fraction prepared from apical hooks of etiolated pea seedlings, but we have not yet characterized it.

Data which we have accumulated suggest that involvement of $sABP_1$ and $sABP_2$ in the transduction of the auxin signal may be associated with transcriptional events, since both binding sites stimulated transcription in isolated pea nuclei only in the presence of $1o^{-8}$M IAA. This stimulation was not due to the presence of RNA-polymerase in the respective cytosol fractions (Mayerbacher, unpublished results). Similar data were reported by van der Linde et al. (1984) and Bailey et al. (1985) for the soluble auxin receptor partially purified and characterized from tobacco tissues and cell suspensions.-

Since tissues consist of a number of different cell types, the data obtained with the pea epicotyls give only a rough estimate of the real situation in particular cells. Therefore we have conducted experiments using soybean cell suspensions, to elucidate the characteristics of sABP in established cell cultures, which are unlikely to undergo gross genetic changes and thus representing a rather homogenous cell population. In protein fractions from the cell suspensions, which correspond to the fractions obtained from pea epicotyls, we found two sABPs ($sABP_{1-soy}$ and $sABP_{2-soy}$), and these had characteristics nearly identical to those of $sABP_1$ and $sABP_2$ from pea epicotyls (Herber, unpublished results). The cell suspension system enabled us also to study correlations between the growth of the culture and the kinetic data of sABP (K_d and number of binding sites, R_T): During the growth cycle, the number of binding sites per cell decreases for $sABP_{2-soy}$, while for $sABP_{1-soy}$ this parameter peaks at the onset of the log-phase of growth. This can be an indication of a correlation between auxin-binding to this site and cell division (Fig. 1). Another interesting feature of these experiments was the change observed in auxin affinity of both binding sites during the growth cycle (Tab. 1).

Tab. 1: Change in the affinities of the auxin-binding to $sABP_{1-soy}$ and $sABP_{2-soy}$ during the growth cycle of soybean cell suspensions (*:K_d-values are mean of 3 replications)

sABP	pI	K_d*	day of the growth cycle
$sABP_{1-soy}$	5.3-5.5	2×10^{-8}M	0
		2×10^{-8}M	1
		2×10^{-9}M	2
		2×10^{-9}M	3
		2×10^{-9}M	4
$sABP_{2-soy}$	6.3-6.4	4×10^{-8}M	0
		4×10^{-8}M	1
		6×10^{-8}M	2
		1×10^{-8}M	3
		4×10^{-9}M	4

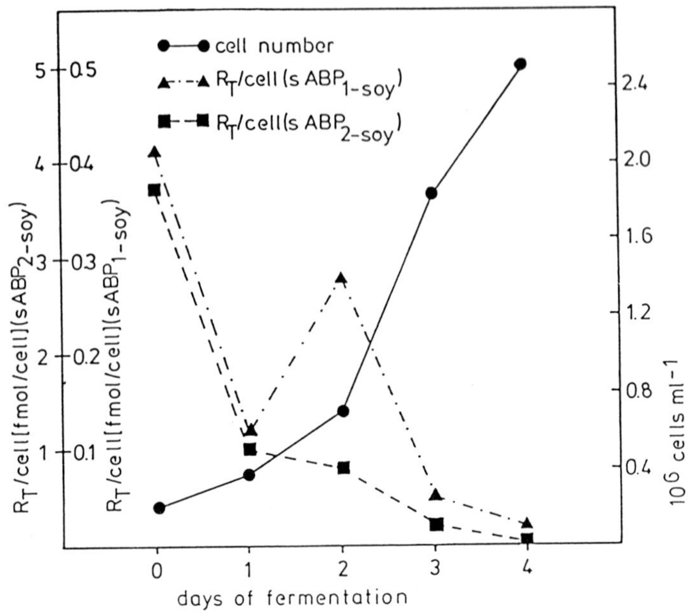

Fig. 1: Growth-curve (●—●), and number of binding sites/cell for $sABP_{1-soy}$ and $sABP_{2-soy}$ in the lag - and log - phase of a batch-cultured soybean cell suspension

Discussion

According to the terminology introduced by Firn (1986), apparent "changes in the sensitivity" of a cell or tissue to respond to phytohormones might be explained in more detail as changes in the following terms:

a) "receptivity"; - changes in the number of hormone receptors,
b) "affinity"; - changes in the affinity of the receptor to the hormone (presumably by covalent modification of the receptor or by some allosteric effect due to the binding of another molecule to the receptor),
c) "response capacity"; - changes in the chain of events subsequent to the primary reaction between hormone and receptor,
d) "uptake efficiency"; - changes in the properties of the growth substance uptake system.

In a recent paper by Bailey et al. (1985) it was demonstrated that sABP from tobacco cell suspensions shows an interesting behaviour with respect to the presence of the receptor in the cytosol. A transfer of sABP from the cytoplasm to the nucleus clearly occurs in the beginning of the growth cycle, while in the late log/early stationary phase sABP again accumulates in in the cytoplasm. Partially purified receptor from tobacco was found to stimulate transcription in isolated nuclei only in the presence of IAA (van der Linde et al. 1984, Bailey et al. 1985), as we report for the pea receptor with preliminary evidence in the present paper. In our study using soybean cell suspensions, we did not examine the possible transfer of the receptors to the nulceus, but assuming it, this could account for the observed decrease of $sABP_{2-soy}$ (Fig. 1), just as the peak of $sABP_{1-soy}$ could be understood as a "change of receptivity".

The observed changes in the affinities of the two binding sites during the growth cycle of soybean cell suspensions are far from being understood. Interestingly, a sharp increase in affinity (tenfold decrease of the K_d, Tab. 1) is observed for $sABP_{1-soy}$ at day 2 of the growth cycle, parallel to the peak in the number of binding sites at that particular day (Fig. 1). Changes in the affinity of an auxin receptor have been reported by Vesper et al., who found a time-course dependent increase in sensitivity to endogenous auxin in maize membranes (Vesper et al. 1978), which subsequently could be correlated to changes in the affinity of the membrane receptor (Vesper 1986). Thus, at the receptor level, data are available indicating both receptivity as well as affinity changes, which occur in correlation to growth effects.-

On the other hand, in our study we separated cytosolic proteins according to their pI, thus having the possibility to identify two different binding sites, which would not have been detected by affinity chromatography. Both sites have similar characteristics in pea and soybean, but each shows, as mentioned above, unique behaviour with respect to the time course of their appearance (pea, Jacobsen, 1984). This may be regarded as a reflection of a "qualitative change in sensitivity", i.e. the initiation of a new chain of events.

Conclusion

Studies on the physiology of phytohormone activity have been conducted for a long time using bioassay systems and exogenous application of phytohormones. The data obtained by these experiments have been necessary for identifying new phytohormones and their range of activities, however, they require a high standardization of the experimental protocols, and dose-response curves are necessarily subjects of interpretation, since they are not fixed entities (Firn 1986). The concept of assuming hormone receptors as targets for the primary action (recognition, transduction of the hormonal signal) tackles the mystery of hormone activity from a more cellular and molecular point of view. Thus the present contribution provides experimental evidence for understanding hormone depending morphogenetic events in plants in a more direct way, avoiding the "black-box-approach" which is an intrinsic prerequisite in bioassay systems, which nevertheless have their own value.

References

Bailey,H.M.,Barker,R.D.J.,Libbenga,K.R.,van der Linde, P.C.G., Mennes,A.M. and Elliott,M.C.(1985) Auxin-binding sites in tobacco cells, Biol.Plant. (Praha) 27(2-3), 1o5-109

Barwale,U.B.,Kerns,H.R. and Widholm,J.M. (1986) Plant regeneration from callus cultures of several soybean genotypes via embryogenesis and organogenesis, Planta 167, 473-481

Bruns,R.,Lawson-Wendling,K. and Pugsley,T.(1983) A rapid filtration assay for soluble receptors using Polyethylenimine-treated filters, Anal.Biochem. 132, 74-81

Firn,R.D.(1986) Growth substance sensitivity: The need for clearer ideas, precise terms and purposeful experiments, Physiol.Plant 67, 267-272

Jacobsen,H.-J. (1981) Soluble auxin-binding proteins in pea, Cell Biol.Intern.Rep. Vol.5(8), 768

Jacobsen,H.-J. (1982) Soluble auxin-binding proteins in pea epicotyls, Physiol.Plant. 56, 161-167

Jacobsen,H.-J. (1984) Two different soluble cytoplasmic auxin-binding sites in etiolated pea epicotyls, Plant&Cell Physiol. 25(6) 867-873

Jacobsen,H.-J. and Hajek,K. (1985) Genotype-specific soluble auxin-binding in etiolated pea epicotyls, Biol.Plant.(Praha) 27(2-3),11o-113

Jacobsen,H.-J. and Hajek, K. (1986) Growth-stage dependent occurrence of soluble auxin-binding proteins in pea, Molecular Biol. of Plant Growth Control, Allan R.Liss,Inc., New York, in press

Kysely,W., Myers,J., Lazzari,P.A., Collins,G.B. and Jacobsen,H.-J. (1986) Plant regeneration via somatic embryogenesis in pea, Plant Cell Rep., in press

Lazzari,P.A., Hildebrand,D.F. and Collins,G.B., (1985) A procedure for plant regeneration from immature cotyledon tissue of soybean, Plant Molec.Biol.Rep., Vol. 3(4) 16o-167

Ranch,J.P., Oglesby,L. and Zielinski,A.C. (1985) Plant regeneration from embryo-derived tissue cultures of soybeans, In Vitro, Vol. 21(11) 653-658

Trewavas,A.J. (1981a) What is the function of growth substances in the intact growing plant? - In: Joint DPGRG and BPGRG Symposium "Aspects and Prospects of Plant Growth Regulators", B.Jeffcoat, (ed.), British Plant Growth Regulator Group, ISBN o-9o6673-o4-6, 197-2o9

Trewavas,A.J. (1981b) How do plant growth substances work? Plant Cell Environm. 4, 2o3-228

Trewavas,A.J. (1982) Growth substances sensitivity:The limiting factor in plant development, Physiol.Plant. 55, 6o-72

Trewavas,A.J.(1983) Is plant development regulated by changes in the concentration of growth substances or by changes in the sensitivity to growth substances? Trends Biochem.Sci.7,354-357

van der Linde,P.C.G.,Bouman,H., Mennes,A.M. and Libbenga,K.R., (1984) A soluble auxin-binding protein from cultured tobacco stimulates RNA synthesis in vitro, Planta 16o, 1o2-1o8

Vesper,M.J. and Evans,M.L.(1978) Time-dependent changes in the auxin sensitivity of coleoptile segments: Apparent sensory adaptation. Plant Physiol. 61, 2o4-2o8

Vesper,M.J. (1985) Detection of a change in apparent K_d of auxin binding sites that correlates with a change in auxin sensitivity in Zea mays coleoptiles, Abstract 12th"International Conference on Plant Growth Substances", Heidelberg 26-31.8. 1985, 54

Wardrop,A.J. and Polya,G.M.(1977) Properties of a soluble auxin-binding protein from dwarf bean seedlings, Plant Sci.Lett. 8, 155-163

PHYTOHORMONE-RECEPTORS FROM TOBACCO CROWN GALL TISSUES

P. Rüdelsheim*, M. De Loose**, D. Inzé**, M. Van Montagu**, J.A. De Greef* & H.A. Van Onckelen*

*Dept. Biology
University of Antwerpen, UIA
B-2610 Wilrijk, Belgium

**Lab. voor Genetica
Rijksuniversiteit Gent
B-9000 Gent, Belgium

Plant cell and tissue cultures are useful tools for the study of plant hormone receptors as it is clearly illustrated by the work of the 'Leiden-group' on different aspects of auxin binding in tobacco callus tissue. In this contribution we will review some physiological and genetical aspects of the tobacco crown gall system, illustrating not only the possibilities offered by this model to investigate the early mode of action of phytohormones, but also the necessity to integrate the study of plant hormone receptors as an essential complementation of data on the endogenous phytohormonal levels.

1. The crown gall system

Upon an Agrobacterium tumefaciens infection a well defined segment called T-DNA of a bacterial plasmid, the Ti-plasmid, is transferred, stably integrated and expressed as a part of the plant nucleus genome, resulting in the formation of a crown gall tumour (for a review see Gheysen et al., 1985).

The combined effects of the T-DNA genes 1, 2 and 4 enables the transformed tissue to grow in culture on synthetic media lacking phytohormones, whereas the nontransformed tissue requires the external addition of auxins and cytokinins for in vitro culture.

NATO ASI Series, Vol. H10
Plant Hormone Receptors. Edited by D. Klämbt

Due to the analogy between the effects of exogenously applied phytohormones on non-transformed plant tissues (Skoog & Miller, 1957) and the effect of mutations in these genes on the morphology of the transformed tissues, the activities were predicted to be "auxin-like" for genes 1 and 2 and "cytokinin-like" for gene 4 (Garfunkel et al., 1981 ; Ooms et al., 1981 ; Leemans et al., 1982 ; Joos et al., 1983).

Recently the gene 4 product has been identified as being an isopentenyl-transferase, an initial step in the cytokinin biosynthesis (Fig. 1,a) (Akiyoshi et al., 1984 ; Barry et al., 1984 ; Buchmann et al., 1985).

Similar the gene 1 and 2 products have been shown to account for a T-DNA encoded 'tryptophan - indole-3-acetamide - IAA' pathway (Fig. 1,b). In the first step a gene 1 (iaaM) encoded

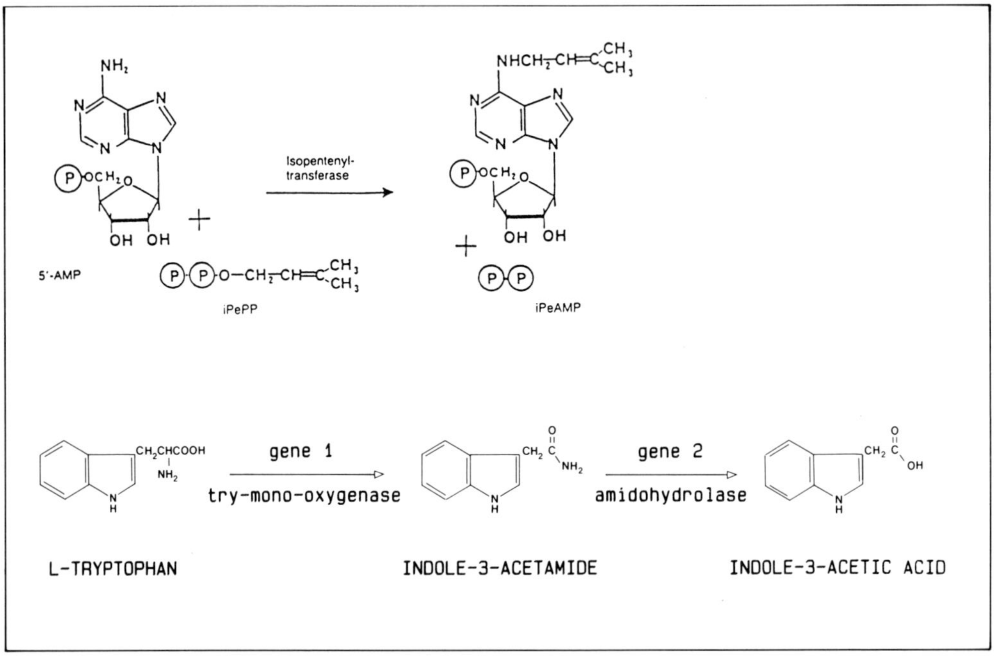

Fig. 1 : The activities of T-DNA encoded proteins are involved in the cytokinin metabolism in the case of gene 4 and in the production of IAA for gene 1 and 2.

tryptophan-2-monooxygenase activity is involved (Inzé et al., 1984 ; Follin et al., 1985 ; Van Onckelen et al., 1985, 1986 ; Tomashow et al., 1986), whereas the second step is catalysed by a gene 2 (iaaH) encoded hydrolase (Schröder et al., 1984 ; Tomashow et al., 1984 ; Kemper et al., 1985).

From the genetical and physiological data thusfar it seems that the crown gall system offers a specific way to manipulate the growth requirements and the morphology of the tissue by changing the endogenous levels of phytohormones regulated by T-DNA encoded enzymes.

2. T-DNA genes and the endogenous phytohormonal levels

Several reports indicated that the phytohormone autotrophous growth of the transformed tissues is due to a T-DNA induced enhancement of the endogenous levels of auxins and cytokinins (for a review see Nester et al., 1984), although the increased phytohormonal levels seem to be not the only factors involved (Nakajama et al., 1979 ; Weiler & Spanier, 1981 ; Pengelly & Meins, 1982).

Table 1 gives a summary of the endogenous auxin and cytokinin levels found in different tissue lines.

The results obtained on the octopine crown gall of Nicotiana tabacum L. cv. Wisconsin 38 confirmed the T-DNA dependent enhancement of the endogenous IAA and cytokinin levels. Yet this is not the case for the nopaline type tumour tissue of Nicotiana tabacum cv. petit Havana SR1, where cytokinin levels comparable to those of the non-transformed tissue were detected.

Soybean crown gall induced by the same nopaline type Agrobacterium tumefaciens, proliferated into a green and a pale cell line (Pedersen et al., 1983).

Both tumour lines showed a significantly higher endogenous cytokinin content than the non-transformed soybean callus in

Table 1 : Summary of the minimal and maximal levels of the endogenous auxin (IAA, pmol/g fr.wt) and cytokinin (pmol ZR equivalents/g fr.wt), as they were determined over several growth periods of transformed and non-transformed tissue lines. The specifications of the tissues and most of the kinetics of the endogenous levels were published earlier. 1: Van Onckelen et al. 1984 ; 2: Rüdelsheim et al. submitted ; 3: Wyndaele et al. 1985 ; 4: p GV2250 described by Leemans et al. 1983 ; 5: p GV4025 described by Van Lijsebettens et al. 1985. Materials and methods were described in these references.

Host Plant	Tissue Specification	IAA	cytokinin
Nicotiana tabacum L.	non-transformed callus[1]	< 5	< 1
cv. Wisconsin 38	octopine wild type[1]	10-300	500-6000
	octopine(-2)mutant[1]	5-100	500-16000
	octopine(-4)mutant[4]	50-750	5-100
Nicotiana tabacum L.	non-transormed callus[2]	5-15	15-60
cv. petit Havana SR1	nopaline wild type[2]	40-250	5-70
	nopaline(-1)mutant[2]	5-80	10-500
	nopaline(-2)mutant[2]	10-200	70-1600
	nopaline(-4)mutant[5]		
	-non supplemented	5-20	5-60
	-suppl. BAP (0.3mg/l)	20-100	5-20
Glycine max L.	non-transformed callus[3]	5-25	< 10
Merr. cv. Mandarin	nopaline wild type		
	-green line[3]	5-25	50-150
	-pale line[3]	5-25	10-50

which hardly any significant cytokinin levels could be detected. On the other hand, no difference in the endogenous IAA concentration was found between the two cloned soybean crown gall types and the untransformed callus.

These results indicate that the introduction of the T-DNA encoded auxin and cytokinin system not always results in an enhancement of the endogenous levels of both phytohormones.

However, all these wild type tumour tissues showed the ability to grow on medium which was not exogenously supplemented with phytohormones, illustrating that this phytohormone independent growth is not necessarily correlated with a T-DNA induced enhancement of both endogenous levels.

In tissues, mutated in gene 1 and/or 2 the T-DNA encoded IAA pathway is at most only partial active and cannot account for the production of IAA. This is illustrated by the accumulation of IAM (up to 50,000 pmol/g fr.wt) in -2 mutant tissues, whereas in a wild type crown gall tissue the IAM level is about 90 pmol/g fr.wt (Van Onckelen et al., 1986 ; Rüdelsheim et al., submitted). Yet all tissues mutated in gene 1 and/or gene 2 show an endogenous IAA level which is considerable higher than the one found in the non-transformed tissue and which, in some cases, only differs slightly from what is detected in the wild type crown gall tissue.

The cytokinin level in these mutant tissues is remarkably higher than what is found in the wild type crown gall tissue, although gene 4 remained unchanged. Similar observations on the effect of mutations of gene 1 and/or 2 on the endogenous cytokinin level have been made earlier (Amasino & Miller, 1982 ; Akiyoshi et al., 1983).

Mutants of gene 4 of an octopine type plasmid induce tumours that proliferate on phytohormone-free medium as unorganised callus. It has been argued that this lack of the "cytokinin" gene could be overcome by high endogenous IAA levels, sufficiently high to sustain the proliferation of the tissue (Nester et al., 1984). Indeed mutation of gene 4 resulted in a lower cytokinin level compared with the wild type. At the same time however the endogenous IAA level in the -4 mutant was higher than in the octopine wild type tissue. Gene 4 mutants of the nopaline type plasmid induce cytokinin requiring tumours (Joos et al., 1983 ; Binns, 1983). This requirement was however not as explicite since we were able to subculture the nopaline type gene 4 mutant tissue on a medium lacking exogenously applied cytokinins, although the tissue was growing more vigourously on the cytokinin supplemented medium (+ BAP 0.3 mg/l).

The omission of the nopaline type plasmid gene 4 had no effect on the endogous cytokinin level compared with the level found in the nopaline wild type tumour which, however, by itself was already very low. The endogenous IAA level of the mutant tissue growing on a non-supplemented medium was surprisingly low, since in these tissues the full T-DNA encoded auxin pathway is present (comparable to the wild type tumour !). If the tissue was cultured on a cytokinin supplemented medium, a significantly higher endogenous IAA level was observed.

These results indicate that besides the well defined T-DNA system other, probably host plant specific, metabolic pathways and their control mechanisms are involved in the production of phytohormones in transformed tobacco cells.

Furthermore a apparent mutual interaction between auxin and cytokinin was observed. Whether this was a direct interaction or a consequence of specific selection during the rigourous cloning procedure, remains uncertain.

Anyhow, these results point out that any attempt to correlate phytohormonal activities with physiological responses should include the rigourous analysis of the actual endogenous levels.

On the other hand, the divergent responsiveness of the different plant tissues to the incited endogenous levels demonstrates that the "sensitivity" of the tissue, which eventually may be expressed in terms of phytohormone receptor sites, is involved in tumour growth and morphology.

3. Phytohormone modulated gene expression

In this workshop the Leiden group reported an auxin mediated increase of the level of at least three translocatable mRNA's in Nicotiana tabacum tissue.

We have chosen a similar approach in Nicotiana plumbaginifolia as a model to study cytokinin modulated gene expression.

When N. plumbaginifolia cell suspensions are cultured in a minimal medium supplemented with 0.5 mg/l NAA and 0.1 mg/l BAP, and subsequently transferred to a medium with only 0.1 mg/l NAA, most cells become enlarged and do not divide anymore after approximately three weeks. A prolonged culture (5 weeks) in this medium results in extensive root formation. The addition of benzyladenine (1 mg/l) to a cell suspension, which has been depleted for cytokinin during 2 weeks, completely inhibits root formation and stimulates cell division.

Comparison of one- and two-dimensional polyacrylamide gel electrophoretic maps of proteins isolated from benzyladenine treated and untreated cells, previously labelled with ^{35}S-methionine for 4 hours, showed that the synthesis of certain proteins is enhanced, induced or suppressed by the cytokinin treatment. Already 72 hours after the benzyladenine addition to a cytokinin depleted cell suspension, clear differences can be observed. These proteins and their enzymatic activities are further identified.

Currently, the minimal time necessary to induce or repress gene expression upon cytokinin treatment is established by in vitro translation of poly(A)$^+$ RNA.

The ability to study fast changes in the protein and mRNA composition in response to phytohormones will make it possible to correlate receptor sites with the early mode of action of phytohormones.

Acknowledgements

This work was supported by grants of the "ASLK-Kankerfonds", the "Fonds voor Wetenschappelijk Geneeskundig Onderzoek (F.G.W.O. 3.001.82)" and the "Services of the Prime Minister (O.O.A. 12.0561.84)" to M.V.M. M.D.L was supported by a grant of the "Instituut voor Wetenschappelijk Onderzoek in Nijverheid en Landbouw (I.W.O.N.L.)". P.R was a Research Assistent, D.I a Research Associate and H.V.O a Senior Research Associate of the "Nationaal Fonds Wetenschappelijk Onderzoek (N.F.W.O.)".

References

Akiyoshi, D.E., R.O. Morris, R. Hinz, B.S. Mischke, T. Kosuge, D.I. Garfinkel, M.P. Gordon & E.W. Nester. Cytokinin/auxin balance in crown gall tumors is regulated by specific loci in the T-DNA. Proc. Natl. Acad. Sci. USA, 80: 407-411, 1983.

Akiyoshi, D.E., H. Klee, R.M. Amasino, E.W. Nester & M.P. Gordon. T-DNA of Agrobacterium tumefaciens encodes an enzyme of cytokinin biosynthesis. Proc. Natl. Acad. Sci. USA, 81: 5994-5998, 1984.

Amasino, R.M. & C.O. Miller. Hormonal control of tobacco crown gall tumor morphology. Plant Physiol. 69: 389-392, 1982.

Barry, G.F., S.G. Rogers, R.T. Fealey & L. Brand. Identification of a cloned cytokinin biosynthetic gene. Proc. Natl. Acad. Sci. USA, 81: 4776-4780, 1984.

Binn, A.N.. Host and T-DNA determinants of cytokinin autonomy in tobacco cells transformed by Agrobacterium tumefaciens. Planta 158: 272-279, 1983.

Buchmann, I., F.J. Mamer, G. Schröder, S. Waffenschidt & J. Schröder. Tumour genes in plants : T-DNA encoded cytokinin biosynthesis. EMBO J. 4(4): 853-859, 1985.

Follin, A., D. Inzé, F. Budar, C. Genetello, M. Van Montagu & J. Schell. Genetic evidence that the tryptophan-2-mono-oxygenase gene of Pseudomonas savastanoi is functionally equivalent to one of the T-DNA genes involved in plant tumour formation by Agrobacterium tumefaciens. Mol. Gen. Gent. 201: 178-185, 1985.

Garfinkel, D.I., R.B. Simpson, L.W. Reaur, F.F. White, M.P. Gordon & E.W. Nester. Genetic analysis of crown gall : fine structure of the map of the T-DNA by site-directed mutagenesis. Cell 27: 147-153, 1981.

Gheysen, G., P. Dhaese, M. Van Montagu & J. Schell. Genetic flux in plants. In : "Advances in Plant Gene Research", vol. 2 (Hohn, B. & E.S. Dennis, eds.) Springer Verlag, Wien, 1985.

Inzé, D., A. Follin, M. Van Lijsebettens, C. Simoens, C. Genetello, M. Van Montagu & J. Schell. Genetic analysis of the individual T-DNA genes of Agrobacterium tumefaciens : further evidence that two genes are involved in indole-3-acetic acid synthesis. Mol. Gen. Gent. 194: 265-274, 1984.

Joos, H., D. Inzé, A. Caplan, M. Sormann, M. Van Montagu & J. Schell. Genetic analysis of T-DNA transcripts in nopaline crown galls. Cell 32: 1057-1067, 1983.

Kemper, E., S. Waffenschmidt, E.W. Weiler, T. Rausch & J. Schröder. T-DNA encoded auxin formation in crown gall cells. Planta 163: 257-262, 1985.

Leemans, J., J.P. Hernalsteens, R. Deblaere, H. De Greve, L. Thia-Toong, M. Van Montagu & J. Schell. Genetic analysis of T-DNA and regeneration of transformed plants. In : "Molecular Genetics of the Bacteria Plant Interaction" (A. Pühler, ed.) Springer Verlag, Berlin, Heidelberg, 1983.

Leemans, J., R. Deblaere, L. Willnïtzer, H. De Greve, J.P. Hernalsteens, M. Van Montagu & J. Schell. Genetic identification of functions of TL-DNA transcripts in octopine crown galls. EMBO J. 1: 147-152, 1982.

Nakajima, H.T., T. Yokota, T. Matsumoto, N. Noguchi & N. Takahashi. Relationship between hormone content and autonomy in various autonomous tobacco cells cultured in suspension. Plant & Cell Physiol. 20: 1489-1499, 1979.

Nester, E.W., M.P. Gordon, R.M. Amasino & M.F. Yanofsky. Crown gall, a molecular and physiological analysis. Ann. Rev. Plant Physiol. 35: 387-413, 1984.

Ooms, G., P.J. Hooykaas, G. Moleman & R.A. Schilperoort. Crown gall plant tumour of abnormal morphology induced by Agrobacterium tumefaciens carrying mutated octopine Ti plasmids : analysis of T-DNA functions. Gene 14: 33-50, 1981.

Pedersen, H.C., J. Christiaensen & R. Wyndaele. Induction and in vitro culture of soybean crown gall tumors. Plant Cell Rep. 2: 201-204, 1983.

Pengelly, W.L. & F. Meins, Jr. The relationship of indole-3-acetic acid content and growth of crown gall tumor tissues of tobacco in culture. Differentiation 21: 27-31, 1982.

Rüdelsheim, P., E. Prinsen, M. Van Lijsebettens, D. Inzé, M. Van Montagu, J. De Greef & H. Van Onckelen. The effect of mutations in the T-DNA encoded auxin pathway on the endogenous phytohormone content in cloned Nicotiana tabacum crown gall tissues. Plant & Cell Physiol. in press.

Schröder, G., S. Waffenschmidt, E.W. Weiler & J. Schröder. The T-region of Ti plasmids codes for an enzyme synthesizing indole-3-acetic acid. Eur. J. Biochem. 138(2): 387-391, 1984.

Skoog, F. & C.O. Miller. Chemical regulation of growth and origin formation in plant tissues cultured in vitro. Symp. Soc. Exp. Biol. 11: 118-131, 1957.

Thomashow, L.S., S. Reeves & M.F. Thomashow. Crown gall oncogenesis : Evidence that a T-DNA gene from the Agrobacterium Ti plasmid $pTiA_6$ encodes an enzyme that catalyzes synthesis of indoleacetic acid. Proc. Natl. Acad. Sci. USA 81: 5071-5075, 1984.

Thomashow, M.F., S. Hugly, W.G. Buchholz & L.S. Thomashow. Molecular basis for the auxin-independent phenotype of crown gall tumor tissues. Science 231: 616-618, 1986.

Van Lijsebettens, M., D. Inzé, J. Schell & M. Van Montagu. Transformed cell clones as a tool to study T-DNA integration mediated by Agrobacterium tumefaciens. J. Mol. Biol. 188: 129-145, 1986.

Van Onckelen, H., P. Rüdelsheim, R. Hermans, S. Horemans, E. Messens, J.P. Hernalsteens, M. Van Montagu & J. De Greef. Kinetics of endogenous cytokinin, IAA and ABA levels in relation to the growth and morphology of tobacco crown gall tissue. Plant & Cell Physiol. 25(6): 1017-1025, 1984.

Van Onckelen, H., P. Rüdelsheim, D. Inzé, A. Follin, E. Messens, S. Horemans, J. Schell, M. Van Montagu & J. De Greef. Tobacco plants transformed with the Agrobacterium T-DNA gene 1 contain high amounts of indole-3-acetamide. FEBS Lett. 181(2): 373-376, 1985.

Van Onckelen, H., E. Prinsen, D. Inzé, P. Rüdelsheim, M. Van Lijsebettens, A. Follin, J. Schell, M. Van Montagu & J. De Greef. Agrobacterium T-DNA gene 1 codes for tryptophan-2-monooxygenase activity in tobacco crown gall cells. FEBS Lett. 198(2): 357-360, 1986.

Weiler, E.W. & K. Spanier. Phytohormones in the formation of crown gall tumors. Planta 153: 326-337, 1981.

Wyndaele, R., H.A. Van Onckelen, J. Christiaensen, P. Rüdelsheim, R. Hermans & J.A. De Greef. Dynamics of endogenous IAA and cytokinins during the growth cycle of soybean crown gall and untransformed callus. Plant & Cell Physiol. 26(6): 1147-1154, 1985.

AUXIN TRANSPORT: BINDING OF AUXINS AND PHYTOTROPINS TO THE CARRIERS. ACCUMULATION INTO AND EFFLUX FROM MEMBRANE VESICLES.

R. Hertel
Institut Biologie III, Universität Freiburg, Schänzlestr.1,
D-78 Freiburg i.Br., F.R.G.

This paper concerns the elements of the auxin transport system, elements that probably are very sophisticated carrier proteins in the plasmalemma. These proteins may be related to the auxin receptors relevant for this hormone`s action.

Auxin transport is a polar and active process; it occurs in parenchyma cells as shown e.g. with maize coleoptiles by removal of the bundles (Hertel and Leopold, 1963); transport is "repeated" at each cell with accumulation into and "secretion" from the cytoplasm. Evidence for this cell-to-cell, repetetive process was provided by Cande and Ray (1976): by application of agar side blocks against the abraded cuticle of coleoptiles it was found that IAA in the polar transport stream exchanged rapidly with the free space, and therefore did not move in a symplast.

Older experiments on transport of labelled auxin through tissue segments, and on uptake of auxin into thin sections or cultured cells (e.g. Hertel and Leopold, 1963; Rubery, 1978; Sussman and Goldsmith, 1981) suggested: (a) Net IAA uptake is dependent on a pH gradient and/or on metabolic energy; it is partially saturable for IAA. On the other hand, (b) NPA and TIBA - the phytotropins in general - inhibit the efflux ("secretion").

Reconstruction of the driving force of auxin transport was then achieved by the chemiosmotic theory (Rubery and Sheldrake, 1974). In the presence of a pH-gradient across membranes of cells or sealed vesicles, where the inner compartment is less acidic than the external milieu, IAAH crosses the membrane, dissociates and becomes trapped as a result of low membrane permeability for IAA^- which, in turn, could exit via an anion carrier.

Two specific trans(plasma)membrane processes are postulated: (a) a (partially) saturable uptake of IAA driven by a pH gradient, and (b) an efflux of IAA via polarly distributed, NPA/TIBA inhibitable exit carriers. These two elements should be seen as two binding sites.

NATO ASI Series, Vol. H10
Plant Hormone Receptors. Edited by D. Klämbt

(1) Transport Elements as Binding Sites: Binding of Auxins and Phytotropins to Presumed Carriers.

The uptake carrier (a) has been characterized as IAA-(site III-) binding in plasma membrane fractions. The (complex) efflux carrier (b) has been studied extensively in terms of phytotropin (NPA) binding.

(a) "Site III" is the uptake carrier.

After having found two types of auxin binding (I, II) at inner membranes from maize, possessing high affinity towards 1-NAA, a third type of auxin association ("site III") - with very different characteristics - was observed in homogenates from *Cucurbita* hypocotyles (Jacobs and Hertel, 1978). A high specificity was measured with several auxins, weak auxins, and anti-auxins. The striking distinction - compared to site I/II - is the fact that 1-NAA has a very low affinity compared with IAA, 2-NAA, 2,4-D. Subsequently, major parts of this "binding" had to be reinterpreted as accumulation into sealed vesicles (Hertel et al, 1983).

"Site III" activity was found to reside on the plasmalemma (Jacobs and Hertel, 1978). Evidence came from differential centrifugation in absence and presence of metrizamide, and from sucrose or metrizamide density gradient centrifugation.

Assays for site III were performed with maize membranes (Dohrmann et al., 1978) where there are no (or very few) pH-tight vesicles. Binding of labelled 2,4-D in presence of cold 1-NAA (= site III) was found in gradients at the position of plasma membrane markers.

The uptake carrier was further characterized as a IAA-(site III-) binding in *Cucurbita* membrane fractions (Benning 1986b). ß-Octylglucoside (detergent) and nigericin (ionophore) were added to permeabilize the membrane particles that could still be pelleted; under such conditions, any saturable association should be binding. A saturation curve with increasing IAA concentration is shown in Fig.1. The K_D for IAA was determined as 0.3 µM, and the amount of binding sites was ca. 8 pmoles/gr fr.wt. In maize, the amount of such sites had been found at ca. 20 pmoles/gr fr.wt. (Dohrmann et al, 1978).

Analog specificity was tested; as expected for site III, 1-NAA did compete much less effectively than did 2,4-D (Fig.2).

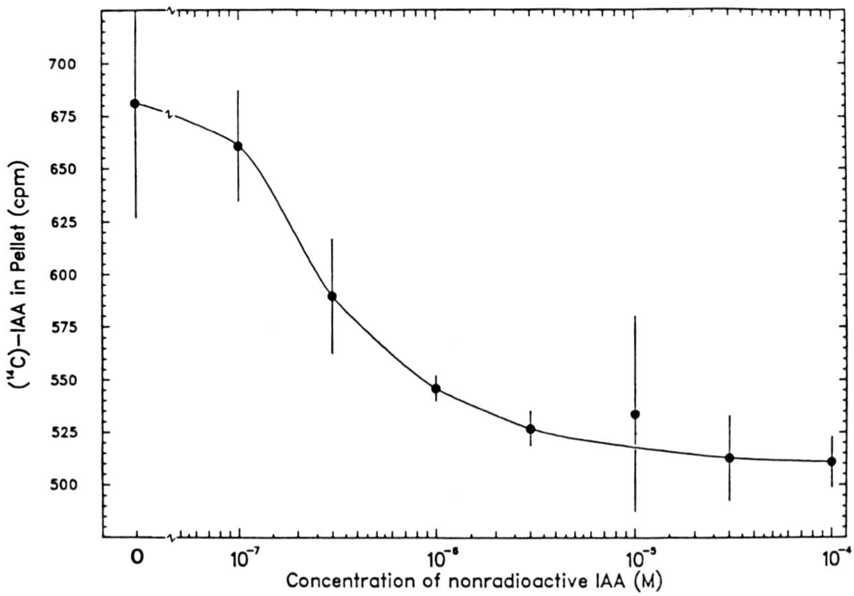

Fig.1. Binding of ^{14}C-IAA (0.03 μM) to permeabilized membranes in presence of increasing concentrations of unlabelled IAA. Data from Benning (1986b). Microsomes from Cucurbita hypocotyls were tested at pH 6 with the centrifuge assay in presence of 8,000 cpm in 1 ml of test solution containing 0.3 % ß-octylglucoside and 1 μM nigericin.

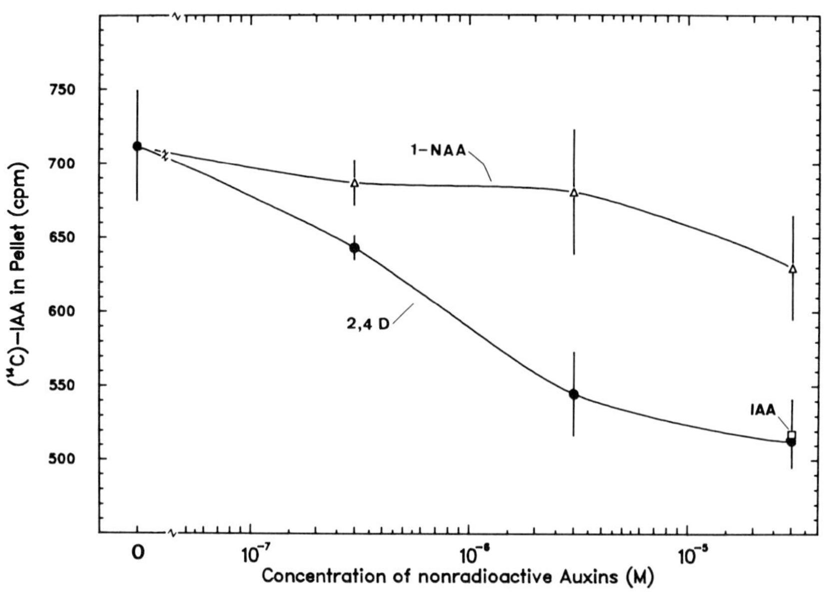

Fig.2. Binding of ^{14}C-IAA in presence of increasing concentrations of unlabelled 2,4,D or 1-NAA. Data from Benning (1986b). Other conditions were as in Fig.1.

(b) Phytotropin binding occurs at the efflux carrier or an associated protein.

Depta et al (1983) studying binding of phytotropins and cross competition with NPA and 2,3,5-TIBA, suggested that the efflux carrier complex bears three different but interacting areas: one site for auxin to be translocated and two regulatory sites for inhibitor binding. The latter appears to be non-competitive with auxins since inhibition of transport (auxin efflux) by NPA, 2,3,5-TIBA and 3,4,5-TIBA appears to be independent of the IAA donor concentration.

Of the regulatory sites, one has a high affinity for NPA and a low affinity of 2,3,5-TIBA, while the other one has intermediate affinities for both. Some kinetic evidence supports the suggestion of two different, interacting NPA binding sites. Higher concentrations of unlabelled NPA stimulate the off-reaction of bound labelled NPA (Michalke, unpubl.).

In line with the idea of two different regulatory sites at the efflux carrier are indications that NPA stimulates IAA transport (efflux) at lower concentrations (e.g. 0.01 µM, see Hertel et al, 1983, their Fig.5A) while at higher concentrations (1 µM), NPA is inhibitory as are other phytotropins that inhibit however over the entire range of effective concentrations.

It was also noted that transport inhibition by high 3,4,5-TIBA concentrations was only partial, whereas with NPA and 2,3,5-TIBA it was almost total. If these three compounds act via common sites in vivo, then 3,4,5-TIBA might displace the other inhibitors. Total inhibition by NPA or 2,3,5-TIBA was partially alleviated by increasing concentrations of the "soft" inhibitor 3,4,5-TIBA (Depta et al, 1983). Furthermore, a time-dependent change in sensitivity (= adaptation) of IAA transport towards the inhibitor 3,4,5-TIBA was shown (W.Petersen, B.G.Kang, unpubl.; see in Hertel, 1986).

Thein and Michalke (unpubl.) solubilized the NPA binding sites from <u>Cucurbita</u> with 0.3% CHAPS or 1.5% octylglucoside. Binding tests were performed in binding buffer pH 5; the material was quickly diluted into pH 7.8 buffer and filtered through Whatman glass filters GF-B coated with polyethylenimine (method of Bruns et al, 1983). Complicated kinetic features were observed.

An accidental finding by Thein (unpubl.) during characterization of the NPA binding site warrents some comment. Binding of NPA to membrane particles as well as to solubilized material from

Cucurbita is reduced by low concentrations of bisulfite (K_i = 30 uM). Fig.3 presents a concentration-effect curve. _Cucurbita_ microsomes were fractionated with PEG according to Michalke (1982). Fraction I membranes were treated with different concentrations of bisulfite at pH 7 for 30 min; then the ^{3}H-NPA binding test was performed at pH 5.

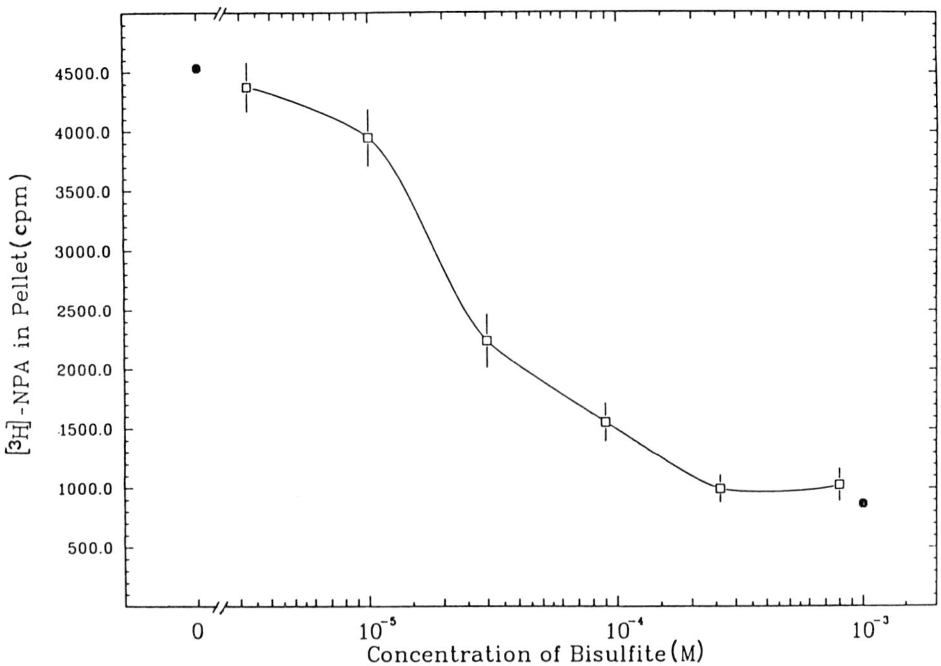

Fig.3. Effect of Bisulfite on saturable ^{3}H-NPA Binding to Zucchini Membranes. Data from M. Thein.

The inhibition by bisulfite is reversible and not due to the reducing potential or to cleavage of S-S bonds. Hence an addition of HSO_3^-, which is the effective form of bisulfite, to a modified amino acid at the binding site is suggested.

(2) <u>Accumulation and Efflux of Auxin. Studies With Vesicles.</u>

(a) Uptake of IAAH+H^+ is saturable and voltage-dependent.

(b) Is the efflux carrier an anion carrier?

Auxin transport studies using membrane vesicles from hypocotyls of _Cucurbita_ have been introduced by Hertel et al (1983). Important quantitative analyses of this in vitro system came from Clark and Goldsmith (1986a and b) and Lomax et al (1985). Sabater and Sabater (1986) extended the study to vesicles from lupin.

(a) With Cucurbita membranes, Benning (1986a) studied the accumulation of IAA, of a membrane potential probe, and of butyric acid as a probe for pH-gradients. Ion gradients (K^+, H^+) were applied in the presence and absence of ionophores e.g. valinomycin. In all cases tested, the accumulation of IAA equals neither potential probe nor pH-probe accumulation, but represents an intermediate between the two. Auxin molecules seem to be taken up as positively charged ions and a pH-gradient is required for accumulation. The uptake mechanism thus appears to be a specific, cotransport of IAA^- and no less than two protons.

This voltage-dependent symport ($IAA^- + H^+ + H^+$) was postulated (Hertel, 1983; Lomax et al, 1985) to account for the high IAA accumulation to values several times higher than to be expected from pH gradients alone, and for the increase in uptake observed with certain valinomycin/K^+ combinations. It was not implied that in the vesicle system a transmembrane potential difference was always required for high accumulations; in its absence, the two protons per transfer would suffice.

The contribution of IAAH diffusion is not necessarily very small; it just has to be lower than the flow through the carrier.

Fast uptake kinetics of IAA clearly showed the involvement of a saturable component (Benning, 1986a). In order to distinguish between binding at the inside of the vesicles and uptake by carrier, the IAA uptake rate at earliest measurable points was determined (first 20 s after addition of the radioactive auxin; Fig. 4). The time of addition of the vesicles to the 25-fold diluted radioactive test buffer is defined as 0 s. The other points give the time span elapsing from addition of the vesicles to the radioactive test buffer until the 25-fold dilution with non-radioactive test buffer. In the presence of 10^{-5}M cold IAA the association rate of ^{14}C-IAA with the vesicles is lower from the very beginning. If no binding on the outside is involved, these association kinetics are only understandable by postulating carrier-mediated uptake of auxin. Binding from the inside as postulated by Clark and Goldsmith (1986) can not be reconciled with the result shown in Fig. 4.

On the other hand, however, there is one important finding of Clark and Goldsmith (1986, 1987)) that shrinking the vesicles with external osmotica led to a corresponding reduction of accumulation

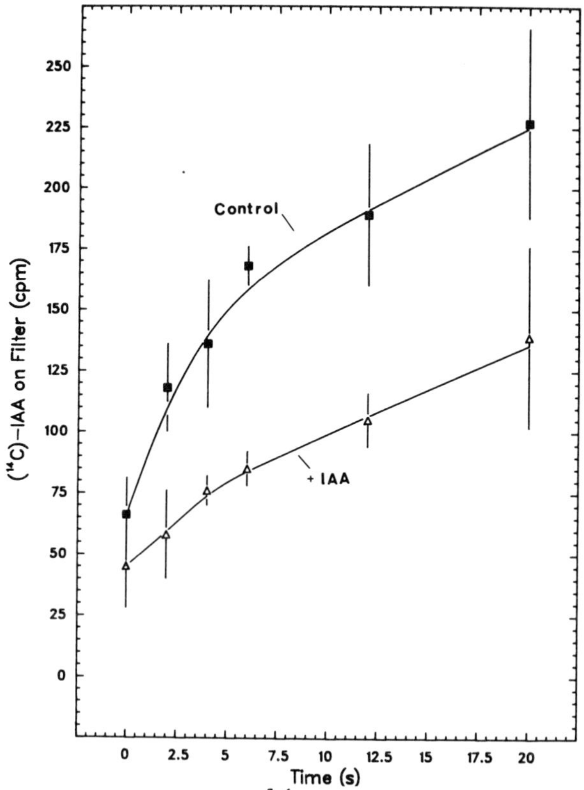

Fig. 4. Initial Uptake of ^{14}C-IAA (0.1 µM) in Absence and Presence (+ IAA) of 10 µM unlabeled IAA. Data from Benning (1986a).

of butyric acid and of IAA at high concentrations while there was much less effect of shrinking on the association of low IAA concentrations. This can not easily be explained with the idea of IAA accumulation into the vesicular space, and it did suggest to the authors involvement of some saturable, special binding.

(b) The question whether the efflux carrier is transporting auxin anions still remains open. In the original chemiosmotic scheme (Rubery and Sheldrake, 1974), the IAA^- anion was leaving the cell (at the basal end). However, Hasenstein and Rayle (1984) showed that auxin transport - with efflux assumed to be limiting - is not affected by hyperpolarization in vivo, and Benning (1986a) did not find any large effect of transmembrane potential on the NPA sensitive portion (= efflux) of IAA accumulation in vesicles. These results led Hertel (1985, 1986) to propose that IAA^- was exported electroneutrally, together with an H^+ or another cation.

Two points, however, strongly argue in favor of the original anion efflux scheme: no evidence for any H^+, K^+ etc. involvement could be obtained (Benning and Hertel, unpubl.), and 1-NAA which is mostly taken up by the 1-NAAH ion trap mechanism, and not via the accumulation uptake carrier, is well transported in the tissue, This 1-NAA transport could hardly proceed with an efflux that was thermodynamically equivalent (symport with 1-NAA^- plus H^+); it would easily operate, however, with an anion NAA^- carrier.

The auxin accumulation as well as the control (e.g. a block) of the efflux carrier may play a "physiological" role for the cell concerned. IAA which is highly accumulated when at low external concentration, may be the more effective auxin even if the affinity of the internal receptor was higher towards 1-NAA (see Ray et al, 1977). Concerning another aspect, Larkin et al (1982) found that certain phytotropins reduced the net auxin efflux from suspension-cultured cells. These phytotropins stimulated divisions of protoplasts of tobacco and *Petunia* at otherwise suboptimal concentrations of exogenous auxins.

A possible connection of auxin transport with Ca^{2+} fluxes has often been discussed; such a link is supported by data showing correlation between oppositely moving Ca^{2+} and IAA (see dela Fuente et al, 1986). With *Cucurbita* in vitro transport it was found to be essential to remove Ca^{2+} with EDTA or EGTA from the extraction medium in order to obtain a significant phytotropin effect (Hertel et al, 1983, and unpubl. results); this indicates that the efflux carrier is blocked by Ca^{2+} on the cytoplasmic side.

Furthermore, there is evidence that Ca^{2+} channels or related proteins do exist in plant membranes. Andrejauskas et al (1985) reported saturable, specific and reversible binding of labeled verapamil - a Ca-channel blocker known to work in animal tissues. In *Cucurbita* and in maize the apparent K_D is 0.1 μM, and the number of sites is 60 pmol/mg protein; these numbers coincide with the corresponding data from animals - too well to be a chance convergence. After fractionation on density gradients and with free flow electrophoresis (Andrejauskas, unpubl.), it could be shown that in *Cucurbita* and maize a significant part of the verapamil binding is localized at the plasmalemma and another fraction at the tonoplast.

Surprisingly, it was detected that verapamil binding in plant membranes increased affinity in the presence of low concentrations (EC_{50} = 1 μM) of 3,4,5-TIBA, the "soft" inhibitor - described above - of auxin transport. The doubling in binding is due to a decrease of the apparent K_D for verapamil from 0.06 to 0.03 μM. 3,4,5-TIBA, with an EC_{50} of 20 μM, also increases verapamil binding to rabbit muscle membranes, and with an IC_{50} of 8 μM, interferes with contraction in isolated rabbit muscle preparations.

This Ca^{2+} related verapamil site is probably not identical with the efflux carrier because the stronger inhibitors of auxin transport, the phytotropins , are ineffective, and because 2,3,5-TIBA acts only at higher concentrations. On the other hand, the effects of 3,4,5-TIBA on both IAA transport and on a Ca^{2+} related process may indicate functional similarities and/or homologies.

(3) The Problem of Auxin Receptors: Relation Between Sites for Transport and Sites for Action.

Some concluding comments and speculations shall be added concerning possible functional and phylogenetic relations between auxin transport and action, and the possible role of the (too) many published binding sites (Hertel, 1983). The arguments are partly similar to the critical appraisal of binding sites by Firn (1987).

Hormone research moves on in a hermeneutic circle: receptors can not be studied and recognized without some preconception about the hormone`s mechanism of action, and on the other hand, one needs to know the receptor in order to understand this mechanism. The question of auxin`s primary mechanism of action is still open. Three hypotheses are discussed (see Evans, 1985; Hertel, 1983, 1985).

The first one is a gene activation model, analogous to steroid action in animals, i.e. via a cytosolic receptor and subsequent stimulation of transcription in the nucleus.

A second hypothesis explained auxin action as proceeding mainly via a stimulation of proton secretion with subsequent acid-induced wall loosening. This widely accepted hypothesis has been disproved by Kutschera and Schopfer (1985) who measured auxin effects in maize coleoptiles with fusicoccin controls.

Finally, fast effects on membranes as Ca^{2+} release or phosphoinositol metabolism still need more experimental support.

Addressing the problem in a more abstract way one may ask whether the various auxin effects are derived from one "master reaction". Auxin provokes not one but many different responses. In the pea epicotyl, for example, elongation is promoted, protons are secreted, ethylene production is stimulated, and the synthesis of new RNA and proteins is induced. Comparing different organs, it is well established that, while in roots elongation is inhibited, coleoptile elongation is promoted by auxin.

Several types of hypothetical networks - and mixtures thereof - can describe the multitude of auxin effects: (a) Each response has its own receptor (= specific binding site) and its own transduction chain; (b) all reactions branch off at one point ("master reaction") from one common chain with one input receptor; (c) there is a sequence of reactions ("primary", "secondary" etc.) each with or without a "little branch".

It is conceivable that IAA "latches on" to several different regulator proteins, preadapted in evolution. These would be convergent and not homologous receptors. In addition to a fast, membrane /Ca^{2+}-linked action, a second - steroid-like action of auxin could then be responsible for the early messenger RNAs - perhaps via one of the reported soluble, high affinity IAA binding sites.

A compromise was proposed "between" one or many receptors (Hertel, 1983): Auxin action and auxin transport are two separate processes since phytotropins can inhibit transport while allowing auxin stimulation of growth. Nevertheless many similarities exist between the transport, auxin`s action in shoots and coleoptiles, and action in roots: the initial kinetics, adaptation, optimum curves as well as the analog specificity pattern. A phylogenetic homology was postulated: Three different, but homologous proteins are thought to be involved: (1) the efflux carrier of transport, (2) the action-related IAA-receptor of stem and coleoptile cells, and (3) a similar IAA-binding protein in roots. All three are auxin binding proteins coded for by different genes that are likely to be derived from a common ancestor gene coding for one membrane protein. (4) The uptake carrier has different specificity.

In Tab. 1, the homologous elements are tentatively assigned to some of published membrane-associated auxin binding sites (see Rubery, 1981; cytosolic sites were extensively discussed at this workshop).

Table 1. Relation between membrane-associated auxin binding sites and physiological functions (from Hertel, 1983, 1985).

Element of model	In vitro binding site	Intracellular localization	Proposed function
Efflux carrier regulatory portion	? NPA-site(s)	Basal plasmalemma	Transport, efflux
Symport	site III	Total plasmalemma	Uptake of IAA and H^+
internal Efflux carrier	site II(+I)	Tonoplast?	Auxin action in shoot
Efflux carrier of root	$site_{root}$	Plasmalemma?	Action on root growth

The two sites involved in transport - site III and the NPA receptor - have been discussed above.

For action in roots, auxin binding sites, localized in heavy maize microsomes, have been decribed (Moloney and Pilet, 1982).

The key element for auxin action in coleoptiles and shoots might be auxin binding site II and/or I, probably localized at an inner compartment (Dohrmann et al, 1978). These sites show a fitting specificity pattern compared to elongation growth (Ray et al, 1977). Furthermore, Löbler and Klämbt (1985) argue that both this relevant auxin receptor and the response system are concentrated in the epidermis.

Unfortunately, these sites, originally reported for maize, were not yet found in any dicotyledoneous plant. Thus, the reasonable postulate (see Firn, 1987) is not satisfied that "all cells capable of showing the same response - here e.g. the maize coleoptile and the pea epicotyl - should also contain a closely related receptor."

Acknowledgements. The work was supported by the "Deutsche Forschungsgemeinschaft" (He 418/8-1). I thank E. Andrejauskas, C. Benning, W. Michalke and M. Thein for the permission to quote unpublished data. Their commments and criticisms as well as discussions with K.A. Clark, M.H.M. Goldsmith, T.L. Lomax, and P.H.Rubery are gratefully acknowledged.

REFERENCES

Andrejauskas E, Marme' D, Hertel R (1985) J biol Chem 260:5411
Andrejauskas E, Marme' D, Hertel R (1986) BBRC, in press
Benning C (1986a) Planta, in press
Benning C (1986b) Diplomarbeit Freiburg i.Br. F.R.G.
Bruns RF, Lawson-Wendling K, Pugsley TA (1983) Anal Biochem 132:74
Cande WZ, Ray PM (1976) Planta 129:43
Clark KA, Goldsmith MHM (1986) In: Bopp M (ed) Plant Growth Substances 1985. Springer-Verl, Heidelberg, p 203
Clark KA, Goldsmith MHM (1987) This volume
Dela Fuente RK, Tang PM, De Guzman CC (1986) In: Bopp M (ed) Plant Growth Substances 1985. Springer-Verl, Heidelberg, p 224
Depta H, Eisele KH, Hertel R (1983) Plant Sci Letts 31:181
Dohrmann U, Hertel R, Kowalik H (1978) Planta 140:97
Evans ML (1985) CRC Crit Rev Plant Sci 2:317
Firn RD (1987) This volume
Hasenstein KH, Rayle D (1984) Plant Physiol 76:65
Hertel R (1983) Z Pflanzenphysiol 112:53
Hertel R (1985) Gior Bot Ital 119:s 46
Hertel R (1986) In: Bopp M (ed) Plant Growth Substances 1985. Springer-Verl, Heidelberg, p 214
Hertel R, Leopold AC (1963) Planta 59:535
Hertel R, Lomax TL, Briggs WR (1983) Planta 157:193
Jacobs M, Hertel R (1978) Planta 141:1
Kutschera U, Schopfer P (1985) Planta 163:483 and 163:494
Larkin PJ, Scowcraft WR, Geissler AE, Katekar GF (1982) Aust J Plant Physiol 9:297
Löbler M, Klämbt HD (1985) J Biol Chem 260:9848 and 260:9854
Lomax TL, Mehlhorn RJ, Briggs WR (1985) Proc Natl Acad Sci USA 82:6541
Michalke W (1982) In: Marme' D, Marre` E, Hertel R (eds) Plasmalemma and tonoplast. Elsevier, Amsterdam p 129
Moloney MM, Pilet PE (1982) In: Marme' D, Marre` E, Hertel R (eds) Plasmalemma and tonoplast. Elsevier, Amsterdam p 293
Ray PM, Dohrmann U, Hertel R (1977) Plant Physiol 60:585
Rubery PH (1981) Annu Rev Plant Physiol 32:569
Rubery PH, Sheldrake AR (1974) Planta 118:101
Sabater M, Sabater F (1986) Planta 167:76
Sussman MR, Goldsmith MHM (1981) Planta 151:15

PRELIMINARY SEPARATION OF PEA STEM NPA RECEPTORS BY HIGH PERFORMANCE ION EXCHANGE CHROMATOGRAPHY

Mark Jacobs and Timothy W. Short
Department of Biology, Swarthmore College
Swarthmore, PA, USA 19081

Introduction

The auxin anion carrier (AAC) is one of the most interesting and important proteins in the transport and response system that allows higher plants to utilize the hormone auxin. It is this efflux carrier, hypothesized to occur in differentially greater numbers at the basal ends of auxin-transporting cells (8),that mediates and confers polarity upon the process of basipetally polar auxin transport (4,8). That process, in turn, is responsible for the distribution of auxin from its sites of synthesis in the shoot apex and young leaves to its target tissues below. As evidence mounts that auxin action depends upon concentrations of free auxin in the cell cytoplasm as opposed to otherwise compartmentalized or conjugated forms of the hormone (1,5), the potential importance of the AAC as a regulatory point not only for transport but also for action has become clear. Fortunately,more is known about the regulation of the AAC's activity than is the case for other auxin binding proteins and carriers, and the greatest amount of research has been done with the phytotropin naphthylphthalamic acid (NPA) (see discussion in 3). NPA can specifically inhibit polar auxin transport, and apparently does so by dramatically reducing basal efflux of the auxin anion from transporting cells (2,9). It is thus likely that the NPA receptor protein is the same as, or closely interactin with, the ACC.

In previous work, we have made monoclonal antibodies (mAb's) against the NPA receptor in etiolated pea stem tissue and used the antibodies in indirect immunofluorescence experiments to identify the location of the presumed AAC's in that tissue (6). The preponderance of sites appears to be at the basal ends of cells sheathing the vascular bundles. We have also checked the species specificity of the mAb's, examined their effects on polar auxin transport *in vivo*, and used them to immunoprecipitate the NPA receptor (7). As a parallel approach to our immunological studies of the AAC, we have recently been exploring non-immunological techniques for protein separation and purification. Here, we report research done using high performance ion exchange chromatography (HPIEC) to begin to resolve the NPA receptor from a Triton-solubilized pea stem cell membrane protein preparation.

Material and Methods

Plant Material - Pea seedlings (Pisum sativum, cv Alaska) were grown 7 days in the dark, then a Triton-solubilized membrane protein preparation was obtained from third internodes as described previously (6, Table 1), except that the buffer in the resuspension medium was 10 mM Na citrate instead of MOPS and both grind medium and resuspension medium contained the protease inhibitor PMSF at 0.5 mM.

Chemicals - Chemicals were purchased from Sigma Chemical Co. (St.

NATO ASI Series, Vol. H10
Plant Hormone Receptors. Edited by D. Klämbt

Louis,MO,USA) ^{3}H-NPA was purchased from Research Products International Corp. (Elk Grove, IL. USA) and unlabeled NPA was a gift from R. Hertel.

HPIEC - High performance ion exchange chromatography was carried out using a Gilson/Rainin HPLC controlled by an Apple IIe computer. An Altex DEAE-5 PW anion exchange column (Spherogel TSK, Beckman), 7.5 cm x 7.5 mm I.D., was used with a 1 ml injection loop. Protein peaks were measured at 254 nm.The HPLC was equilibrated with a solution of 10 mM Na citrate, pH 6.5, for several hr, then 1 ml of solubilized pea membrane protein, prepared as described above, was injected. This was eluted at a flow rate of 1 ml/min subject to the following program: 0-5 min 10 mM Na citrate, pH 6.5 (solution "A"); 5-25 min, a linear gradient of 0-0.5 M NaCl in "A"; 25-50 min, constant 0.5 M NaCl in "A"; 50-60 min, a linear gradient of 0.5 M - 0 M NaCl in "A". Fractions were collected either at set times (every 5 min) or by collecting particular peaks in the 254 nm absorption trace.

NPA Binding - Collected HPIEC fractions were placed immediately on ice and tested for NPA binding activity using the Amicon micropartition system described previously (6). Here, to 2.2 ml of each fraction was added ^{3}H-NPA to a final concentration of 10^{-9}M. This mixture was vortexted, and 500 ul of it placed into each of two Amicon micropartition assemblies ("hot" duplicate samples).To the remaining 1.2 ml of radioactive solution from each fraction was added unlabeled NPA to a final concentration of 10^{-5} M. This was also vortexted and 500 ul of it added to each of two separate Amicon assemblies ("hot + cold" duplicates). After centrifugation for 2.5 h, 1600 x g, 4° C, the dry filter membranes, retaining the solubilized receptor-ligand complex, were removed and counted in Scintisol (Isolab) in a liquid scintillation counter. "Specific binding" is equal to "hot" - "hot + cold" counts.

SDS Gel Electrophoresis - Aliquots of each HPIEC fraction (100 ul) were mixed with 100 ul of SDS application buffer (0.125 M Tris-HCl, pH 6.8, 4 % SDS, 20 % glycerol, 10 % B-mercaptoethanol), boiled, then 150 ul of the mixture applied to a 10 % discontinous 1.5 mm polyacrylamide gel. Gels were run at 130 V through the stacking gel, 275 V through the running gel, and stained with a silver reagent.

Results and Discussion

We wished to use an anion exchange column on a HPLC to attempt to separate NPA binding activity from other detergent-solubilized pea stem cell membrane proteins. Earlier attempts to measure NPA binding in fractions collected from HPLC sizing columns indicated the need for a large volume (1-2 ml) injection loop when using the protein concentrations we obtained in the pea stem extracts. A preliminary run under the conditions described in Material and Methods indicated at least some NPA binding activity in 4 of the 12 eluted fractions, but several further checks of those fractions showed consistent binding only in 2 fractions, one eluting at 15-20 min, the other at 55-60 min (Fig.1). In these runs, the complex of peaks eluting at 55-60 min had significantly higher binding activity than the shoulder eluting at 15-20 min, and this late peak was further fractionated to attempt to resolve more precisely the NPA receptor protein. Fig.2 shows a HPIEC separation

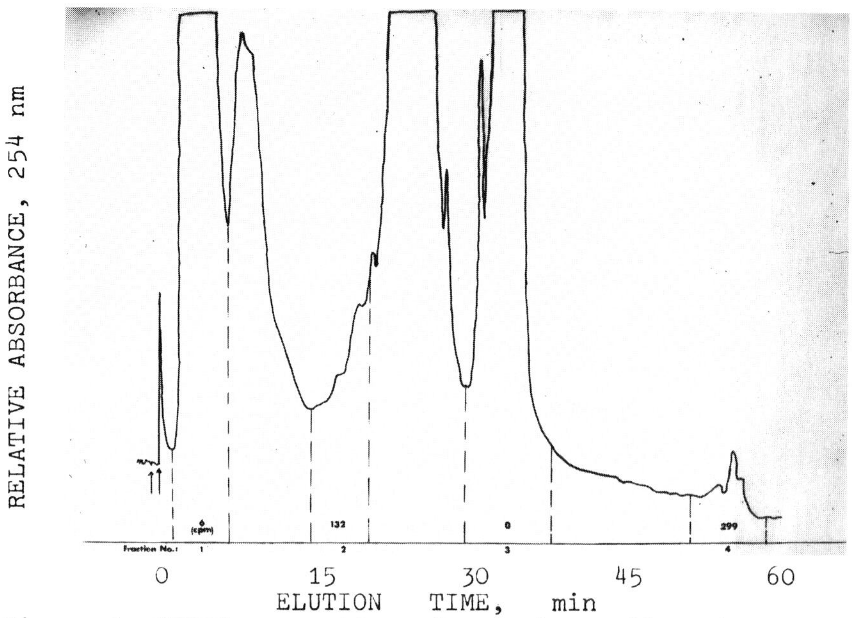

Figure 1. HPIEC separation of pea stem cell membrane proteins using a DEAE column. Run conditions as in Materials and Methods. Injection given at second upward arrow. Specific NPA binding for each fraction is indicated in cpm above the fraction number.

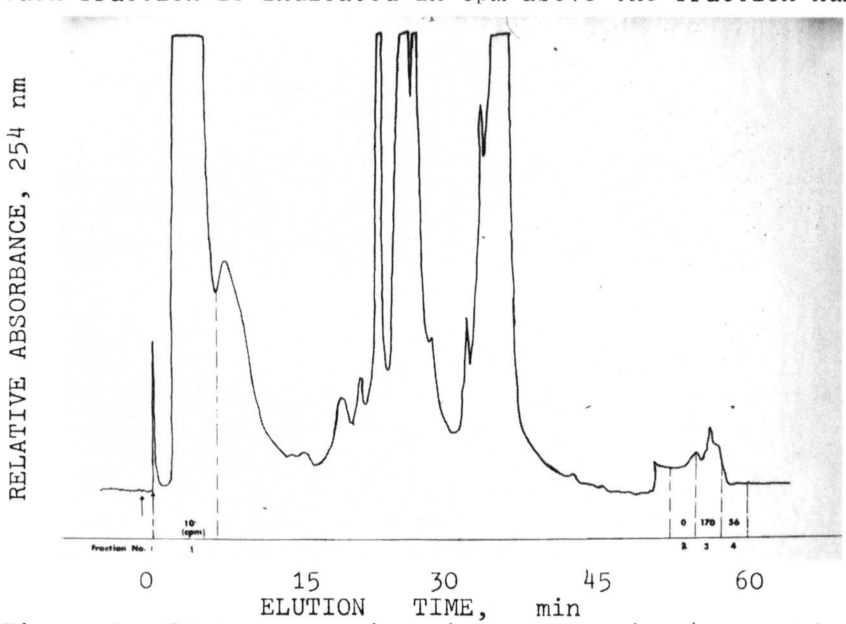

Figure 2. HPIEC subfractionation of fraction 4 from Figure 1. Conditions and methods as in Figure 1.

in which fractions were collected only from 0-6 min, 52-54.5min, 54.5-57 min, and 57-59.5 min. NPA binding measured in those 4 fractions is shown, and the third (54.5-57 min) fraction shows the greatest binding activity.

When the fractions numbered 2, 3 and 4 from an experiment such as that shown in Fig.2 were applied to a 10 % SDS polyacrylamide gel, proteins present could be visualized by silver staining (Fig.3). When one examines fraction 3 - the peak of NPA binding

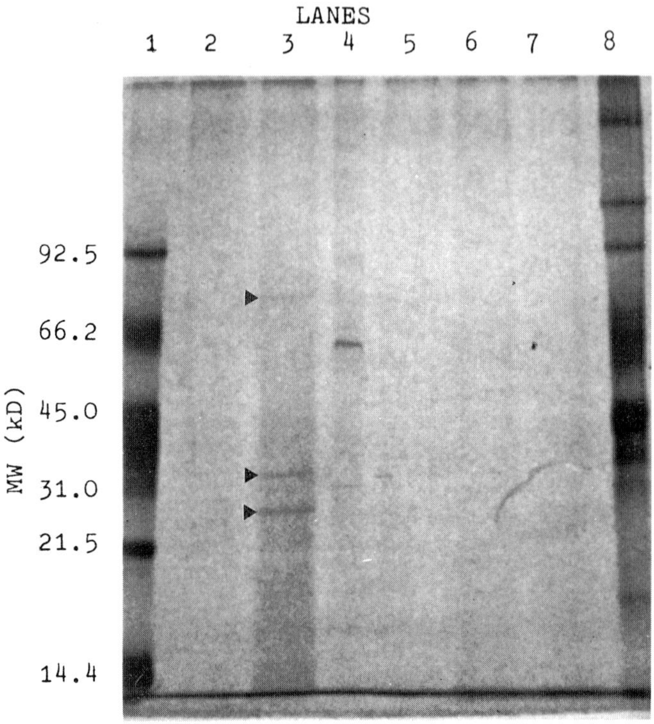

Figure 3. SDS Polyacrylamide gel electrophoresis of proteins from fractions 2-4, Figure 2. Lane 1: low MW standards. Lane 2: fraction 2. Lane 3: fraction 3. Lane 4: fraction 4. Lanes 5-7: fractions from a more dilute run. Lane 8: high MW standards. arrows in Lane 3 point to the three protein bands stained with silver reagent.

activity from HPIEC - one sees three protein bands. The highest MW band is ca. 77 kD, the lower MW bands are at ca. 34 kD and 28 kD. The band at 77 kD corresponds to the protein we have visualized on SDS PAGE after using our mouse anti-pea NPA receptor mAb's to immunoprecipitate recognized antigens from a Triton-solubilized pea stem cell membrane protein preparation (7). We do not yet know whether the lower MW bands are related to the 77 kD band (subunits or breakdown products) or are unrelated proteins eluting at the same time from the column. Experiments testing the effects of various protease inhibitors during pea protein preparation, and experiments utilizing non-reducing gels, are underway to check their possible effects on the band pattern in fraction 3. It would be also interesting to perform an immunoprecipitation experiment using only the proteins in fraction 3 from Fig. 2 and our anti-NPA receptor mAb's to see whether any of the three HPIEC-separated proteins (but particulary that at 77 kD) are recognized by the antibodies.

Conclusions

High performance anion exchange chromatography has been used to carry out a preliminary separation of NPA receptor from a pea stem cell membrane protein fraction. Of the three protein bands visible in the HPIEC fraction with the highest NPA binding activity, one is of the same MW (77 kD) as a protein from the same pea stem cell preparation recognized by anti-NPA receptor mAb's (7). We are encouraged to purify and further investigate the 77 kD protein in the hopes of better characterizing the pea NPA receptor and testing the hypothesis that it is identical to the AAC of the polar auxin transport system.

Acknowledgements - We are grateful to the National Science Foundation for a grant (PCM 8314844) that supported this research and to Caitlin Kennedy and Phillip Kloeckner for assistance with many of the experiments.

References

1. Bandurski, R.S. (1980) Homeostatic control of concentrations of indole-3-acetic acid. In: Plant Growth Substances 1979, pp.37-49, Skoog,F., ed. Springer, Heidelberg.
2. Cande, W.Z., Ray, P.M. (1976) Nature of cell-to-cell transfer of auxin in polar transport. Planta 129, 43-52.
3. Depta, H., Eisele, K.-H., Hertel, R. (1983) Specific inhibitors of auxin transport: action on tissue segments and in vitro binding to membranes from maize coleoptiles. Plant Sci. Letters 31, 181-192.
4. Goldsmith, M.H.M. (1977) The polar transport of auxin. Annu. Rev. Plant Physiol. 28, 439-478.
5. Hertel, R. (1983) The mechanism of auxin transport as a model for auxin action. Z. Pflanzenphysiol. 112, 53-67.
6. Jacobs, M., Gilbert, S.F. (1983) Basal localization of the presumptive auxin transport carrier in pea stem cells. Science 220, 1297-1300.
7. Jacobs, M., Short, T.W. (1986) Further characterization of the presumptive auxin transport carrier using monoclonal antibodies. In: Plant Growth Substances 1985, pp. 218-226, Bopp, M., ed. Springer, Heidelberg.
8. Rubery, P.H., Sheldrake, A.R. (1974) Carrier-mediated auxin transport. Planta 118, 101-121.
9. Sussman, M.R., Goldsmith, M.H.M. (1981) The action of specific inhibitors of auxin transport on uptake of auxin and binding of N-1-naphthylphthalamic acid to a membrane site in maize coleoptiles. Planta 152, 13-18.

EFFECT OF SURFACE AND MEMBRANE POTENTIALS ON IAA UPTAKE AND BINDING BY ZUCCHINI MEMBRANE VESICLES.

Kathleen A. Clark and Mary Helen M. Goldsmith
Biology Department, Kline Biology Tower, Yale University
New Haven CT 06511 USA

The polar transport of the endogenous hormone controlling extension growth of plant cells, indoleacetic acid (IAA), is thought to depend on transmembrane pH and electrical gradients resulting in part from the action of proton ATPases in the plasma membrane. According to a recent hypothesis (see Goldsmith, 1977 for review), elements of this transport process are: (1) permeation of the membrane by the undissociated lipophilic indoleacetic acid (IAAH) from the acidic apoplast, followed by dissociation of the weak acid and accumulation of the IAA anion (IAA^-) in the alkaline cytoplasm; (2) a saturable symport of IAA^- with one or more protons; (3) a carrier-mediated efflux of IAA^- down a considerable electrochemical gradient. The efflux is greater from the basal than the apical end of cells and is thought to be responsible for the overall polarity of the process. This step is also the site of action of napthylphthalamic acid (NPA) and herbicides that inhibit polar transport but stimulate net accumulation of auxin by tissues and cells (Sussman & Goldsmith, 1981a&b; Thomson et al, 1973).

We are using membrane vesicles as a simplified system for studying the mechanisms involved in the transport and accumulation of auxin. In particular, we are interested in determining the involvement of the transmembrane pH ($pH_o < pH_i$) and voltage gradients (K^+ diffusion potential, $[K^+]_{in} > [K^+]_{out}$) in IAA uptake.

Previously it has been noted that the uptake of IAA in zucchini vesicles is 3-5x more than expected from the pH gradient, as measured by the uptake of ^{14}C-butyric acid (BA) (Clark and Goldsmith, 1986) or electron spin resonance probes for ΔpH (Lomax et al 1985). In order to explain this extra accumulation of IAA, several hypotheses have been advanced. Recently we (Clark and Goldsmith, 1986) reported evidence suggesting that much of the intravesicular IAA may be associated with the

NATO ASI Series, Vol. H10
Plant Hormone Receptors. Edited by D. Klämbt

membranes, rather than free within the vesicles. When the vesicular volume was reduced by resuspending the vesicles in more concentrated sorbitol solutions (e. g. 800 mOs instead of 350 mOs), the amount of butyric acid varied in proportion to the volume, as expected of a substance in the intravesicular solution. By contrast, the amount of ^{3}H-IAA was nearly the same when the intravesicular volume was osmotically reduced three-fold. Of particular interest in this experiment was the observation that when a saturating concentration (5 µM; Clark and Goldsmith, 1986) of unlabelled IAA was present, ^{3}H-IAA (10 nM) accumulated in accord with the size of the vesicles and the imposed pH gradient. Taken together these observations suggest that much of the IAA accumulated by the vesicles is not freely distributed in the intravesicular space, but rather may be bound to or intercalated into the membrane.

A different hypothesis to explain the increased accumulation of IAA in zucchini vesicles is that, in addition to the transmembrane pH gradient, a voltage gradient may drive IAA accumulation. Both Lomax et al (1985) and Hertel (1983, 1986) observed that valinomycin stimulates auxin accumulation by vesicles in the presence of an outwardly directed gradient of K^{+} ions. This led them both to propose that if influx of IAA involves a positively charged combination such as [IAAH + H^{+}] or [IAA^{-} + $2H^{+}$], a membrane potential (inside negative) could provide an additional driving force for IAA accumulation. This suggestion represents a considerable departure from the original hypothesis (see above) as it necessarily assumes that the membrane must be relatively impermeable to IAAH. This would seem to be a physically implausible assumption given the intrinsic permeability of IAA both in artificial phospholipid bilayers (Gutknecht and Walter, 1980) and in the plasma membrane of <u>Hydrodictyon africanum</u> (Raven, 1975) of about 10^{-3} cm·sec^{-2}.

Phospholipid bilayers and hence the microsomal membrane vesicles isolated from plant cells bear a net negative charge in aqueous media. The extent to which this charge is screened by counterions in the solution affects the magnitude of the surface potential and the distance that it extends out from the bilayer (for review, see McLaughlin, 1977). This means that the ionic strength of the medium may influence the physical properties of the membrane as well as the partitioning of ions and intercalation of amphipathic molecules, such as IAA, into the membrane (Andersen et al, 1978; McLaughlin, 1977; Weigl, 1969; Zimmermann et al, 1977).

The original experiments suggesting that a K^+ diffusion potential stimulates auxin accumulation by vesicles (Lomax et al, 1985; Hertel, 1983, 1986) did not consider the possibility that variations in ionic strength might affect the association of IAA with the membrane. Furthermore, no attempt was made either to monitor the voltage gradient or to define the conditions under which a transmembrane potential was present. Therefore, we felt it worthwhile to explore the relation between K^+ concentration gradients, membrane and surface potentials, and accumulation of IAA into zucchini membrane vesicles.

To generate negative membrane potentials, we set up K^+ gradients by preparing the vesicles in 100 mM K^+ salt solutions and varying the outside $[K^+]$ (0.5 - 100 mM). Since the ionic strength of the medium will alter the surface potential of the membranes, which in turn may affect either insertion of IAA into the lipid bilayer or interaction of IAA with membrane proteins, ionic strength and also anion concentration were maintained constant by substituting either Cs^+, Na^+ or choline chloride for KCl. In some experiments, the chlorides were replaced by the less permeable sulfates of K^+ or Na^+. These results were compared with the situation in which ionic strength was varied along with $[K^+]$. If IAA uptake depends in part on a negative transmembrane potential (Lomax et al, 1985; Hertel, 1983, 1986), accumulation should be enhanced by an outwardly directed potassium gradient, assuming that the vesicles are selectively permeable to K^+. The K^+ ionophore, valinomycin, was added to insure that this was the case.

Materials and Methods.

We have modified the protocol of Hertel's group (Hertel et al, 1983) in order to prepare microsomal vesicles which have a pH gradient, a K^+ gradient or both. Hypocotyls (subapical 2 cm) from 6 d old, dark-grown zucchini (*Cucurbita pepo* L. cv. Black

Beauty) seedlings were homogenized in isotonic medium (350 mOs) buffered at pH 8 (150-300 mM sorbitol, 25 mM HEPES/BTP, 3 mM EGTA, 1 mM PMSF, 1 mM DTT, 0.1% BSA and 0 - 50 mM K_2SO_4/Na_2SO_4 or 100 mM KCl). The temperature was 0 - 4 °C for this and most subsequent procedures; a consistent exception was that when valinomycin was present the temperature was 10 °C. Microsomal membrane vesicles were isolated by differential centrifugation (13000xg supernatant centrifuged 45 min at 50000xg). The 50000xg pellet was washed and resuspended in a weakly buffered pH 8 solution containing the same salt concentration as present in the homogenization medium plus 2.5 mM HEPES/BTP, 1 mM DTT and enough sorbitol to maintain the osmolarity at 350 mOs. Sometimes the vesicles were frozen (-70 °C) at this point and used at a later date.

We used ^{14}C-butyric acid as a ΔpH-probe and measured its uptake simultaneously with that of ^3H-IAA. Since butyric acid has nearly the same pK_a as IAA, accumulation of BA and IAA due to a pH gradient should be the same (Rottenberg, 1979). A trans-membrane pH gradient was established and uptake of IAA and BA initiated when the vesicles were resuspended in a pH 6 solution that consisted of 10 mM MES, ^3H-IAA (10 nM), ^{14}C-BA (5 µM), the desired concentration of salt and enough sorbitol to bring the solution to 350 mOs. The net uptake was monitored after IAA and BA had reached steady state (20 min at 0 °C; 10 min at 10 °C) by sampling aliquots (50 µl; 10 - 20 µg protein) in triplicate. Vesicles were separated from the incubation medium by filtration (Millipore HAWP 0.45 µm). Radioactivity was determined by scintillation counting in Optifluor; tritium counts were corrected for spillover of ^{14}C and counts of both isotopes were corrected for efficiency of the scintillation counter (Beckman LS 7000). The ΔpH-dependent uptake of IAA and BA was the difference in radioactivity associated with the vesicles in the presence or absence of 5 µM carbonyl cyanide m-chlorophenyl hydrazone (CCCP) and valinomycin. The combination of the protonophore, CCCP, and the K^+-ionophore, valinomycin, collapsed the pH gradient more completely, by dissipating both pH and electrical gradients, than CCCP alone. Vesicle volume expressed on a protein basis was equated with the volume of the vesicles available to ^3H-water but excluding ^{14}C-dextran (Rottenberg, 1979). This permitted the ΔpH-driven accumulation of IAA and BA to be expressed on a concentration basis, C_i/C_o; i. e., the concentration of IAA or BA accumulated by the vesicles (C_i) relative to that outside (C_o). Protein was routinely assayed (Bradford, 1976).

Fluorescent quenching of the positively charged cyanine dye $diSC_3(5)$ (3,3'-dipropylthiadicarbocyanine iodine) was used as a membrane potential probe (Beeler et al, 1981; Rottenberg, 1979; Waggoner,1979). A Perkin-Elmer 650-40 spectrofluorimeter, modified to permit constant stirring and dark addition of substances through an injection port, was used to detect fluorescent emission of $diSC_3(5)$ (2 μM) at 665 nm after excitation at 640 nm. Temperature was maintained at 10 °C; vesicle concentration was 5 μg/ml. The ratio of dye concentration to vesicles was optimized in preliminary experiments. Quenching of the dye occurs both because it associates with membranes and accumulates within vesicles.

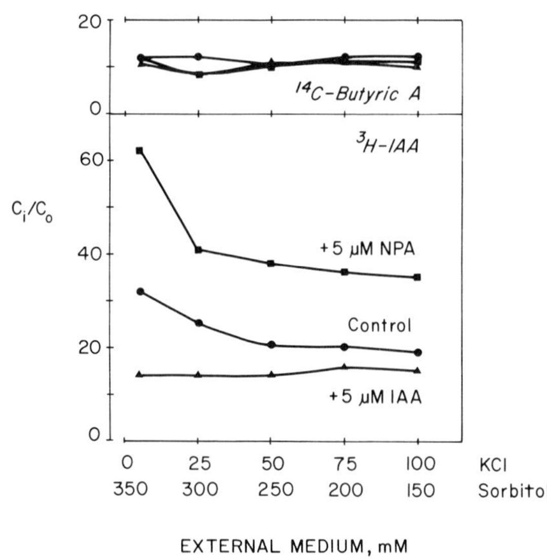

Figure 1. A K^+ gradient may enhance accumulation of ^{3}H-IAA but does not affect ^{14}C-BA uptake. Vesicles were prepared in a pH 8 buffer that contained 50 mM KCl and resuspended in a pH 6 buffer which included 5 - 100 mM KCl, and 150 - 335 mM sorbitol. Accumulation of ^{14}C-BA (5 μM) and ^{3}H-IAA (10 nM) was determined simultaneously after 20 min of incubation at 0 °C. Symbols: circles = controls; squares = + 5 μM NPA; triangles = + 5 μM IAA. See Materials and Methods for further details.

Results and Discussion.

The results shown in Fig. 1 (see above) suggest that the presence of a K^+ concentration gradient ($[K^+]_{in} > [K^+]_{out}$), may enhance accumulation of IAA. Vesicles prepared in 50 mM KCl accumulated more ^{3}H-IAA than ^{14}C-BA when the outward K^+ gradient was steep. At a saturating IAA concentration (5 µM; Clark and Goldsmith, 1986), however, IAA accumulation was unaffected by decreasing the K^+ gradient and not significantly different from that of BA. NPA (5 µM) caused a 2-fold stimulation of IAA uptake regardless of the K^+ gradient. Butyric acid uptake was unaffected by the salt composition of the medium, indicating no change in the pH gradient or size of the vesicles.

When high ionic strength was maintained by ionic substitution, IAA accumulation was not influenced by the K^+ gradient (Fig. 2).

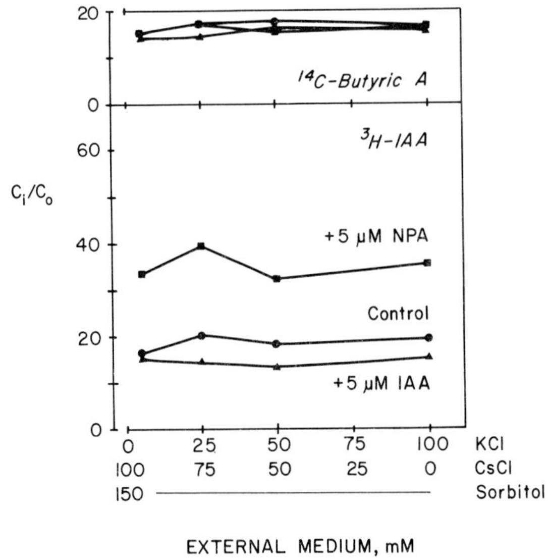

Figure 2. IAA accumulation is not influenced by the K^+ concentration gradient when constant external ionic strength is maintained by ionic substitution. Vesicles were prepared as in Figure 1 but were resuspended in a pH 6 buffer which consisted of 150 mM sorbitol and 5 - 100 mM KCl counterbalanced with 0 - 95 mM CsCl. Same symbol code and assay conditions as described for Figure 1.

Instead, vesicles prepared in 50 mM KCl and resuspended in 100 mM salt, whether KCl, CsCl/KCl, cholineCl/KCl or NaCl/KCl, accumulated IAA similarly, regardless of the imposed K^+ gradient. These results suggest that the effect of ionic strength on net IAA uptake may be related to the magnitude of the negative surface potential of the membranes. The observation that 10 mM $CaCl_2$ depressed IAA uptake to a similar extent as 100 mM KCl (data not shown) supports this conclusion. The divalent cation Ca^{2+} is known to be an order of magnitude more effective as a counterion than monovalent cations (McLaughlin, 1977). Notice also that despite the reduction in uptake at higher external ionic strength (compare with Fig. 1), the uptake of IAA in the presence of NPA remained at least twice that of BA, suggesting that a potassium diffusion potential can not be responsible for the NPA-enhanced accumulation of IAA.

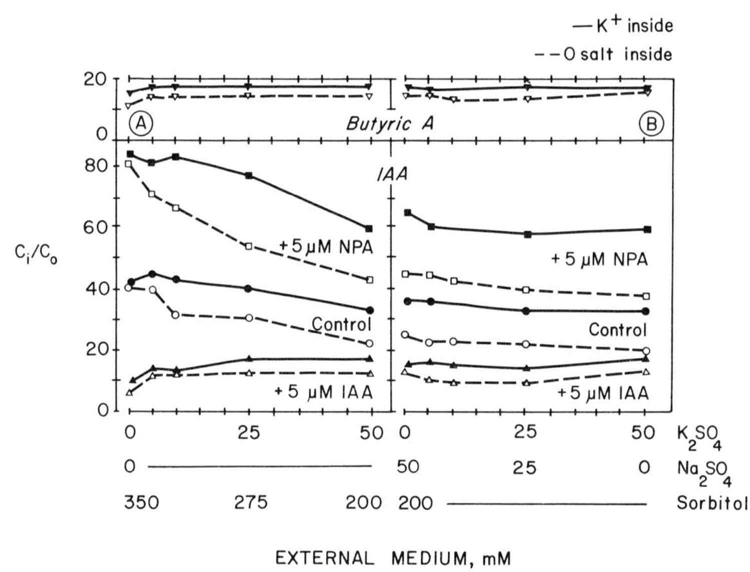

Figure 3. <u>High external ionic strength depresses IAA accumulation by vesicles regardless of whether or not a K^+ concentration gradient exists.</u> Vesicles were prepared in a pH 8 buffer which contained 50 mM K_2SO_4 (__) or no inorganic salt at all (- -). Resuspension media consisted of a pH 6 buffer, 0.5 - 50 mM K_2SO_4 and: (A): 200 - 350 mM sorbitol. (B): 0 - 49.5 mM Na_2SO_4 and 200 mM sorbitol. Same symbol code and assay conditions as described for Figure 1.

Surprisingly, high external ionic strength depressed IAA accumulation by vesicles regardless of either the magnitude or direction of the K^+ gradient. Vesicles prepared in a homogenization medium that contained either 50 mM K_2SO_4 (or 50 mM Na_2SO_4, not shown) or no inorganic salt at all, responded similarly when the K^+ composition of the resuspension medium was varied from zero to 50 mM K_2SO_4 (Figs. 3A and 3B). As seen in Figure 3A, maximum accumulation of IAA did not depend on the presence of a K^+ gradient but occurred when the salt concentration in the resuspending medium was lowest. As the external ionic strength was increased by the addition of K_2SO_4, IAA accumulation was depressed similarly in both vesicle preparations (Fig. 3A). An external concentration of 10 mM $CaCl_2$ was as effective as 50 mM K_2SO_4. When the ionic strength was constant (Fig. 3B), IAA accumulation did not vary despite the decreasing K^+ concentration gradient. These results strongly suggest to us that an outwardly directed K^+ diffusion potential is not responsible for the enhanced accumulation of IAA when the external K^+ concentration is low (Figs. 1 and 3A).

Of course, the idea that a transmembrane K^+ gradient ($[K^+]_{in} > [K^+]_{out}$) induces a negative membrane potential rests upon the supposition that the membranes are selectively permeable to K^+. To verify that a negative membrane potential was present, we monitored the fluorescent response of the positive cyanine dye, $diSC_3(5)$ under conditions similar to those in which we studied IAA uptake. Vesicles which had been prepared in 50 mM K_2SO_4 were resuspended in a pH 6 buffer that contained 2 μM $diSC_3(5)$ and 50 mM concentrations of Na_2SO_4/K_2SO_4. (Ionic strength was held constant in all fluorescent experiments, since it affects binding of the dye to the vesicles.) Upon addition of the vesicles to the dye mixture, immediate quenching of fluorescence was observed (Fig. 4). Since this quenching of fluorescence did not differ significantly when the K^+-loaded vesicles were resuspended in different external K^+ concentrations, it is probably due to hydrophobic interactions of the dye with the membrane rather than distribution of the dye cations between the internal and external solutions in response to a membrane potential. The lack of correlation between dye uptake and the

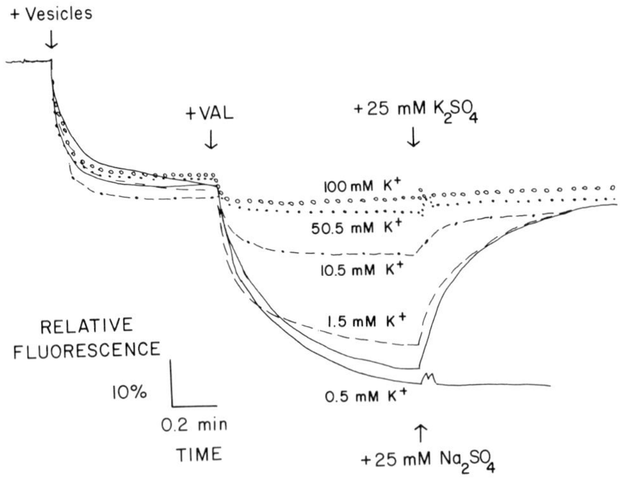

Figure 4. As determined from quenching of the fluorescent dye diSC$_3$(5), a negative membrane potential is generated in vesicles with a transmembrane K^+ gradient, $[K^+]_i > [K^+]_o$, when valinomycin is added. Vesicles prepared in 50 mM K_2SO_4 were diluted 100 fold into a pH 6 buffer that contained 2 μM diSC$_3$(5), 200 mM sorbitol and 0 - 50 mM K_2SO_4 counterbalanced with 0 - 49.5 mM Na_2SO_4. Final external $[K^+]$ was (mM): 0.5 (____), 1.5 (- -), 10.5 (-.-), 50.5 (...) or 100 (ooo). Final concentrations of additions: 5 μg/ml vesicles, 0.3 μM valinomycin and 25 mM K_2SO_4 or 25 mM Na_2SO_4. Temperature was maintained at 10 °C. For other details, see Materials and Methods.

imposed K^+ concentration gradient may indicate that at 10 °C, the selective permeability of these vesicles to K^+ is low. In any event, the fact that the dye uptake fails to respond to changes in the K^+ gradient suggests that under the experimental conditions used in Figures 1 and 3A, a diffusion potential is not present, and therefore can not be evoked to explain the excess IAA accumulated.

One way to insure an increase in membrane permeability to K^+ is through the addition of the K^+ ionophore, valinomycin. (Attention must be paid to the [valinomycin]/membrane ratio as high valinomycin concentrations may also increase membrane permeability to H^+.) When valinomycin was added to vesicles prepared in 50 mM K_2SO_4 and resuspended in low external $[K^+]$ (e.g., 0.5 mM), quenching of fluorescence occurred (Fig. 4). Since this quenching could be reversed by raising the external concentration of K^+ but not Na^+, it was due to accumulation of dye within the vesicles. As the outside K^+ concentration was raised, an inverse relationship was observed between the amount of quenching caused by the addition of valinomycin and the external $[K^+]$. This indicates that the distribution of dye cations was affected by the K^+ diffusion potential. Furthermore, when the K_2SO_4-loaded vesicles were resuspended in 50 mM K_2SO_4, little quenching of the dye occurred upon addition of valinomycin.

Having determined the conditions necessary to generate a K^+-diffusion potential, we proceeded to look at the effect of a membrane potential on IAA and BA accumulation (Fig. 5). Vesicles which had been prepared in 50 mM K_2SO_4 but resuspended in 50 mM Na_2SO_4 (final external $[K_2SO_4]$ was 0.5 mM) were allowed to accumulate ^3H-IAA and ^{14}C-BA for 5 min at 10 °C.

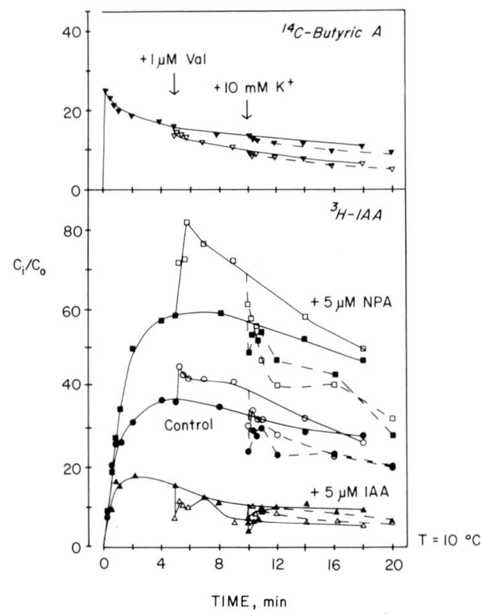

Figure 5. A negative membrane potential, generated by the addition of valinomycin to vesicles in which $[K^+]_i > [K^+]_o$, can cause a transient increase in IAA accumulation. Accumulation of ^{14}C-BA (5 μM) and ^3H-IAA (10 nM) into vesicles (0.17 mg/ml) prepared in 50 mM K_2SO_4 but diluted 100 fold into 50 mM Na_2SO_4 was determined. Additions: 1 μM valinomycin; 5 mM K_2SO_4. Closed symbols, - Val; open symbols, + Val; solid lines, - 5 mM K_2SO_4; dashed lines, + 5 mM K_2SO_4; same symbol code as described for Figure 1.

Generating a negative membrane potential by addition of valinomycin caused an immediate, transient increase in IAA accumulation which was most pronounced if NPA was present. When saturating IAA was included, addition of valinomycin did not cause a significant change in IAA uptake. Addition of 5 mM K_2SO_4 to the solution reversed the uptake induced by valinomycin presumably because the diffusion potential decreased. A somewhat greater rate of decline in BA accumulation indicated an increased decay rate of the pH gradient when valinomycin was added (Fig. 5A). It has not been established whether a negative membrane potential directly facilitates IAA transport or secondarily increases accumulation, for example, by modifying the conformation of a saturable binding site.

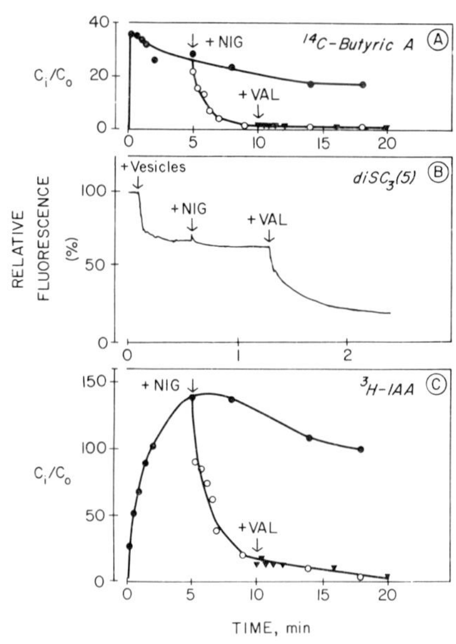

Figure 6. <u>When the pH gradient is eliminated, generation of a membrane potential (negative inside) does not enhance IAA accumulation.</u> Vesicles were prepared in 50 mM K_2SO_4 (pH 8) and resuspended in 50 mM Na_2SO_4 (pH 6); final external [K_2SO_4] was 0.5 mM. (A) and (C): Accumulation of ^{14}C-BA (5 µM) and ^{3}H-IAA (10 nM) into vesicles (0.11 mg/ml). [Nigericin], 1.0 µM; [valinomycin], 1.0 µM; [NPA], 5 µM. (B): Quenching of the fluorescent dye, diSC$_3$(5). Concentration of additions: 2 µM diSC$_3$(5); 5 µg/ml vesicles; 0.1 µM nigericin; 0.3 µM valinomycin. Temperature was maintained at 10 °C.

According to chemiosmotic theory, pH and voltage gradients should be interchangeable forces driving transmembrane transport. Assuming, as postulated by Lomax et al (1985) and Hertel (1983, 1986), that the mechanism of IAA uptake by zucchini vesicles involves a positively charged species, then if either the pH gradient or membrane potential were eliminated, the other gradient should be capable of sustaining some level of IAA accumulation. We have seen that under conditions that should negate the membrane potential (i. e., the same high concentration of K^+ on both sides of the membrane), IAA accumulation remained higher than predicted from the pH gradient, especially when NPA was present (Figs. 2, 3 and 5). That is, a pH gradient alone can support high levels of IAA accumulation. The reverse situation does not appear to be true. When the pH gradient was eliminated, a valinomycin-induced membrane potential did not sustain IAA uptake (Fig. 6). Initially in this experiment, $[K^+]_{in}$ was greater than $[K^+]_{out}$. Adding nigericin, which exchanges protons for potassium ions electroneutrally, effectively collapsed the pH gradient (butyric acid uptake, Fig. 6A). Nigericin did not affect the K^+ concentration gradient since a negative membrane potential could still be generated upon later addition of valinomycin (diSC$_3$(5) fluorescence, Fig. 6B). Under these conditions, however, valinomycin did not stimulate IAA uptake (Fig. 6C). In this experiment, the external proton concentration, which might be important for a proton/IAA symport mechanism, was maintained constant ($pH_o = 6$); nevertheless, in the absence of a pH gradient, a K^+ diffusion potential failed to sustain IAA uptake.

CONCLUSIONS

When vesicles loaded with K^+ are resuspended in increasing concentrations of KCl, the saturable uptake of IAA is reduced. At the same time, accumulation of the pH probe BA indicates that the pH gradient is not. According to the hypothesis suggested by Hertel (1983) and Lomax et al (1985), the reduction in IAA uptake might be attributed to decreasing the $[K^+]$ gradient and concomitant depolarization of the membrane potential. Several observations, however, suggest that it is best viewed as a result of the increased external ionic strength. (1) When ionic strength is held constant in ion substitution experiments, the magnitude of IAA accumulation is not affected by the K^+ concentration gradient. (2) The decreased uptake at high ionic strength is independent of the particular cation and occurs with Na^+, K^+, Cs^+, and choline salts of both chloride and sulfate. Furthermore, Ca^{2+} salts are an order of magnitude more effective than the monovalent cations. These results demonstrate that in order to avoid ambiguity, experiments using K^+ diffusion potentials to study the effects of voltage gradients on IAA uptake should be performed at constant ionic strength. (3) Outwardly directed K^+ gradients give rise to negative membrane potentials (as indicated by fluorescent quenching of the cyanine dye, $diSC_3(5)$) only when valinomycin is present. This indicates that a voltage-dependent uptake mechanism is not a viable explanation for the elevated accumulation of IAA (in excess of the pH gradient) that occurs in the absence of valinomycin with either NPA or at low ionic strengths. (4) Furthermore, a K^+ diffusion potential is detectable spectrofluorometrically on adding valinomycin to K^+-loaded vesicles even when high ionic strength is present. (5) Although a K^+ diffusion potential (inside negative) stimulates IAA accumulation, this is a transient effect, decaying in about 10 mins (at 10 °C). The transient nature of this response probably reflects the decay of the $[K^+]$ gradient and suggests that a $[K^+]$ diffusion potential is unlikely to be the basis for the elevated accumulations of IAA at steady state. (6) Furthermore, such a membrane potential does not lead to a significant accumulation of IAA in the absence of a pH gradient (i.e., in the presence of nigericin, Fig. 6). The above observations are inconsistent with the proposal that a transmembrane voltage gradient provides the additional driving force needed to account for IAA accumulation by vesicles.

The nonspecific component of the ΔpH-dependent IAA uptake (i.e., the fraction that is nonsaturable by IAA and insensitive to NPA) can be accounted for from the size of the pH gradient and does not appear to be affected by increasing ionic strength.

On the other hand, the magnitude of the saturable, NPA-enhanced component of IAA accumulation exceeds that which can be accounted for on the basis of the pH gradient. Perhaps intercalation of amphipathic IAA into the bilayer or binding at a specific site is responsible for this excess specific accumulation rather than carrier-mediated uptake. According to this interpretation, the sensitivity of the saturable component to the external ionic strength might be due to conformational changes in a membrane protein, perhaps caused by changes in the surface potential. The transient stimulation of IAA uptake by a K^+ diffusion potential suggests that a stable negative membrane potential could contribute to the steady-state IAA accumulation; however, another possibility is that the magnitude of the transmembrane potential could also influence the binding of IAA at a specific, NPA-stimulated site.

This work was supported in part by a grant from the U.S. Department of Energy (DOE 85ER1337).

REFERENCES

Andersen OS, Feldberg S, Nakadomari H, Levy S, McLaughlin S (1978) Biophys J 21:35
Beeler TJ, Farmen RH, Martonosi AN (1981) J Membrane Biol 62:113
Bradford MM (1976) Anal Biochem 72:248
Clark KA, Goldsmith MHM (1986) In: Bopp M (ed) Plant Growth Substances 1985. Springer-Verlag, Berlin/Heidelberg, p 203
Goldsmith MHM (1977) Annu Rev Plant Physiol (Bethesda) 28:439
Gutknecht J, Walter A (1980) J Membrane Biol 56:65
Hertel R (1983) Z Pflanzenphysiol 112:53
Hertel R (1986) In: Bopp M (ed) Plant Growth Substances 1985. Springer-Verlag, Berlin/Heidelberg, p 214
Hertel R, Lomax TL, Briggs WR (1983) Planta (Berl) 157:193
Lomax TL, Mehlhorn RJ, Briggs WR (1985) Proc Natl Acad Sci USA 82:6541
McLaughlin S (1977) Curr Top Membranes Transport 9:71
Raven JA (1975) New Phytol 74:163
Rottenberg H (1979) Methods Enzymol 55:547
Sussman MR, Goldsmith MHM (1981a) Planta (Berl) 151:15
Sussman MR, Goldsmith MHM (1981b) Planta (Berl) 152:13
Thomson K-St, Hertel R, Miller S, Tavares JE (1973) Planta (Berl) 109:337
Waggoner AS (1979) Annu Rev Biophys Bioeng 8:47
Weigl J (1969) Z Naturforsch 24b:365
Zimmermann U, Ashcroft RG, Coster HGL, Smith JR (1977) Biochim Biophys Acta 469:23

MODIFICATIONS OF AUXIN EFFLUX CARRIER IN THE AUXIN TRANSPORT SYSTEM BY DIETHYL ETHER AND ETHYLENE

B.G. Kang
Department of Biology, Yonsei University
Seoul 120, Korea

1. INTRODUCTION

A variety of agents in vapor or in gas phase have attracted the attention of some early investigators for their growth-modifying effects in plants. Among these, Schroeder (17, see also 23) noticed a temporary stimulation of growth by low concentrations of ether vapor. Since Neljubov (12) first showed that growth of pea seedlings was affected by ethylene, ethylene research was carried out by many workers thereafter (see 1). Ether and ethylene (2), however unrelated they may be to each other functionally, are likely to influence plant growth through an interference with the auxin transport system. In an extensive study, van der Weij (21) showed that polar auxin movement in Avena coleoptile segments was inhibited ("suspended") when 40 % ether-saturated water was placed in the test chamber. Several studies (e.g. 2) have demonstrated ethylene interference with auxin movement.

Transport of indoleacetic acid (IAA) as a cell-to-cell transfer involves at least two separate processes associated with the plasma membrane leading to cellular entry and exit of auxin molecules (9,16,18). Although auxin uptake is also thought to be mediated by a saturable carrier, auxin efflux carriers preferentially localized on the cellular basal region of the plasma membrane (10) have drawn special attention partly because the efflux process is a rate limiting component and is sensitive to such auxin transport inhibitors as 1-N-naphthylphthalamic acid (NPA) and 2,3,5-triiodobenzoic acid (TIBA) (3,7,20). It was proposed (3) that the efflux carrier complex has regulatory sites that bind NPA and TIBA.

In the present work, effects of ether vapor and ethylene on auxin transport were compared with changes of binding or accumulation of auxin and NPA to isolated membranes, induced by these agents, either in vitro (ether) or in vivo (ethylene) in an effort to support the physiological relevance of the possible carriers.

NATO ASI Series, Vol. H10
Plant Hormone Receptors. Edited by D. Klämbt

2. MATERIALS AND METHODS

2.1. Plant Materials

Seeds of zucchini squash (Cucurbita pepo L. cv. Black Beauty) were planted in wet vermiculite and grown for 6 days. Seeds of mung bean (Vigna radiata L. Wiltzeck) were soaked in running water for about 6 h, planted on 0.4% agar-gel bed in a plastic box, and grown for 3 days. Presoaked pea (Pisum sativum cv. Alaska) seeds were planted in wet vermiculite and grown for 7 days. All plants were grown in complete darkness at 24°C.

2.2. Radiochemicals

(2,3,4,5-^{3}H)-NPA (16.4 Ci/mmole) and (5-^{3}H)-IAA (26.6 Ci/mmole) were obtained from CEA, Gif-sur-Yvette, France.

2.3. In Vivo Treatments

For treatment with ether vapor during an auxin transport period, ether-saturated water was diluted with water to various concentrations, and 10 ml of this solution was placed into a separate well in a glass container (volume, 760 ml) with a glass lid to produce appropriate concentrations of ether vapor inside the container. Twenty percent ether-saturated water was estimated to give rise to approximately 5% ether vapor (V/V) inside the container where auxin transport tests were performed.

For ethylene treatment of intact pea plants, seedlings in a pot were placed in a large glass desiccator, and ethylene was injected into the desiccator to make an ethylene concentration of 100 ul/l.

For pre-incubation of excised pea internode tissue, 500 g of internode segments were incubated in 250 ml Na-phosphate buffer (0.05 M) at pH 6.8 containing 2% sucrose with or without a test substance in a 500 ml Erlenmeyer flask. They were incubated on a shaker at room temperature in daylight.

2.4. Auxin transport

Auxin transport experiments were carried out at room temperature in daylight using 3 mm segments of either mung bean hypocotyls or pea third internodes. The segments were cut from the tissue subjacent to the apical hook with a multiple-blade cutter. Agar (1.5%) blocks (4 mm x 4 mm x 1 mm) containing 100 nM ^{3}H-IAA in 50 mM Na-phosphatase buffer (pH6) were applied on the apical cut surface of vertically positioned segments as doner, and plain buffered agar blocks applied to the basal end as receiver. At the end of a transport period, radioactivity in the receiver blocks was counted. Data presented are the average from 10 segments.

2.5. Particle Preparation

All procedures were carried out at 0-4°C in daylight according to the method described by Jacobs and Hertel (11). Chopped and ground tissue material in extraction medium (2 x ml/g fr. wt. of tissue) which consisted of 0.25 M sucrose, 50 mM tris-acetate at pH 8, 1 mM EDTA, 0.1 M $MgCl_2$ and 1 mM DTE, were squeezed through a nylon cloth, and the filtrate was centrifuged at a low speed (3,700 g) for 10 min. The resulting supernatant was centrifuged at a high speed (200,000 g) for 20 min in a Beckman Ti 50 rotor. The supernatants were discarded, and the pellets resuspended in binding assay medium (ml/g fr. wt. of tissue) with a glass homogenizer. The assay medium for IAA association was 0.25 M sucrose, 10 mM MOPS at pH 6.5, 0.5 mM $MgCl_2$ and 1 mM DTE, and that for NPA binding was 0.25 M sucrose, 10 mM Na-citrate at pH 5,5, 0.5 mM $MgCl_2$ and 5 mM $MgSO_4$.

2.6. Assays for Binding or Accumulation

NPA binding was tested as described by Ray (14). ^{3}H-NPA was added to the binding mixture at 5 x 10^{-10}M with or without 10^{-5}M or appropriate concentrations of cold NPA.

Either 1 ml or 2 ml of resuspended particles in the appropriate binding assay medium was used for binding assay. ^{3}H-IAA association was assayed according to the method described for ^{14}C-NAA binding (15). Labelled IAA was added to the mixture (2 x 10^{-9}M) with or without 10^{-4}M or appropriate concentrations of unlabelled IAA. In the present experiments, 2, 3, 5-TIBA and phenylacetic acid (PAA) were added to the binding mixture (3 x 10^{-6}M and 10^{-6}M respectively), because TIBA was found to increase site III type IAA association (11) and PAA was known to saturate auxin receptors at site I (5).

"Specific binding" denotes the difference in radioactivity between the pellets, minus and plus, cold IAA or NPA added.

3. RESULTS AND DISCUSSION

3.1. Promotion of Auxin Transport and Modulation of Auxin Carriers by Ether

Two striking effects of ether were found: in mung bean hypocotyl segments auxin transport was strongly promoted by low concentrations of ether vapor during a 2 h transport period (Table 1),

Table 1. Effect of Ether Vapor on Auxin Transport in Mung Bean Hypocotyl Segments.

Ether Vapor (Percent, V/V)	Transport (cpm)
0	652
1.25	1045
6.25	1129
12.50	866

and on the other hand, the affinity of specific NPA binding to mung bean membranes was 4 times increased by adding 0.18 % ether to the test medium (Fig.1, right).

Table 2. In vitro Effect of Ether on saturable association of IAA and NPA to Zucchini Membranes.

Ether (Percent, W/W)	Saturable Association (cpm) ^{3}H-IAA	^{3}H-NPA
0	3029	833
0.06	2989	850
0.18	1963	1010
0.61	1879	1134

Similar effects of ether vapor on auxin transport were observed in zucchini hypocotyl segments (data not shown). At high ether concentrations, transport was inhibited in mung bean and in _Cucurbita_ as had been reported for _Avena_ (21). However, in the two early studies on oat coleoptiles (21,23), a promotion of auxin transport at lower ether concentrations had not been observed.

With membranes from zucchini hypocotyls, saturable binding of ^{3}H-NPA as well as association of ^{3}H-IAA was found to be modified by ether added to the binding mixture. Data presented in Table 2 indicate that the effect on NPA binding is similar to that with mung bean vesicles.

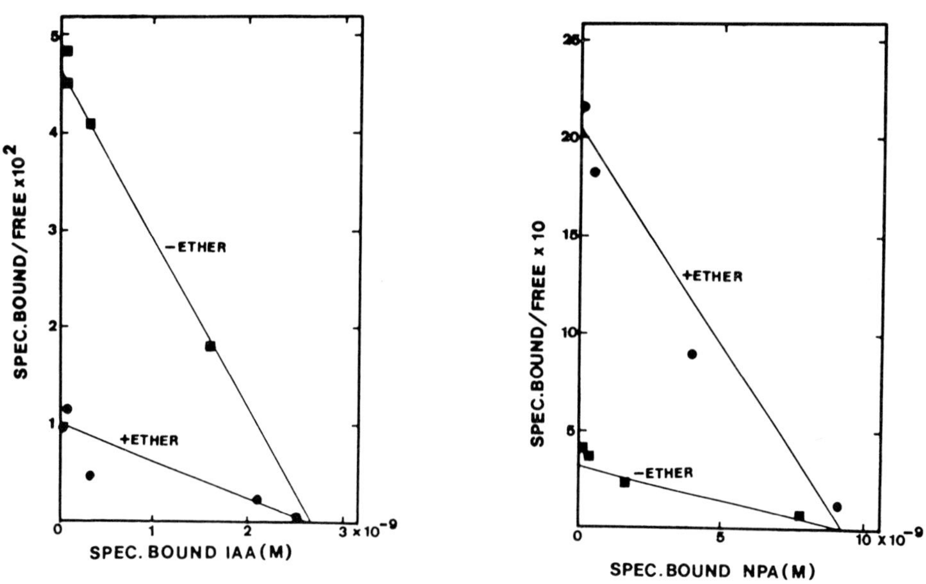

Fig.1. Scatchard plots. ^{3}H-IAA asasociation to zucchini membranes (left); ^{3}H-NPA binding to mung bean membranes (right) in presence (circles) or absence (squares) of ether (0.18%, W/W)

The Scatchard analysis in Fig.1 (right), clearly indicates that membrane treatment with ether results in a decreased K_D value for NPA binding with the number of binding sites being unchanged. In the case of IAA association (Fig.1, left), the previously held notion (11) of a specific site-III-type auxin binding to receptors on the plasma membrane was reinterpreted as being a pH driven accumulation in vesicles (9); thus only an "apparent K_D" - increased by ether - can be given if we are dealing with an IAA uptake into vesicles (9), in case of a binding however, the interpretation would be like that for NPA binding.

Ether could decrease IAA accumulation in membrane vesicles in different ways: it might cause leakage, or block uptake carriers , or else - as the most interesting possibility - it might enhance efflux by modifying the efflux carrier.

The opposite phenomena, at least superficially, caused by phytotropins and TIBA are well documented. The transport inhibitors are known to bring about net accumulation of auxin in tissue segments (19) and in vesicles (9,11) by specifically blocking the auxin exit at the efflux carrier localized on the plasma membrane.

The use of both Cucurbita and mung bean for different aspects of this study on ether effects needs some comment. Mung bean material showed the most pronounced effects of ether on in vivo transport (Table 2) and on NPA binding (Fig.1, right). But probably due to some characteristic feature of the Cucurbitaceae in qualitative chemistry of their membrane lipids (see 6), outside-out vesicles derived from the zucchini plasma membrane seem to have some remarkable behavior in that they can sustain imposed pH gradients to a considerable extent in time (4), and that weak organic acids including auxin can be accumulated in the vesicles with such pH gradients as a driving force (9). For this reason, it happened that zucchini membranes served for site III type auxin association. Although the auxin binding activity was detected in membranes from mung bean hypocotyls, it was nowhere near comparable to that for zucchini hypocotyls (data not shown). Therefore zucchini membranes were employed for kinetic studies of auxin association with regard to ether, and mung bean membranes for NPA binding respectively. Nevertheless the most significant ether effects - promotion of transport (Table 1) and increasing NPA binding (Fig.1, right) were compared with the same mung bean hypocotyls.

Ether is known for long as an anesthetic agent, and its anesthetic action is believed to be related to its possible modification of the plasma membrane with its lipid solubility (22). Ether could possibly modify either the fluidity or the volume of the membrane in such a way to bring about a more "exposed" state of the carrier protein. This idea is consistent with the fact that ether increased NPA binding affinity, while it decreased IAA association.

3.2 Inhibition of Auxin Transport and Reduction of Auxin Efflux Carrier by Ethylene Treatment

Auxin transport capacity in pea third internode segments is progressively decreasing as the period of ethylene treatment of intact seedlings prior to the excision of segments is increasing (Table 3). It is also indicated, however, that when ethylene is applied to excised segments during incubation of the tissue in buffer, no significant effect of ethylene can be detected on subsequent auxin transport (Table 3, lower half). Lack of ethylene effect on auxin transport in isolated pea stem segments was reported earlier (2). The data also show that auxin transport capacity is significantly reduced by merely incubating excised tissue segments in buffer.

Table 3. Effect of Ethylene Pretreaments of Intact Pea Seedlings or Excised Segments on Subsequent Auxin Transport. 3 mm segments were cut for the transport test from intact seedlings. In the case of excised tissue, 1 cm segments were preincubated, and at the end of the appropriate preincubation period, 3 mm segments were cut from their mid portion for auxin transport tests.

Modes of Treatment	Transport (cpm)	
Pretreatment (h) of Intact Seedings	0-2 h	2-4 h
0	467	1805
2	316	1348
4	149	1042
8	143	871
Preincubation (h) of Excised Segments with (+) or without (-) Ethylene	0-1	
0	207	
2 -	154	
2 +	160	
4 -	159	
4 +	139	
6 -	170	
6 +	146	

NPA binding activity in membranes from pea seedlings pre-treated with ethylene for various periods as shown in Table 4 indicates that, in parallel with <u>in vivo</u> auxin transport, NPA

Table 4. NPA binding to Membranes from Pea Seedlings Pretreated with Ethylene for Various Periods.

Ethylene Pretreatment (h)	Specific Binding (cpm)
0	301
2	288
8	229
16	153

binding activity is likewise decrasing with increasing ethylene pretreatment periods. Data in Fig.2 (left) indicate that ethylene pretreatment leads to a decrease in the number of NPA binding sites

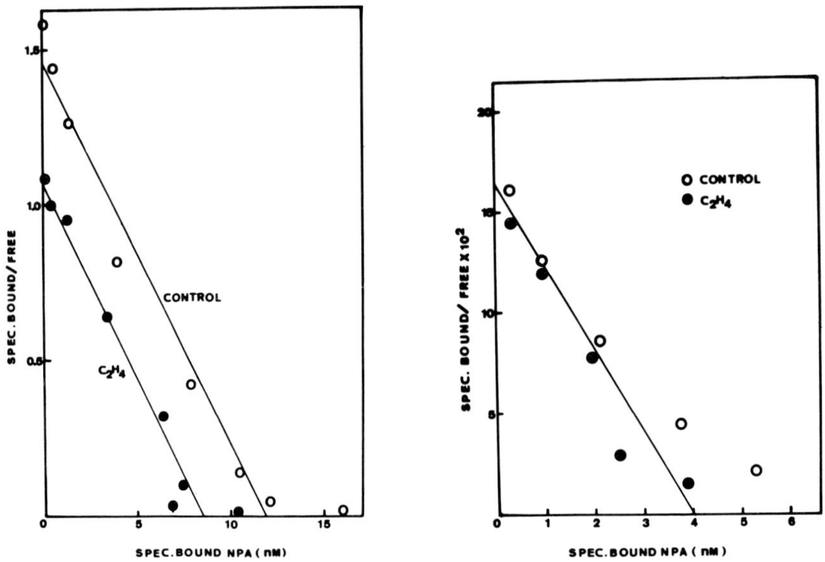

Fig. 2. Scatchard plots of ^{3}H-NPA binding to membranes. Left: from intact pea seedlings pretreated with or without ethylene for 16 h. Right: From excised segments preincubated in buffer with or without ethylene for 8 h.

without affecting the affinity.

These results establish a close, positive correlation between auxin transport on one hand, and NPA binding on the other hand. It should be pointed out, however, that the decrease in transport is seen earlier (at 4 h) than the large difference in NPA sites (at 16 h). This could be explained if small differences of carrier number (at early time) resulted in an amplified difference expressed by the repetetive cell-to-cell transport.

If the gradual decline in the number of NPA binding sites by ethylene reflects loss of the auxin efflux carrier protein, then outside-out membrane vesicles from ethylene pretreated seedlings would have a reduced number of the exit sites for auxin, and thus should accumulate more auxin (increased ^{3}H-IAA association) than vesicles from non-treated seedlings. This has been verified in zucchini membranes (data not shown).

The ethylene effect cannot be manifested if the gas is applied to isolated segments (Fig. 2, right, and Table 3 , lower part). This correlates well to the lack of ethylene effect on auxin transport under similar conditions. The exception is a report on ethylene promotion of IAA transport in excised petiole segments of the semi-aquatic Ranunculus sceleratus (13).

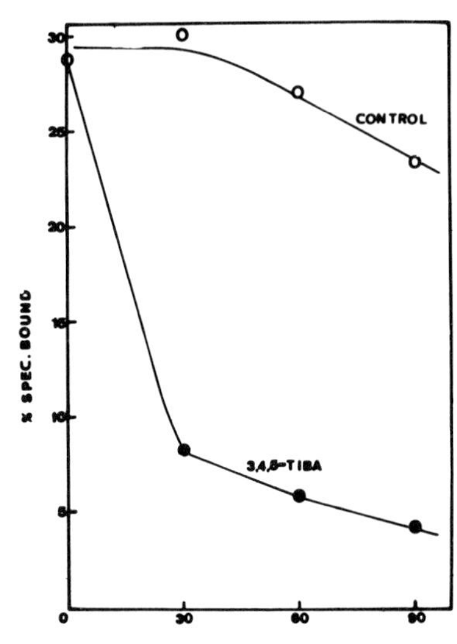

Fig. 3. Time course for decreases in NPA binding activity in membranes from pea tissue preincubated with or without 0.1 mM 3,4,5-TIBA.

3.3 Effects of Pretreatment With Inhibitors

2,3,5-TIBA and NPA both inhibit auxin transport strongly and rapidly (20). The TIBA analog , 3,4,5-TIBA, is less potent even at saturating concentrations (3). In this regard, it is more like "Kartoffelfaktor" (=KF), a naturally occuring ligand from potato tubers competing for NPA receptors (W. Petersen, unpubl., see in 8). Time course data in Fig.3 show that NPA binding activity in pea segments is decreasing with time during incubation of the tissue in buffer, and that presence of 3,4,5-TIBA greatly and rapidly abolishes binding activity. In preliminary experiments, this decrease appeared to be due to a reduced number of NPA binding sites. KF treatment produced a similar effect (data not shown).

Table 5. Partial Recovery of NPA binding from 3,4,5-TIBA Pretreated Pea Segments by Successive Washing and Repelleting

Number of Washing	Specific Binding (cpm)
0	371
1	505
2	774
3	778

Since NPA binding is known to be inhibited by 3,4,5-TIBA with a K_i 10^{-6}M (see 3), particles prepared from 3,4,5-TIBA pretreated tissue were washed and repelleted successively with the test medium to check the possibility that reduction of NPA sites was due to persistent 3,4,5-TIBA on the NPA sites during extraction and particle preparation. Data in Table 5 show partial recovery of binding activity by washing the pellets twice with no further recovery thereafter. However, the extent of recovery was only fractional (see Fig.3), and moreover, NPA binding activity was gradually decreasing during preincubation in non-treated segments as well. Probably this inherent decrease in NPA binding activity upon excision might explain the insensitivity of excised tissue segments to ethylene.

Further research is warranted to elucidate a mechanism by

which ethylene reduces NPA binding sites i.e. auxin carriers in the plasma membrane. Pea tissue extracts as well as those from corn coleoptiles, were shown to have KF-type activity; but this activity was not found to be influenced by ethylene treatment.

Acknowledgements. I thank R.Hertel, Institut Biologie III, Univ. Freiburg i.Br. FRG, in whose laboratory this work, supported by the "Alexander-von-Humboldt-Stiftung", was initiated.

4. REFERENCES

1. Abeles FB (1973) Ethylene in Plant Biology. Academic Press, New York
2. Burg SP, Burg EA (1967) Plant Physiol 42:1224
3. Depta H, Eisele KH, Hertel R (1983) Plant Sci Letts 31:181
4. Depta H, Hertel R (1982) In: Marme' D, Marre` E, Hertel R (eds) Plasmalemma and Tonoplast. Elsevier, Amsterdam p137
5. Dohrmann U, Hertel R, Kowalik H (1978) Planta 140:97
6. Goad LJ (1977) In: Telvini M, Lichtenthaler HK (eds) Lipids and Lipid Polymers in Higher Plants. Spinger-Verl, Heidelberg, p 146
7. Goldsmith MHM (1982) Planta 155:68
8. Hertel R (1981) Biochem Physiol Pflanzen 176:495
9. Hertel R, Lomax TL, Briggs WR (1983) Planta 157:193
10. Jacobs M, Gilbert SF (1983) Science 220:1297
11. Jacobs M, Hertel R (1978) Planta 141:1
12. Neljubov D (1901) Beih Bot Zentralbl 10:128
13. Musgrave A, Walters J (1973) New Phytol 72:783
14. Ray PM (1977) Plant Physiol 59:594
15. Ray PM, Dohrmann U, Hertel R (1977) Plant Physiol 59:357
16. Rubery PH (1977) Planta 152:74
17. Schroeder H (1908) Flora 99:156
18. Sussman MR, Goldsmith MHM (1981) Planta 151:15
19. Sussman MR, Goldsmith MHM (1981) Planta 152:13
20. Thomson KS, Hertel R, Müller S, Tavares JE (1973) Planta 109:337
21. Van der Weij HG (1934) Rev trav bot neerl 31:810
22. Winter PM, Miller JN (1985) Sci Amer 252:94
23. Witt J, Söding H (1965) Planta 65:232

SEARCH FOR ENDOGENOUS LIGANDS TO FUSICOCCIN BINDING SITES.

A. Ballio and P. Aducci.
Department of Biochemical Sciences, University of Rome "La Sapienza", Piazzale Aldo Moro 5, 00185 Rome, Italy.

The search for endogenous ligands to fusicoccin (FC) binding sites has been stimulated by an apparent paradox. In fact, while FC binding sites have been detected in all tissues of higher plants so far examined, in agreement with the largely aspecific _in vivo_ activities of FC, the distribution of the compound in Nature is very restricted (Aducci et al., 1982). To the best of our knowledge FC is only present in almond and peach trees infected by _Fusicoccum amygdali_ Del.(Ballio et al., 1976), a fungus which in culture produces FC and a number of metabolites structurally related to it (Ballio, 1981). Furthermore, cultures of an unclassified fungus produce a family of slightly modified fusicoccins, called cotylenins (Sassa, 1970), which have biological properties identical to those of FC. A recent claim that immature cobs from a soviet maize cultivar contain a biologically inactive derivative of FC yielding FC on acid hydrolysis (Muromtzev et al., 1980) could not be confirmed with a western cultivar (Aducci et al., 1985).

This contradictory situation can be rationalized on the assumption that higher plants contain one or more endogenous ligands for FC binding sites having the plausible role of modulators of the H^+/K^+ pump known to be stimulated by FC. This hypothesis was supported by the demonstration that immersion of maize root segments in a large volume of water increases H^+-excretion, K^+-uptake, hyperpolarization of the cell membrane potential and ATP-ase activity (Leonard and Hanson, 1972 a, b; Lin and Hanson, 1974); these effects, all strongly reminiscent of a FC treatment, might well be due to the removal of a substance capable of interacting with FC binding sites (Gronewald et al., 1979).

NATO ASI Series, Vol. H10
Plant Hormone Receptors. Edited by D. Klämbt

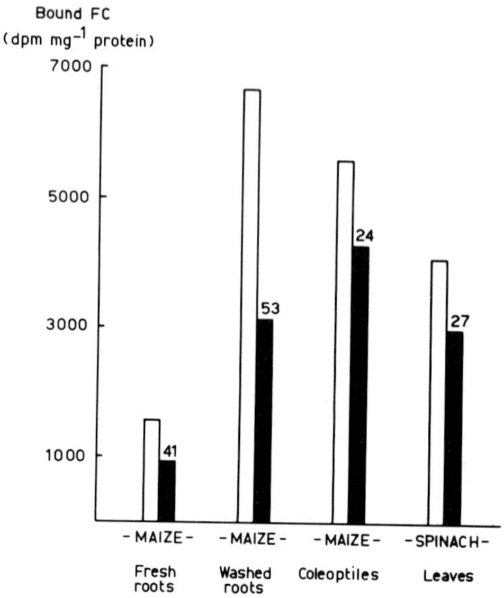

Fig.1. Effect of maize root "wash" on the binding of FC to microsomal preparations of maize roots, maize coleoptiles and spinach leaves. Open columns, no addition; black columns, maize root "wash" added.

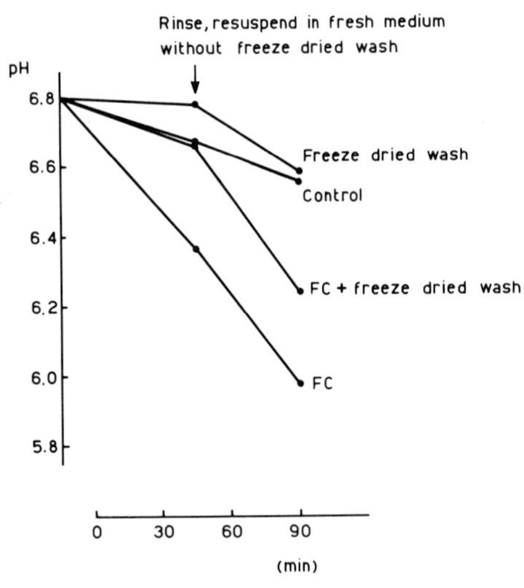

Fig.2. Effect of maize root "wash" on H^+-excretion from maize root segments. The arrow indicates the time at which tissues are removed from incubation medium, rinsed and incubated in the absence of "wash".

In order to obtain more direct evidence in favour of the above hypothesis we tested (Aducci et al., 1980) the binding activity of microsomal fractions prepared from maize roots "washed" according to Hanson and coworkers, and compared it with that of similar fractions prepared from untreated roots. As shown in Fig. 1, 5-6 times more FC is bound after "washing", this increment being strongly reduced on adding back the "wash" to the binding assay mixture. This effect is not limited to maize roots, since "washing" also reduces the amount of FC bound to microsomal fractions of maize coleoptiles and of spinach leaves. It was furthermore observed that the crude "wash" inhibited the FC-promoted H^+- excretion from maize roots. This inhibition could be reversed by transferring the tissue to a fresh incubation medium without added "wash" (Fig. 2).

The above results represented a heavy support to the hypothesis that endogenous ligand(s) for FC binding sites exist in higher plants. The obvious next step was the isolation and structure determination of these ligands. We were aware that the isolation of substantial amounts of these products would involve considerable difficulties, since it is predictable that these ligands, unless their affinity is much lower than that of FC, will be present in a tissue at a concentration similar to that of FC binding sites, namely picomoles g^{-1} fresh weight. Since the molecular weight of the endogenous ligand is lower than 1,000 daltons its content in the fresh tissue should not be higher than 1 $\mu g\ kg^{-1}$. The _in vitro_ FC binding assay, which is capable of detecting these very low quantities in a competition test, has permitted the development of procedures suitable for the purification of the endogenous ligand(s) starting from 0.5 ÷ 1.0 kg batches of maize roots. The specificity of the assay, already demonstrated relative to the plant hormones (Ballio et al., 1981), was further confirmed by the negative results obtained with several phytotoxins and plant growth substances (Table 1).

Table 1. Substances not competing with FC in the binding assay.

Abscisic acid	Benzoxazolinones
Auxins	Brassinolide
Cytokinins	Jasmonic acid
Gibberellic acid	Turgorins
1-N-Naphtylphtalamic acid (NPA)	
Cercospora beticola toxin	
Helminthosporium maydis toxin	
Ophiobolins	

A concentrated solution containing the active substances was prepared from roots of hydroponically grown maize plantlets either by extraction with warm ethanol followed by removal of the solvent, or by expression in a Braun multipress to give a liquid which was depleted of proteins by a short treatment at 80°C. In both cases the extracts were suitable for fractionation by reversed phase chromatography. This fractionation was carried out in three steps, a coarse one, followed by a semipreparative and a high performance run, which eventually afforded a tiny symmetrical peak. One kilo batches of maize roots yielded a barely visible but highly active residue which, besides competing efficiently with FC in the binding assay, discharged very rapidly all radioactivity from solubilised FC binding sites which had been saturated with tritiated FC.

Up to now the amount of this purified and apparently homogeneous product made available by the above procedures is too small for its complete chemical and biological characterization. A Fourier transform infra red spectrum and an electron impact mass spectrum gave structural informations of very little help. At variance with the crude maize extract the purified product, at least at the low concentrations tested, was unable to inhibit *in vivo* the FC stimulated H^+-excretion; rather it slightly stimulated this process. Therefore we examined all other fractions eluted from the first reversed phase column for their effect on FC promoted H^+-excretion and found that this process is inhibited by fractions well separated from the endogenous ligand for FC binding sites and eluted in the region of highly polar substances.

In conclusion we presently know that extracts of maize roots contain a ligand for FC binding sites and a product which inhibits *in vivo* FC stimulated H^+-excretion without competing *in vitro* with FC binding. The two substances can be extracted, separated and purified, but their minute concentration in the tissue has hampered chemical and biological characterization. Further efforts are required for preparing the amounts necessary to achieve this characterization.

Acknowledgements. The work was supported by the Italian National Research Council (C.N.R.) and the Ministry of Public Education (Ministero Pubblica Istruzione). We thank Dr. A. Evidente, Prof. G. Randazzo (University of Naples), Prof. F. Gasparrini, Prof. D. Misiti and Dr. M. Pierini (University of Rome "La Sapienza") for skilful collaboration in the preparation of the purified endogenous ligand, and Ms. F. Maffei for her excellent assistance in plant growth, tissue extraction and bioassays.

REFERENCES.

Aducci P, Ballio A, Evidente A, Federico R, Iasiello I, Marra M, Randazzo G (1985) Phytochem 24:1097

Aducci P, Ballio A, Federico R, Montesano L (1982) In: Wareing PF (ed) Plant Growth Substances 1982. Academic Press, London, p 395

Aducci P, Crosetti G, Federico R, Ballio A (1980) Planta 148:208

Ballio A (1981) In: Durbin RD (ed) Toxins in Plant Disease. Academic Press, New York, p 395

Ballio A, D'Alessio V, Randazzo G, Bottalico A, Graniti A, Sparapano L, Bosnar B, Casinovi CG, Gribanovski-Sassu O (1976) Physiol Plant Pathol 8:163

Ballio A, Federico R, Scalorbi D (1981) Physiol Plant 52:476

Gronewald JW, Cheeseman JM, Hanson JB (1979) 63:255

Leonard RT, Hanson JB (1972a) Plant Physiol 49:430
Leonard RT, Hanson JB (1972b) Plant Physiol 49:436
Lin W, Hanson JB (1974) Plant Physiol 54:799
Muromtzev GS, Kobrina NS, Voblikova VD, Koreneva YM (1980) Izv Akad Nauk SSSR, Ser Biol 2:897
Sassa T (1970) Agric Biol Chem 34:1588

THE INTERACTION OF FUSICOCCIN WITH SPECIFIC BINDING SITES.

P. Aducci and A. Ballio.

Department of Biochemical Sciences, University of Rome "La Sapienza", Piazzale Aldo Moro 5, 00185 Rome, Italy.

Introduction.

Fusicoccin (FC) is the major phytotoxic metabolite of the fungus *Fusicoccum amygdali* Del. (Ballio et al., 1964), a pathogen of peach and almond trees. Its structure, elucidated by Ballio et al., (1968), is reported in Fig. 1.

FC is present in host plants infected by the fungus, but as far as we know is foreign to other plants. It stimulates growth by cell extension in a number of tissues of higher plants, as well as movement of stomata and seed germination, and has become a useful tool for investigating several important physiological processes controlled by phytohormones (Marrè, 1979). Both the pathological and physiological responses may be considered as a consequence of the activation of a H^+/K^+ exchange system driven by a plasma membrane ATPase. The direct *in vitro* activation of the plasmalemma H^+-ATPase by FC is still an open question. Recently, a stimulating effect of FC on radish plasmalemma H^+-ATPase has been reported (Rasi-Caldogno and Pugliarello, 1985), but other experiments (Cleland, 1985)

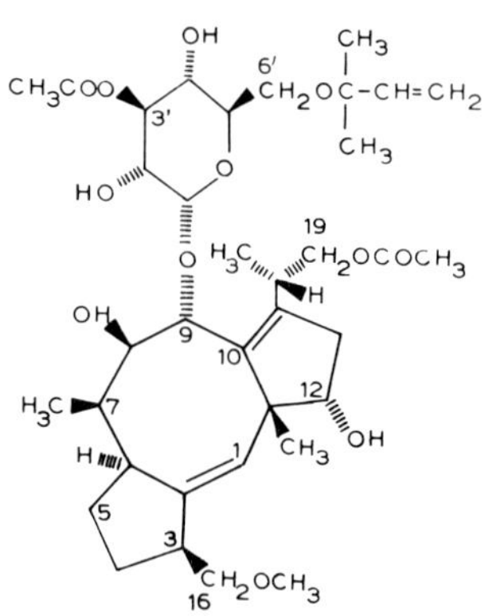

Fig. 1. Structure of fusicoccin.

NATO ASI Series, Vol. H10
Plant Hormone Receptors. Edited by D. Klämbt

in reconstituted pea root vesicles indicate that FC lowers the apparent Km for ATP of H^+ pumping without influencing the rate of hydrolysis of ATP.

The primary activation of the plasmalemma ATPase is supported, among other findings, by the occurrence of high affinity binding sites able to interact specifically with FC in plasmalemma-enriched fractions of all tissues so far examined (Dohrmann et al., 1977; Ballio, 1979; Pesci et al., 1979; Ballio et al., 1980; Stout and Cleland, 1980). It is reasonable to think that these sites are responsible for the first event in the mechanism of action of FC.

In the present paper results dealing with various aspects of their interaction with FC are surveyed and more recent work carried out in our laboratory is also discussed.

Characterization of FC binding sites.

For the studies on FC binding sites tritiated dihydro-FC has been used, taking advantage of the observation that the biological activities of FC are unchanged after hydrogenation of the t-pentenyl group on C-6' of the FC glucosyl moiety (Aducci et al., 1982b).

The hypothesis that binding sites are receptors that mediate the physiological action of FC is also supported by the parallelism of physiological activities of a wide range of FC derivatives with their ability to compete with tritiated FC in binding studies (Ballio et al., 1981). The choice of a suitable tissue for binding studies has not created difficulties since FC binding sites have been detected in a wide number of tissues of different plants. Most of the studies have been performed with maize roots and shoots and with spinach leaves, but binding sites were also found in oat roots, pea internodes, zucchini hypocotyls, almond leaves and barley and tobacco leaf protoplasts (Aducci et al., 1982b and references therein). In all tissues the radioactive ligand is maximally associated with plasmalemma-enriched microsomal fractions and in various separation systems the radioactivity coincides with plasmalemma markers.

The direct proof for a surface membrane localisation of FC binding sites is supported by different pieces of evidence. The first comes from the observation that tobacco leaf protoplasts, briefly exposed to a non-penetrating FC derivative, rapidly agglutinate when treated with anti-FC antibodies capable of recognizing structural features of the ligand not involved with the binding site (Aducci et al., 1980). The other evidence derives from the observation that proteins released after rehydration of osmotically shocked oat leaf protoplasts contain a component which specifically bind FC (Rubinstein, 1982).

FC binding sites present in microsomal preparations of maize coleoptiles were the first to be characterised. Their binding is a slow process, dependent on temperature and pH. The protein nature of the binding sites is suggested by their sensitivity to agents such as trypsin, heat and mercurials (Dohrmann et al., 1977; Pesci et al., 1979). The apparent molecular weight is approximately 80,000 daltons. The thermodynamic properties of FC binding sites have been studied, mainly by Scatchard analysis which showed high binding affinities and low concentration of sites (Pesci et al., 1979; Ballio, 1979; Aducci et al., 1984b). The high affinity of the binding sites represents a reasonable explanation for the persistence of the phytotoxic symptoms in plants affected by the fungus (Fusicoccum amygdali Del.). The occurrence in maize coleoptiles of at least two classes of binding sites was indicated by biphasic Scatchard plots and better by treatment of the binding data with a non-linear least squares fitting program (Aducci et al., 1984b). The confidence on the absolute values of K_d (K_{d1}=3.3 ± 1.6 x 10^{-10}M; K_{d2}=1.3 ± 0.53 x 10^{-8}M) estimated by this computerized analysis is very high and is strenghtened by their independence on protein concentration. This fitting procedure provides reliable data and its use should be recommended in studies aimed to compare the thermodynamic properties of binding sites from different tissues and plants; in fact attention has been frequently drawn to the easy misinterpretations of biphasic Scatchard plots (Norby et al., 1980; Klotz, 1982). The measurement of the number of FC binding sites by the fitting procedure, which cannot be as accurate as that of the dissociation constants, confirms that minute amounts of them are present in plants, thus suggesting that their purification will likely meet with practical difficulties.

Solubilisation of FC binding sites.

The purification and biochemical characterization of FC binding sites required the solubilisation of membrane-bound FC binding proteins. FC receptors or complexes of FC with its receptors have been solubilised by a variety of methods and from different tissues.

Table 1. Solubilisation of membrane-bound FC binding sites.

TISSUE	METHOD	REFERENCE
Maize coleoptiles	chaotropic agents, DOC, Triton X-100	Pesci et al., 1979
Maize mesocotyls	acetone precipitation	Aducci et al., 1986, unpublished
Maize roots	acetone precipitation	Aducci et al., 1986, unpublished
Oat roots	Triton X-100	Stout and Cleland, 1980
Spinach leaves	acetone precipitation	Aducci et al., 1982a

The acetone precipitation method used by us for spinach receptor solubilisation (Aducci et al., 1984a) and, more recently, for binding sites present in maize mesocotyls and roots (unpublished results) has the great advantage to provide material capable to bind FC in buffer-soluble form and free of detergent or chaotropic agents. All preparations of solubilised receptors investigated so far had the apparent molecular weight of 80,000 daltons; the isoelectric point was 4.9 for both spinach and maize.

Most of the work of our group used soluble preparations of spinach leaves, a material particularly suitable to biochemical studies because of its availability in large amounts. When the procedure developed for the purification of FC binding proteins (Aducci et al., 1982a) was scaled up, the binding activity was lost. Studies directed to overcome this difficulty showed that phosphatase and/or α-mannosidase were responsible of this effect and that fluoride or molybdate had a good stabilizing effect. Thus, it appears that both phosphate group and α-mannosyl residues are features of the binding protein necessary for the interaction with the ligand. Consequently the hypothesis that a phosphorylation-dephosphorylation process may modulate the activity of FC receptors was considered. Such a modulation has been observed in several mammalian systems. It has recently been noted that auxin receptors display an improved binding activity in conditions that minimize phosphatase activity and encourage phosphorylation (Van der Linde et al., 1985; Mennes et al., this book). A similar mechanism might be shared by the wheat germ cytokinin binding protein, which has been shown to become phosphorylated by an endogenous kinase (Polya and Davies, 1983).

In order to investigate whether the inactivated FC receptors could be reactivated upon acting of endogenous protein kinases (PK), the PK activities present in spinach leaves have been characterized. Ca^{2+}-dependent PK and a C-type PK were detected in different cell compartments. Conditions suitable for enzyme activation were used in the attempt to reactivate phosphatase-inactivated FC binding sites. Unfortunately these attempts have been fruitless, thus suggesting that the phosphate groups present in the FC binding glycoprotein may be linked to sugar residues. Such a structural feature has been found in some animal glycoproteins (Neufeld and Ashwell, 1980), but has never been observed in plants. Further investigations are necessary to clarify this point.

Proteoliposomes as a model system for transduction of FC signalling.

As a first approach to the study of the molecular mechanism involved in the transduction of the primary FC signal to the cell machinery a model system consisting of solubilised FC binding sites entrapped into phospholipid vesicles has been developed. Proteoliposomes able to bind FC specifically were obtained by entrapping stabilised soluble preparations of spinach leaf microsomes into phosphatidylcholine vesicles by a "freeze-thaw" procedure (Aducci et al., 1986). It appears that the entrapped binding protein is preferentially outside-oriented since binding is lost on trypsin treatment. Detergent permeabilization experiments show that only 20% of the binding sites are inside-oriented.

The time-course of FC binding to proteoliposomes has been compared with those to membranes and to soluble sites (Fig. 2, left). At 27°C saturation is roughly similar for membranes and proteoliposomes, whereas soluble sites reach saturation 30 min before. The displacement rate of FC bound to proteoliposomes is intermediate between that of the other two systems. Furthermore, a minimal estimate of association and dissociation rate constants k_1 and k_2 for the FC-proteoliposomes complex, based on simple assumptions, gives a value of the equilibrium constant corresponding to 5×10^{-8}M, namely in the same range reported for microsomal preparations.

The pH-dependence of FC binding to proteoliposomes has also been compared to that of membranes and soluble sites (Fig. 2, right).

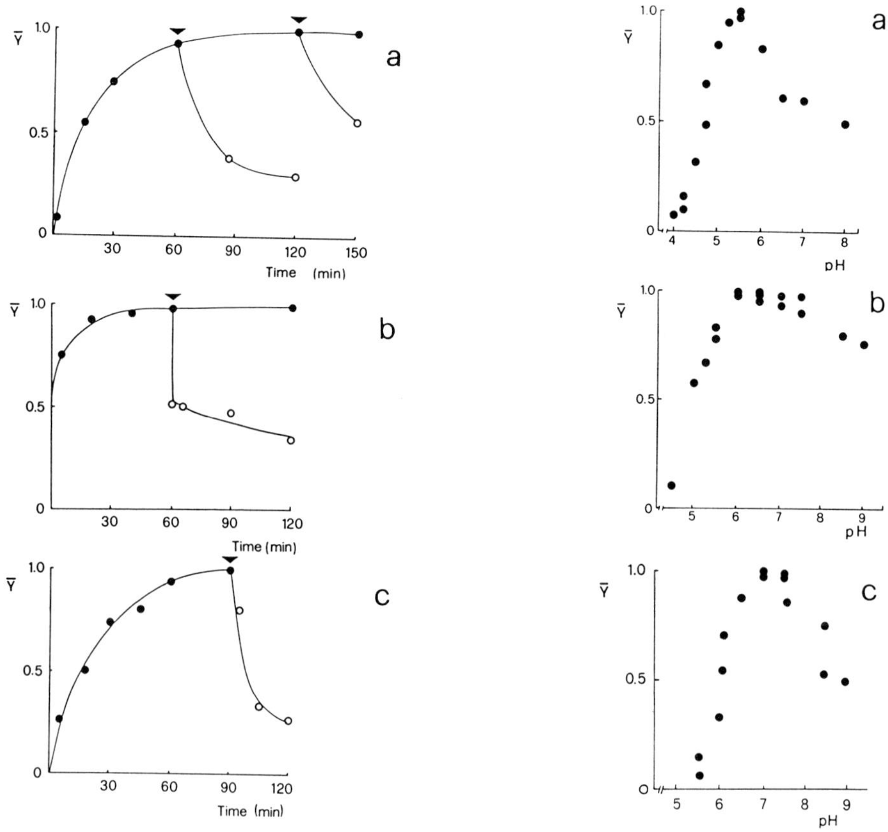

Fig. 2. Time-course (left) and pH-dependence (right) of FC binding to spinach leaf membranes (a) soluble sites (b) and proteoliposomes (c). Arrows in the time profiles indicate the addition of cold 10^{-5}M FC.

The proteoliposome pattern is comparable to that of microsomal preparations, but is shifted 1.5 units towards more alkaline values. The soluble sites instead have a broad pH optimum between 5.5 and 8.5.

Analysis of pH data according to Hill equation allows to calculate for the three systems the number of residues n responsible for the pH-dependent modulation of FC binding; these values are n=2.00, n=1.36, n=2.00 for the membrane, the soluble sites and the proteoliposomes, respectively. Moreover, the pK values corresponded to

4.65 for the membranes, 5.05 for the soluble sites and 5.95 for the proteoliposomes. These data indicate a higher hydrophobicity in the proteoliposomes than in the membranes, a likely consequence of a different phospholipid composition of the two systems. As for the soluble sites, their n and pK values might be influenced by the probable conformational changes following the removal of lipid components.

Acknowledgements. The work was supported by Italian Research Council (C.N.R.) and the Ministry of Public Education (Ministero Pubblica Istruzione). The protein kinase experiments were carried out with Dr. M. Marra and Dr. J.D. Walton, and those with liposomes with Dr. F. Persichetti and Ms. M.R. Fullone.

REFERENCES

Aducci P, Federico R, Ballio A (1980) Phytopath Medit 19:187

Aducci P, Ballio A, Federico R (1982a) In: Marmé D, Marrè E, Hertel R (eds) Plasmalemma and tonoplast. Elsevier, Amsterdam p 279

Aducci P, Ballio A, Federico R, Montesano L (1982b) In: Wareing PF (ed) Plant growth substances 1982. Academic Press, London p 395

Aducci P, Ballio A, Fiorucci L, Simonetti E (1984a) Planta 160:422

Aducci P, Coletta M, Marra M (1984b) Plant Sci Letts 33:187

Aducci P, Ballio A, Fullone MR, Persichetti F (1986) Plant Sci Letts in press

Ballio A (1979) In: Geissbühler H (ed) Advances in pesticide science pt 2. Pergamon Press, Oxford, New York p 336

Ballio A, Chain EB, De Leo P, Erlanger BF, Mauri M, Tonolo A (1964) Nature 203:267

Ballio A, Brufani M, Casinovi CG, Cerrini S, Fedeli W, Pellicciari R, Santurbano B, Vaciago A (1968) Experientia 24:631

Ballio A, Federico R, Pessi A, Scalorbi D (1980) Plant Sci Letts 18:39

Ballio A, Federico R, Scalorbi D (1981) Physiol Pl 52:476

Cleland RE (1985) Plant Physiol 77:S-87

Dohrmann G, Hertel R, Pesci P, Cocucci S, Marrè E, Randazzo G, Ballio A (1977) Plant Sci Letts 9:291

Klotz IM (1982) Science 217:1247

Marrè E (1979) Ann Rev Plant Physiol 30:273

Neufeld EF, Ashwell G (1980) In: Lennarz WJ (ed) The biochemistry of glycoproteins and proteoglycans. Plenum Press, New York p 252

Norby JG, Ottolenghi P, Jorgen J (1980) Anal Biochem 102:318
Pesci P, Cocucci SM, Randazzo G (1979) Plant Cell and Environ 2:205
Polya GM, Davies JR (1983) Plant Physiol 71:482
Rasi-Caldogno F, Pugliarello MC (1985) Biochem Biophys Res Comm 133:280
Stout RG, Cleland RE (1980) Plant Physiol 66:353
Van der Linde PLG, Maan AC, Mennes AM, Libbenga KR (1985) Proc 16th FEBS Congress VNU Science Press

ABOUT THE SEARCH FOR THE MOLECULAR ACTION OF HIGH-AFFINITY AUXIN-BINDING SITES ON MEMBRANE-LOCALIZED RAPID PHOSPHOINOSITIDE METABOLISM IN PLANT CELLS

B. ZBELL AND **C. WALTER**

BOTANISCHES INSTITUT, RUPRECHT-KARLS-UNIVERSITÄT HEIDELBERG
IM NEUENHEIMER FELD 360, D-6900 HEIDELBERG, FEDERAL REPUBLIC OF GERMANY

INTRODUCTION

One possible strategy for the clarification of the molecular action of auxin is founded on the so-called auxin receptor hypothesis. This hypothesis means that auxin acts only after its binding to a specific subcellular site, the auxin receptor. In other words, the auxin binding reaction should induce a hormone-specific effect. Though the detection of numerous auxin binding sites on membranes and soluble proteins from various plants upto now functions other than those of auxin transport sites are surely the most difficult to establish. As consequence the molecular mechanism of auxin action is one unsolved problem of plant cell biology concerning the mechanism of signal recognition and signal processing in the single plant cell (FIRN,1983; RUBERY,1984).

For the research on the molecular action of auxin suspensions of *in vitro* cultured cells of *Daucus carota* are certainly a suitable experimental system since their growth and differentiation are affected by the available auxin. In previous investigations (ZBELL,1983,1985) using membrane fractions prepared from carrot suspension cells high-affinity auxin-binding sites could be detected with an apparent dissociation constant $K_D = 5 \times 10^{-7}$ mol l^{-1} and a density of 1 pmol g^{-1} cell fresh weight or 100000 per cell, respectively (ZBELL,1983). Due to their enzymatic avtivities and buoyant density the membranes were identified as right-side-out vesicles of intracellular membranes probably containing those of the endoplasmic reticulum, Golgi apparatus or tonoplast. An ATP-dependent proton translocation leading to an intravesicular acidification could be detected as a property of the isolated membrane vesicles. The activity of the proton translocation *in vitro* could be stimulated either by auxins or by particular inhibitors of protein dephosphorylation. These findings were interpreted as an indication for an auxin-mediated protein phosphorylation. In connection with other findings, i.e. the auxin-mediated release of membrane-bound orthophosphate and the occurrence of a phospholipase C-like activity, the following hypothesis concerning the mechanism of the primary action of auxin was postulated (ZBELL,1983). Auxin does stimulate the activity of a membrane-associated phospholipase C (EC 3.1.4.10) leading to the hydrolysis of membrane-bound phosphoinositides (PI) resulting in the release of inositolphosphate and diacylglycerol . The inositolphosphate may be dephosphorylated by a phosphatase activity leading to the release of phosphate.

NATO ASI Series, Vol. H10
Plant Hormone Receptors. Edited by D. Klämbt

Diacylglycerol (1,2-DG) may act as a membrane-bound second messenger for the stimulation of a C-type similar protein kinase. Such a mechanism in plant cells may be similar to that occurring in animal cells, and it is well-known as the hormone-mediated phosphoinositide response (PI response) as postulated by MICHELL (1975) with a simultaneous protein phosphorylation (NISHIZUKA,1984) and Ca^{2+} mobilization from intracellular stores (BERRIDGE and IRVINE,1984). The complex view of the metabolic reactions involved in the PI response as it is established to occur in numerous animal cells as a consequence of the action of various hormones or transmitters is shown in Fig. 1 .

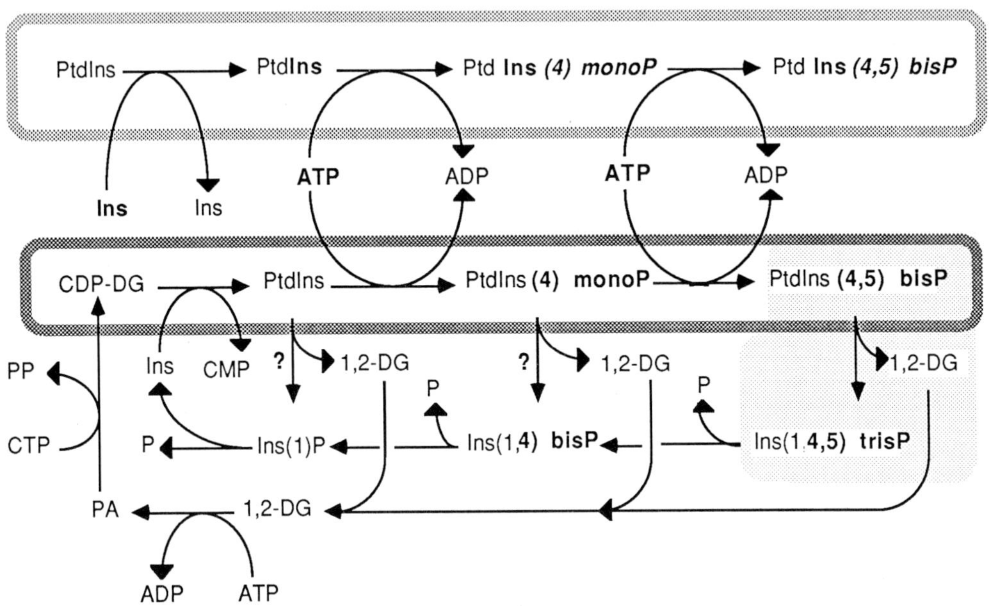

Fig.1. Pathways of phosphoinositide metabolism in an eukaryotic cell. Two disinct pools of PtdIns exist which derive either by an exchange reaction or by *de novo* synthesis. PtdIns of both pools can be converted by subsequent phosphorylations to generate PtdIns(4)monoP and PtdIns(4,5)bisP, but only the polyphosphoinositides generated by the*de novo* synthesis are the target substrates of the signal-dependent enzymatic cleavage and the release of Ins(1,4,5)trisP and 1,2-DG with second messenger functions in the cell (dotted area).

On the basis of recent experimental data a similar mechanism of the PI response is suggested to occur in plant cells. In general, membrane-bound phosphatidylinositol (PtdIns) is converted by the action of a PtdIns kinase (EC 2.7.1.67) to phosphatidylinositol (4) monophosphate (PtdIns(4)monoP), which is further phosphorylated to phosphatidylinositol (4,5) bisphosphate (PtdIns(4,5)bisP) by a PtdIns(4)monoP kinase (EC 2.7.1.68). These phosphorylated compounds of PtdIns are also called polyphosphoinosotides. After binding of an agonist to its receptor GTP is utilized by a so-called N-protein (LITOSCH and FAIN, 1986). As a consequence the receptor is functionally coupled to a phosphodiesterase, the phospholipase C (EC 3.1.4.10), and PtdIns(4,5)bisP is hydrolyzed to generate 1,2-DG and Ins(1,4,5) trisP, which act as second messengers for the stimulation of the membrane-associated protein kinase C (NISHIZUKA,1984) and the release of Ca^{2+} from intracellular stores (BERRIDGE and IRVINE,1984), respectively.

Evidence for the occurrence of a similar mechanism of the PI metabolism in plant cells was first presented by BOSS and MASSEL (1985) as well as HEIM and WAGNER (1985) showing the presence of all polyphosphoinositides in phospholipid extracts from suspension cells of *Daucus carota* and *Catharanthus roseus* , respectively, which were prelabelled *in vivo* with *myo* -[2-^{3}H]-inositol. However, WHEELER and BOSS (1986) reported that membranes prepared from prelabelled carrot suspension cells were found to contain PtdIns and PtdIns(4)monoP but not PtdIns(4,5)bisP. The latter compound was obviously degraded during the preparation of the membranes. In order to overcome these problems another experimental strategy was chosen, i.e. the *in vitro* labelling of PtdIns by [γ-^{32}P]ATP leading only to labelled PtdIns(4)monoP and PtdIns(4,5)bisP. In deed, a low endogenous activity of PtdIns phosphorylation reactions was recently found to occur in membranes isolated from *Triticum aestivum* (SANDELIUS and SOMMARIN,1986). In comparison to *myo* -[2-^{3}H]-inositol labelling the *in vitro* phosphorylation of phosphatidylinositol via [γ-^{32}P]ATP offers the following advantages: Only PtdIns(4)monoP and PtdIns(4,5)bisP are labelled, that means, only the substances being converted by a stimulus response action are labelled leading to a high signal noise ratio for the analysis.

Several other investigations on plants demonstrate also the occurrence of the particular compounds being known to be metabolized in the signal-mediated PI response in animals. However, until yet reports are missing to prove the cooperation of the single elements of the PI metabolism in a process of signal transduction in plants. For instance, though the presence of polyphosphoinositides in cell suspension cultures of *Glycine max* and *Petroselinum crispum* no evidence was recently found for the involvement of the PI response for the elicitor-induced phytoalexin synthesis (STRASSER et al., 1986). Until yet the only indication for an auxin action on membrane-bound PI turnover was found in isolated soybean membranes by MORRÉ et al. (1984). On the basis of these data an alternative view of the original hypothesis was formulated by ZBELL and MORRÉ (1986). It was suggested that the primary action of auxin in plant cells does not involve the action of a phospholipase C but the coordinated actions of a phospholipase D (EC 3.1.4.4) and a phosphatidate phosphatase (EC 3.1.3.4). The former enzymatic activity leads to the cleavage of PtdIns generating inositol and phosphatidate which is processed further to 1,2-DG and orthophosphate by the latter enzyme. In order to enlight the mechanism of an auxin action on membrane-localized PI turnover a set of experiments was performed using *myo* -[2-^{3}H]-inositol and [γ-^{32}P]ATP as substrates for the labelling of membranes prepared from carrot cells. First, but at the moment also preliminary results are presented here which point to, but not yet prove the occurrence of an auxin-mediated control of PI turnover in plant cells.

MATERIAL AND METHODS

Radiochemicals. *myo* -[2-^{3}H]-inositol (577 GBq mmol^{-1}) and [γ-^{32}P]ATP (111 TBq mmol^{-1}) were purchased from from the Radiochemical Centre Amersham (Buckinghamshire,Great Britain).

Plant material. Cell cultures of *Daucus carota* L. were grown as 60 ml suspensions in 250 ml wide-necked Erlenmeyer flasks on rotary shaking machines (100 rpm) at +28° C in the dark. The liquid medium was prepared from commercially available nutrients (FLOW Laboratories, Meckenheim, FRG) and was supplemented with 20 g l^{-1} sucrose and 4.5 µmol l^{-1} 2,4-D. The suspensions were transferred always after a period of 10 days to a fresh medium, and for the preparation of the subcellular fractions the carrot cells were harvested during the log-phase of the culture.

Preparation of intracellular membranes. Membranes of the carrot cells were prepared by a modification of the method of BOSS and RUESINK (1979). Briefly, the cells (120 g fresh weight) were collected by suction on a coarse fritted-glass filter and homogenized with the same amount of sea sand and 120 ml grinding medium (50 mmol l^{-1} Tris-HCl, pH 8.0, 250 mmol l^{-1} sucrose, 1 mmol l^{-1} EDTA, and 0.1 mmol l^{-1} $MgCl_2$) in an ice-cold mortar. After 15 min of grinding the paste was centrifuged twice at 2000 g for 10 min with a table centrifuge (Beckman TJ-6R) in a swinging bucket rotor to get a supernatant cleared of sea sand, cell debris, and nuclei. The second supernatant was centrifuged at 50000 g for 60 min in a high-speed centrifuge (Beckman J-21 B) with a fixed-angle rotor (JA-20) to sediment the membranes. The membrane pellet was suspended by the use of a glass rod and subsequent strokes with a teflon-fitted glass potter. Aliquots of the suspended membranes were layered on top of a 2 ml cushion of 270 g l^{-1} Renografin-60 (ρ=1.093 g cm^{-3}) in a gradient buffer (50 mmol l^{-1} Tris, 10 mmol l^{-1} KCl, 1 mmol l^{-1} EDTA, 0.1 mmol l^{-1} $MgCl_2$, pH 7.5) and were centrifuged by an ultracentrifuge (Kontron TGA-45) in a swinging bucket rotor (Spinco SW 41 Ti) at 120000 g for 2 h. The membrane fraction was collected by the use of a pasteur pipette and after its dilution with three volumes of washing buffer (25 mmol l^{-1} Tris-HCl, pH 7.5, 10 mmol l^{-1} KCl, 1 mmol l^{-1} EDTA, 0.1mmol l^{-1} $MgCl_2$) it was sedimented by ultracentrifugation in a fixed-angle rotor (Spinco 50 Ti) for 45 min at 200000 g. The pelleted membranes were suspended in a buffered sucrose solution (25 mmol l^{-1} Hepes-KOH, pH 7.5, 250 mmol l^{-1} sucrose). These steps including centrifugation and resuspension of the membranes were twice repeated, and the final membrane suspension was stored in polyethylene vials until use in liquid nitrogen. All operations were performed at +4° C and with precooled solutions.

Inositol exchange reaction. The assay contained 37 KBq *myo* -[2-^{3}H]-inositol, 500 µmol l^{-1} inositol, 10 mmol l^{-1} $MnSO_4$, 100 µmol l^{-1} CMP, 250-300 µg membrane protein in a final volume of 500 µl buffer (25 mmol l^{-1} Hepes, pH 8.0, and 250 mmol l^{-1} sucrose). The reaction was started by the addition of the membranes, incubated at +25° C for the times specified in the legends and terminated by the addition of 1 ml of an ice-cold stop solution containing 2-propanol/conc. HCl (100/1; v/v).

Phosphatidylinositol phosphorylation. The standard assay contained 93 KBq [γ-^{32}P]ATP, 100 µmol l^{-1} ATP, 10 µmol l^{-1} GTP, 10 mmol l^{-1} $MgSO_4$, 25 mmol^{-1} LiCl , 200-300 µg membrane protein in a final volume of 500 µl buffer (25 mmol l^{-1} Hepes, pH 7.5, and 250 mmol l^{-1} sucrose). The reaction was normally started by the addition of an aliquot of the ATP/GTP mixture, incubated at room temperature for the times specified in the legends and terminated by the addition of 1 ml of an ice-cold stop solution containing 2-propanol/conc. HCl (100/1; v/v).

Phospholipid extraction. The lipids were extracted from the acidified propanolic solutions with the low-toxicity solvent system according to HARA and RADIN (1978) using n-hexane as the organic solvent. After evaporation of the organic solvents and addition of a scintillation mixture the radioactivity of the ^{3}H- or ^{32}P-labelled phospholipids was determined in a liquid scintillation counter.

Anion exchange chromatography of inositolphosphates. After phospholipid extraction the inositolphosphates of the aqueous extract were separated by the anion exchange chromatography on DOWEX AG 1-X2 (200-400 mesh) as described by STRAUB and GERSHENGORN (1986) but modified by the use of a batch mode. ^{32}P-activity of the fractions was measured via the Cerenkov-radiation in a liquid scintillation counter.

RESULTS AND DISCUSSION

Inositol exchange reactions: The pathway of the membrane-bound PtdIns biosynthesis includes two different enzymatic reactions of inositol incorporation into phospholipids. First, inositol can be utilized by the action of the cytidine diphosphate diglyceride:inositol transferase (EC 2.7.8.11) leading to a net synthesis of labelled PtdIns. Secondly, inositol can be incorporated by an inositol exchange reaction resulting only in a labelling of already existing PtdIns (s. Fig.1). As mentioned, the inositol exchange reaction was found in isolated soybean membranes to be stimulated by auxin in longterm as well as in shortterm incubation periods (MORRÉ et al., 1984). However, maintaining the experimental conditions of MORRÉ et al. the auxin effect could not be verified even though the membranes prepared from carrot suspension cells possessed of a high *in vitro* inositol exchange activity. This activity exhibited all the characteristic properties including its stimulation by CMP, its total dependency on manganese ions and the magnitude of the activity as it was first described by SEXTON and MOORE (1981). There was no evidence for any significant longterm effect of auxin on the labelling of phosphatidylinositol via the inositol exchange reaction. Furthermore, after a labelling of the membranes with *myo*-[2-^{3}H]-inositol via the inositol exchange reaction and a subsequent wash out of unincorporated radioactivity any shortterm effect by auxin on the prelabelled membranes could not be found. In other words, using prelabelled membrane fractions of carrot cells any significant auxin-mediated loss of inositol labelling of the phospholipids could not be detected as it was reported by MORRÉ and coworkers for the soybean membranes. In accordance with these findings there was no evidence for any dose effect of auxin on the inositol exchange activity of prelabelled membranes. Furthermore, any significant effect of auxin on the inositol exchange reaction could not be measured even in the presence of its substrate inositol and its cofactors CMP and manganese if the membranes were prelabelled with *myo*-[2-^{3}H]-inositol (Fig. 2).

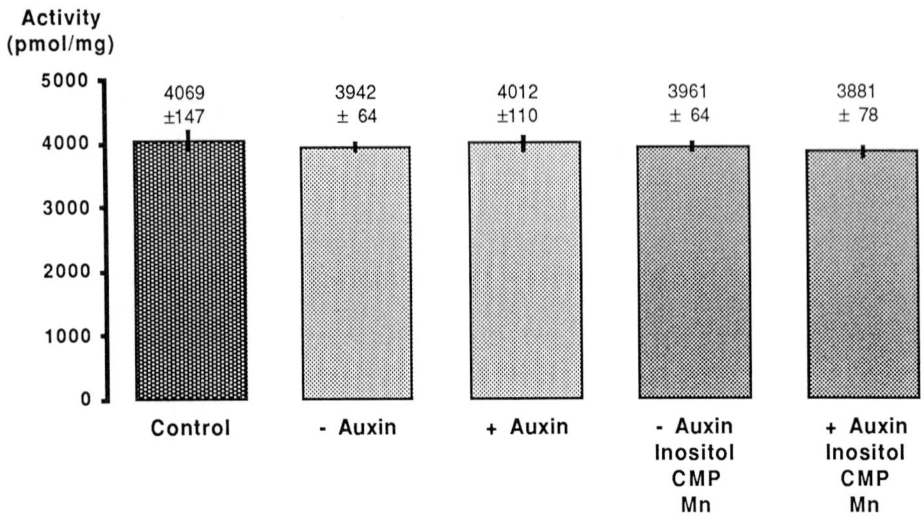

Fig. 2. No auxin effect on the turnover of membrane-bound PI derived from the inositol exchange reaction. Membrane fractions from carrot suspension cells were prelabelled for 1 h with *myo* -[2-^{3}H]-inositol via the inositol exchange reaction. After three washings the labelled membranes were incubated for 2 min in the absence or presence of the substrate and the cofactors of the inositol exchange reaction and, as indicated, with 1 µmol l^{-1} 2,4-D, too. The control shows the radioactivity of unincubated membranes.

There was also no indication of an auxin response on the inositol exchange reaction in the presence of Mg·ATP leading to the phosphorylation of PtdIns and even after the addition of GTP. The failure to detect any effect of auxin on the inositol exchange reaction is not caused by any loss of exchange activity after the prelabelling and the subsequent washes of the membranes. After the addition of CMP and manganese, which are essential cofactors for the inositol exchange reaction, prelabelled and washed membranes were found to increase the membrane-bound radioactivity in the presence of *myo* -[2-^{3}H]-inositol but to reduce the labelling in the presence of excess amounts of unlabelled inositol (Fig. 3). This incompetence of auxin to control the inositol exchange reaction in carrot cells corresponds well to the well-known inability of hormones and transmitters to act on this mechanism in animal cells, and it is also in accordance with the recent finding of LABARCA et al. (1985) that the inositol exchange reaction generates an agonist-insensitive pool of PI.

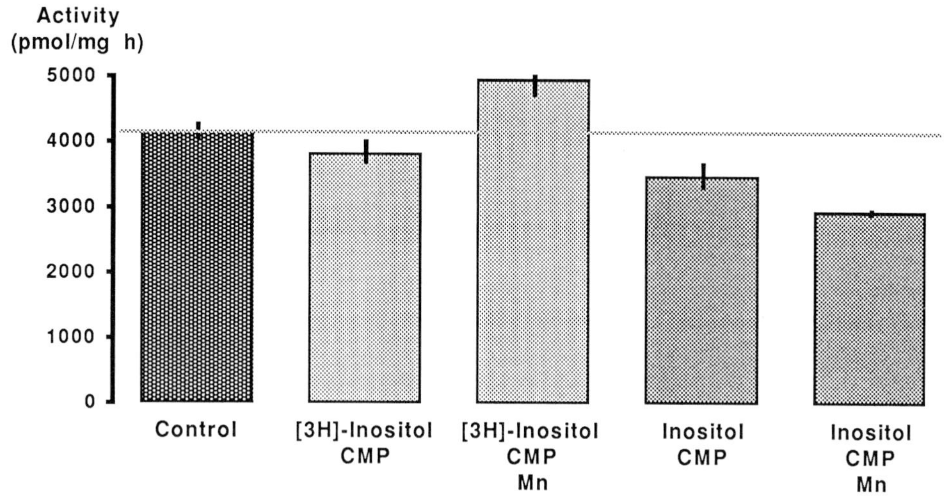

Fig.3. Inositol exchange activity of prelabelled and washed membranes. Membrane fractions from carrot suspension cells were prelabelled for 1 h with *myo* -[2-^{3}H]-inositol via the inositol exchange reaction. After three washings the labelled membranes were incubated for 1 h in the presence of labelled or unlabelled substrate and the cofactors of the inositol exchange reaction. The control shows the radioactivity of unincubated membranes.

Phosphoinositide phosphorylation: Membranes prepared from carrot suspension cells were found to have the capacity for PtdIns phosphorylation *in vitro* . Without the use of phosphatase inhibitors this endogenous activity was found to be more than several times higher as it was reported for the *Triticum* membranes by SANDELIUS and SOMMARIN (1986) and in a comparable magnitude as in animal cells. This PtdIns phosphorylation depends absolutely on the presence of Mg^{2+} and can be saturated with increasing ATP concentrations indicating Mg·ATP as the true enzymatic substrate. With this experimental tool it was possible to answer the main question: Does auxin control the PI turnover in plant cells ? As a first attempt the kinetics of the PtdIns phosphorylation in the absence or presence of 1 μM IAA were analyzed. A very rapid reduction of [^{32}P]-label could be observed in the phospholipid fraction with a minimum at 30 s obviously initiated by the action of auxin (Fig. 4).

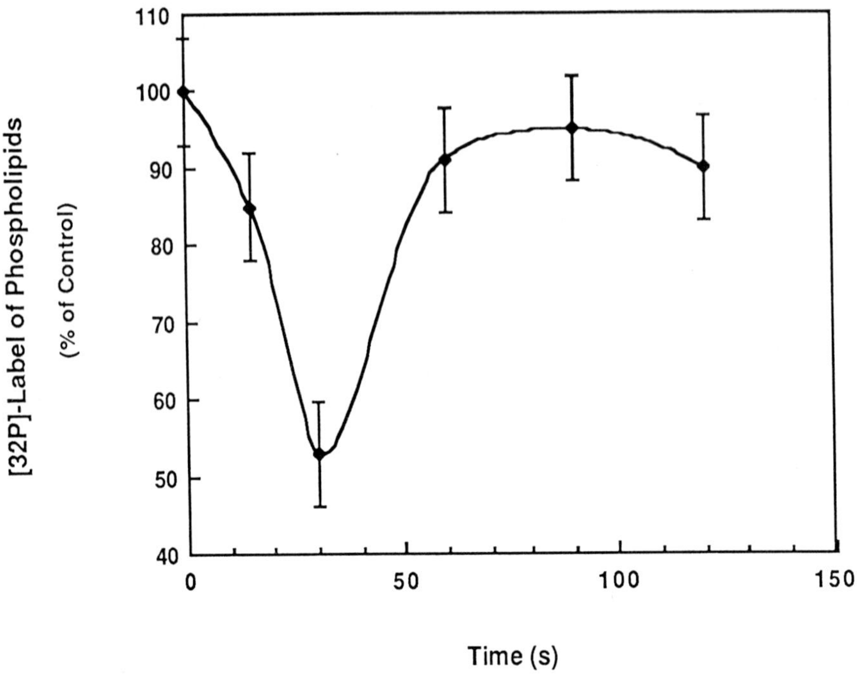

Fig. 4. Kinetics of the IAA-mediated loss of [^{32}P]-label of phosphoinositides. Membrane fractions from carrot suspension cells were preincubated for 5 min in the absence (control) and presence of 1 μmol l^{-1} IAA, and afterwards the reaction for the phosphorylation of PtdIns under standard conditions was started by the addition of [γ-^{32}P]ATP and 10 μmol l^{-1} GTP-γ-S.

This decrease of [^{32}P]-label is not caused by an auxin-mediated inhibition of the PI phosphorylation activity but it is really caused by an auxin-mediated hydrolysis of [^{32}P]-labelled phosphoinositides. This could be clearly ascertained for the same assays by the detection of [^{32}P]-labelled inositolphosphates after the elution of the aqueous extract from the DOWEX anion exchange resin. The corresponding kinetics of the IAA-mediated release of the [^{32}P]-labelled inositol compounds Ins(1,4,5)trisP and Ins(1,4)bisP are presented in Fig. 5. A very rapid increase of the amounts of both inositolpolyphosphates was detected, and the kinetics observed with an maximum activity at 30 s corresponds well with the kinetics of the loss of the [^{32}P]-label in the phospholipid fraction. The results presented derive from assays performed in the presence of 10 μM GTP-γ-S. In the absence of GTP the same type of kinetics of the IAA-mediated loss of [^{32}P]-label of phospholipids did occur, but it appears that the addition of GTP or GTP-γ-S stabilizes the auxin-mediated reaction. Repeating this type of experiments several times it became clear that the auxin-mediated effects on PtdIns phosphorylation and PI hydrolysis are very reproducible and do occur always very fast with nearly identical kinetics.

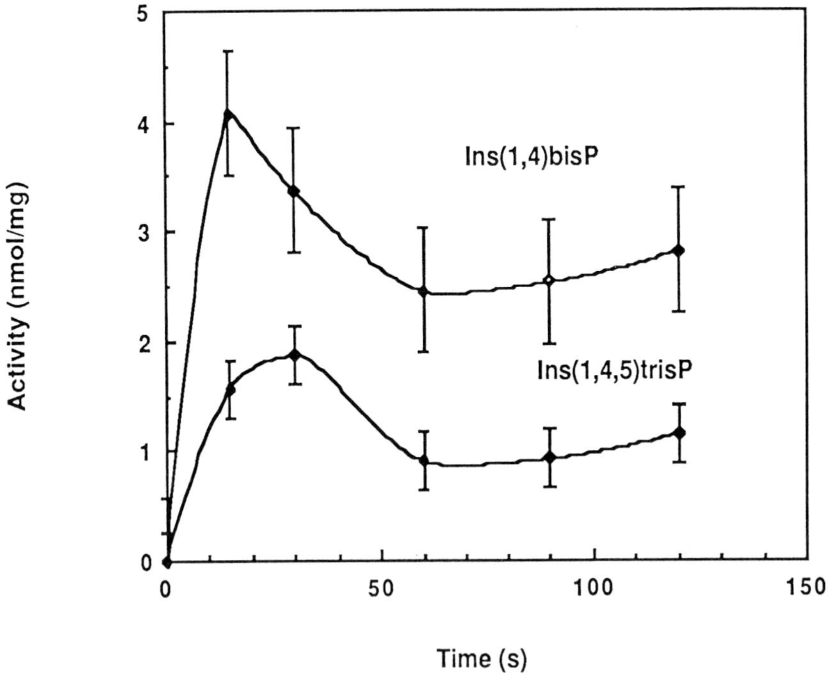

Fig. 5. Kinetics of the IAA-mediated release of [^{32}P]-label of inositolphosphates. After the phospholipid extraction of the assays described in Fig. 4 the inositolphosphates of the corresponding aqueous extracts were separated by anion exchange chromatography. The data represent the corrected values for the radioactivity of the inositolphosphates found in the auxin-containing assays as the label of the control assays was subtracted.

These auxin-dependent reactions exhibit also a dose response relationship. In the presence of 10 μmol l^{-1} GTP a continuous loss of [^{32}P]-label of the phospholipids was found with increasing IAA concentrations. This dose effect relationship is also established in the auxin-mediated release of [^{32}P]-labelled inositolphosphates (Fig. 6). It is remarkable that IAA in dependence to its concentration leads to a progressive release of Ins(1,4,5)trisP upto a saturation level at 10^{-6} mol l^{-1} hormone, but it does affect a linear increase on the release of Ins(1,4)bisP. This phenomenon may reflect the processes related to the rapid turnover of the inositolpolyphosphates (s.Fig.1). Higher IAA levels than 10^{-6} mol l^{-1} were found to evoke a further release of both compounds indicating obviously an unspecific effect of IAA on the membranes. The dose of IAA needed for the apparent half maximal PI response seems to be lower than the apparent dissociation constant of the membrane-bound high-affinity auxin-binding sites which was previously determined as a value of 5 x 10^{-7} mol l^{-1}. Furthermore, it can be calculated from the number of auxin binding sites in the assay that the first messenger auxin leads to the release of the second messenger Ins(1,4,5)trisP with an amplification of 500 - 1000.

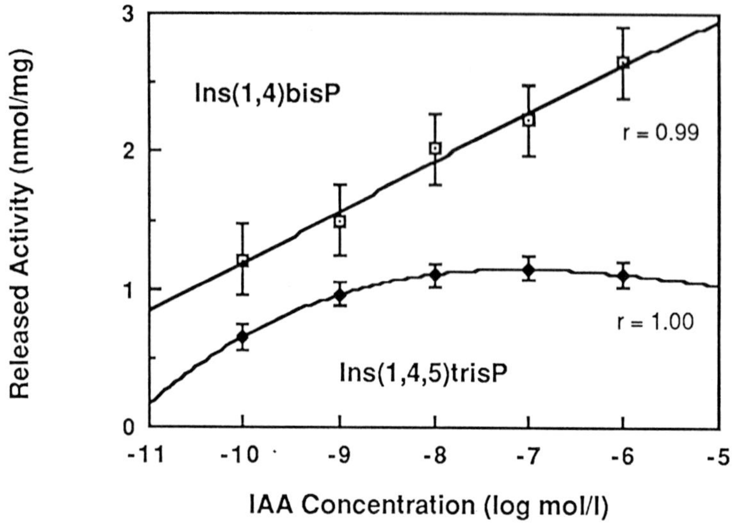

Fig.6. Dose response relationship of the auxin-mediated release of inositolphosphates. Membrane fractions from carrot suspension cells were preincubated for 2 min in the absence (control) and presence of IAA at the indicated concentrations, and afterwards the reaction for the phosphorylation of PtdIns under standard conditions was started by the addition of [γ-^{32}P]ATP and 10 μmol l^{-1} GTP. After the phospholipid extraction of the assays the inositolphosphates of the aqueous extracts were separated by anion exchange chromatography. The data represent the corrected values for the radioactivity of the inositolphosphates found in the auxin-containing assays as the label of the control assays was subtracted.

The dose response relationship of the rapid auxin-mediated reactions on the PI turnover in isolated carrot cell membranes point to a higher affinity of the receptor-like auxin-binding sites than it was determined previously (ZBELL,1983), and the time course of the auxin-mediated hydrolysis of phosphoinositides and the corresponding release of inositolpolyphosphates may reflect complex kinetics of auxin action. What that means in relation to the precise mechanism of signal transduction at the auxin receptor must be clarified in future work as well as all the other open questions. At the moment the data presented can be evaluated only as a first indication for an occurrence of a hormone-controlled PI response in plants, since it is still open if it is a similar or a distinct signal transduction mechanism in comparison to the well-known animal process. Much more data are needed about the precise subcellular localization and the auxin specificity of the reactions, the substrate specificity of the phospholipase C-like activity, the necessity of GTP as a coupling agent pointing to the involvement of an N-protein, the time course and the regulation of the turnover of the particular polyphosphoinositides and inositolphosphates. The results of this research will allow to verify or to denye the hypothesis presented. Only in the case of its verification it may be possible to discern a signal chain of the molecular action of auxin on plant cell growth as it is shown in Fig. 7 .

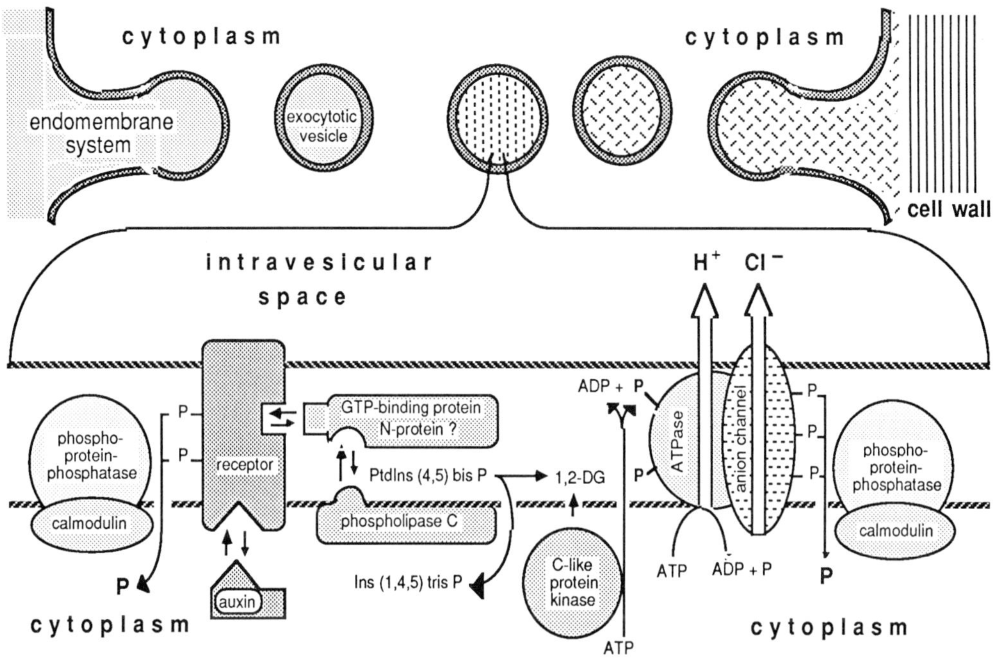

Fig. 7. A model linking the auxin-mediated PI response to the exocytosis in plant cells. The first messenger auxin binds to its receptor on intracellular membranes, and this signal may be transduced to the phospholipase C by the involvement of a N-protein. The activation of the phospholipase leads to the hydrolysis of PtdIns(4,5)bisP and the corresponding release of 1,2-DG and Ins(1,4,5)trisP which act as second messengers for protein phosphorylation and the release of Ca^{2+} from intracellular stores, respectively. A membrane-bound proton pump being activated by the protein phosphorylation generates a proton gradient which is the driving force for the traffic of the exocytotic vesicles. After fusion of the vesicles with the plasma membrane the vesicle content is unloaded into the cell wall space leading to plant cell elongation growth.

The auxin mediated processes may start with the membrane-localized PI response as the primary action of the hormone and continue with the stimulation of the proton translocation on membrane vesicles which are involved in the process of exocytosis and are loaded with materials needed for plant cell growth. Like in animal cells the traffic of the exocytotic vesicles may be sustained by Ca^{2+} which is released from its intracellular stores by Ins(1,4,5)trisP. The possible function of this compound as a second messenger also in plant cells was demonstrated by its action on the release of $^{45}Ca^{2+}$ from preloaded membrane vesicles prepared from *Cucurbita pepo* hypocotyl (DRØBAK and FERGUSON,1985). In its late steps this view of an auxin dependent course of exocytosis resembles the model recently presented by THEOLOGIS (1986), but it differs in the early steps of the direct action of the hormone: Here, the auxin-mediated PI response and there, the auxin-mediated gene regulation are connected to the process of exocytosis in the plant cell. These different views concerning the molecular action of auxin need not be contradictory since upto now the occurrence of multiple auxin binding sites leading to multiple auxin actions cannot be excluded.

LITERATURE

BERRIDGE, M.J. and IRVINE, R.F. (1984) Inositol trisphosphate, a novel second messenger in cellular signal transduction. Nature 312, 315-321.

BOSS, W.F. and MASSEL, M.O. (1985) Polyphosphoinositides are present in plant tissue culture cells. Biochem. Biophys. Res. Commun. 132, 1018-1023.

BOSS, W.F. and RUESINK, A.W. (1979) Isolation and characterization of concanavalin A-labeled plasma membranes of carrot protoplasts. Plant Physiol. 64, 1005-1011.

DRØBAK, B.K. and FERGUSON, I.B. (1985) Release of Ca^{2+} from plant hypocotyl microsomes by inositol-1,4,5-trisphosphate. Biochem. Biophys. Res. Commun. 130, 1241-1246.

FIRN, R.D. and KEARNS, A.W. (1983) The search for the auxin receptor. In: WAREING, P. (ed.) Plant Growth Substances 1982 pp 385-393 Academic Press New York

HARA, A. and RADIN, N.S. (1978) Lipid extraction of tissues with a low-toxicity solvent. Anal. Biochem. 90, 420-426.

HEIM, S. and WAGNER, K.G. (1985) Evidence of phosphorylated phosphatidylinositols in the growth cycle of suspension cultured plant cells. Biochem. Biophys. Res. Commun. 134, 1175-1181.

LABARCA, R., JANOWSKY, A. and PAUL, S.M. (1985) Mangsanese stimulate incorporation of [^{3}H]inositol into an agonist-insensitive pool of phosphatidylinositol in brain membranes. Biochem. Biophys. Res. Commun. 132, 540-547.

LITOSCH, I. and FAIN, J.N. (1986) Regulation of phosphoinositide breakdown by guanine nucleotides. Life Sci. 39, 187-194.

MICHELL, R.H. (1975) Inositol phospholipids and cell surface receptor function. Biochim. Biophys. Acta 415, 81-147.

MORRÉ, D.J., GRIPSHOVER, B., MONROE, A. and MORRÉ, T.J. (1984) Phosphatidylinositol turnover in isolated soybean membranes stimu-lated by the synthetic growth hormone 2,4-dichlorophenoxy-acetic acid. J. Biol. Chem. 259, 15364-15368.

NISHIZUKA, Y. (1984) Turnover of inositol phospholipids and signal transduction. Science225, 1365-1370.

RUBERY, P.H. (1981) Auxin receptors. Ann. Rev. Plant Physiol. 32, 569-596.

SANDELIUS, A.S. and SOMMARIN, M. (1986) Phosphorylation of phosphatidylinositols in isolated plant membranes. FEBS Lett. 201, 282-286.

SEXTON, J.C. and MOORE, T.S., Jr. (1981) Phosphatidylinositol synthesis by a Mn2+-dependent exchange enzyme in castor bean endosperm. Plant Physiol. 68, 18-22.

STRASSER, H., HOFFMANN, C., GRISEBACH, H. and MATERN, U. (1986) Are polyphosphoinositides involved in signal transduction of elicitor-induced phytoalexin synthesis in cultured plant cells? Z. Natursch. 41c, 717-724.

STRAUB, R. and GERSHENGORN, M.C. (1986) Thyrotropin-releasing hormone and GTP activate inositol trisphosphate formation in membranes isolated from rat pituitary cells. J. Biol. Chem. 261, 2712-2717.

THEOLOGIS, A. (1986) Rapid gene regulation by auxin. Ann. Rev. Plant Physiol. 37, 407-438.

WHEELER, J.J. and BOSS, W.F. (1986) Are polyphosphoinositides in the plasma membrane of carrot cells? Plant Physiol. 80 (4), Supplement, Abstract 421.

ZBELL, B. (1983) Über die molekulare Wirkung von Auxin. Biochemische Untersuchungen an isolierten Membranen aus *in vitro* kultivierten Zellen von *Daucus carota* L. Doctoral Thesis, Free University of Berlin.

ZBELL, B. (1985) An auxin mediated control of an intracellular proton pump via reversible protein phosphorylation and its consequence for the primary action of auxin. In: Proc. 12th Internat. Conf. Plant Growth Subst., Heidelberg, Abstract 1003, p 62

ZBELL, B. and MORRÉ, D.J. (1986) The molecular action of auxin. In: Models in Plant Physiology , Biochemistry and Technology (D.W. NEWMAN and K.G. WILSON, eds.). CRC Press, Boca Raton, Florida (in press)

HORMONE RECEPTOR MANIPULATION BY HYDROSTATIC PRESSURE: INTERACTION BETWEEN Ca^{2+}, MEMBRANE COMPONENTS AND PHOSPHATIDYLINOSITOL IN PEA FOLIAGE MICROSOMES

Y.Y. Leshem and G. Bar-Nes
Department of Life Sciences, Bar-Ilan University
Ramat-Gan 52 100, Israel

ABSTRACT

While presenting a Ca^{2+}-mediated bridging hypothesis between membrane phospholipids and protein, a method is described whereby differential compressibilities of membrane components can be utilized to induce gentle ejection of enzymes and/or hormone receptors. Auxin as a proton source may displace Ca^{2+} and enhance electrostatic bridge breaking thus initially causing greater exposure of immersed protein, while upon pressurization ejection of protein receptor ensues. Experimental evidence shows a stoichiometric interaction between Ca^{2+} and protein extrusion on one hand and ethylene production on the other, phosphatidylinositol having a pronounced regulatory effect.

INTRODUCTION

Hormone-receptor interaction constitutes the initial step in a series of events by which many hormones regulate cellular metabolism (7). Binding is not *a priori* evidence that the moiety which recognizes the hormones mediates hormone action. However, in certain cases a receptor role has been demonstrated which is directly associated with hormone action. The magnitude of hormone receptor effect is subject to regulation by rate of receptor production, loss or relocation in membrane domains. Thus in a given situation high endogenous hormone concentrations associated with low receptor availability or *vice versa* are not conducive to optimal physiological response. A typical case illustrating this point is young carnation flower tissue in which production of ethylene binding sites temporally precede actual rise in ethylene

NATO ASI Series, Vol. H10
Plant Hormone Receptors. Edited by D. Klämbt

production (17) thus sensitizing the young flower to exposure to exogenous ethylene.

Current concepts of membrane biophysics (16) assign an important role to divalent cations, especially Ca^{2+} in determination of membrane phospholipid (PL) phase states and hence of PL-membrane protein interaction. Mobility of integral or peripheral proteins (hormone receptors and/or enzymes) may be restricted by electrostatic Ca^{2+} bridging which would markedly alter microviscosity and induce phase transitions from a liquid crystalline to a solid gel configuration. Shinitzky (16) has shown that all negatively charged membrane PL's, especially phosphatidylinositol (PI) and phosphatidylserine by virtue of electrostatic cross-linking with phosphate sections of headgroups induce physiologically meaningful structural rearrangements in membrane architecture. Recent research on Ca^{2+} associated hormone responses in mammals and plants (1,5,12,13,20) has moreover indicated the importance of membrane polyphosphoinositides. In keeping with the contention that the very acid nature of phosphoinositides - phosphatidylinositol 4-phosphate (PIP) and phosphatidylinositol 4,5 bisphosphate (PIP_2) is important in the function of these phospholipids in the cell, NMR spectroscopy has recently shown that PIP and PIP_2 below pH 5 respectively contain two and three negative charges and at pH 8 contain three and five (19).

Regarding specific "anchoring" of putative membrane bound protein hormone receptors it has been shown (8,11) that by means of electrostatic Ca^{2+} crosslinking such bridging may be via a) PL-protein binding between membrane PL and carboxyl tails of proteins, b) protein-cytoskeleton anchoring, c) bridging between PL's themselves to create localized membrane domains possessing greater microviscosity than adjacent domains. Such anchoring besides limiting the proteins' mobility may also cause their deeper insertion within the bilayer and thus less exposure for interaction with effectors on the membrane's exterior.

Isolation techniques for membrane proteins are often faced with the drawback stemming from loss of receptor activity and cell disintegration due to employment of detergents. In the present research we report on a piezobiophysical pressurization technique which allows gentle shedding of membrane proteins without disruption of cells (14) and so doing we endeavoured to obtain further insight on the interactions between auxin and ethylene receptors and their possible relation to Ca^{2+} bridging phenomena outlined above.

METHODS

Microsomal membrane preparation

Pea foliage microsomal membranes were prepared from 70 gr fresh weight tissue as detailed elsewhere (10).

Selective release of integral proteins

The procedure of Deckmann *et al*. (4) was followed (See Fig 1).

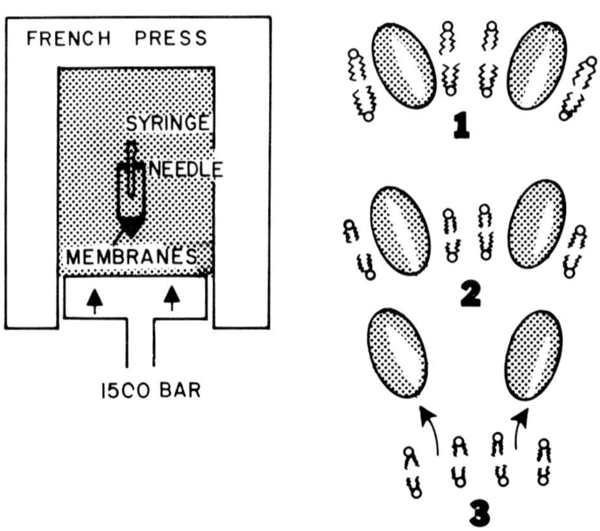

Fig. 1. Hydrostatic pressure procedure for obtaining integral membrane proteins based on differential isothermic compressibilities of membrane constituents (4). Details in text.

The rationale underlying this technique is based on the finding that isothermic compressibilities $[-(1/V). \Delta V/ \Delta P]$ of membrane components as effected by hydrostatic pressure (P) relate to the energetic quantity $P \Delta V$, where ΔV is the corresponding change in specific volume of the investigated entity. Empiric findings have indicated that in water, protein and gel phase phospholipid, compressibility is small, and below the 2000 bar range for practical purposes negligible (6,18). However, liquid crystalline phase PL is endowed with marked compressibility. Thus upon progressive increase of hydrostatic pressure, PL's are compressed, initially resulting in greater exposure of embedded protein and finally, above a threshold value, protein may be completely ejected (Fig. 1-right).

Ca crosslinks supposedly would be progressively broken and Ca^{2+} released to the medium. In the present research it was surmised that indole acetic acid (IAA) serving as a proton donor, in keeping with the "acid effect" interpretation of mode of auxin action (3) could "relax" PL-protein bridging. The initial effect would be one of greater exposure of hormone receptors possibly including ethylene receptors, and finally of complete ejection of the receptor. This mechanism, if verified, could explain concentration effects as for example where different auxin levels either promote or inhibit ethylene production. Released Ca^{2+}, depending on which leaflet of the bilayer it is located, may either inhibit or trigger the "phosphatidyl-linoleyl(-enyl) cascade" (9).

Experimental procedure

Microsomal membrane aliquots comprising of 25 μg protein in 1.7 ml [N-(2-hydroxyethyl)] piperazine-N'-3 propane sulfonic acid (EPPS) buffer 0.002M, pH 8.5 containing 10^{-3}M $CaCl_2$ were placed in capped Eckendorf plastic tubes, following the procedure of Deckman *et al* (4) and after adding buffer to the top, were sealed with a cap through which a short hypodermic syringe needle was placed to allow pressure equilibration. In the auxin treatment prior to pressurization membranes were incubated with 10ppm IAA for 18 hours at 22°C, controls being incubated with buffer without IAA in a like manner. When PI was added, this was achieved by adding

500 μg PI to the initial membrane isolate and sonicated for 60 sec to enhance incorporation into microsomes. To improve protein shedding and prevent its reinsertion into PL bilayers, 1% benzyl alcohol and 2% sucrose were added throughout.

Such preparations were placed in a 45 ml pressure bomb and subjected to 1500 bar hydrostatic pressure in an Aminco French Press for 20 min. After release the samples were centrifuged at 150,000 g for 90 min at 4°C and the supernatant collected. Ethylene production of the microsomal pellets was assessed on a Varian Model 3300 gas chromatograph as detailed elsewhere (10), while protein and Ca^{2+} content of the supernatant were determined by Coomassie Blue and tetramurexide staining, respectively (2,15). All trails employed 4-6 replicates and results represented are typical of several experimental runs.

RESULTS AND DISCUSSION

Figure 2 indicates results of an experiment indicating on a relative scale, the effect of auxin and hydrostatic pressure on ethylene and extrusion of Ca^{2+} and protein.

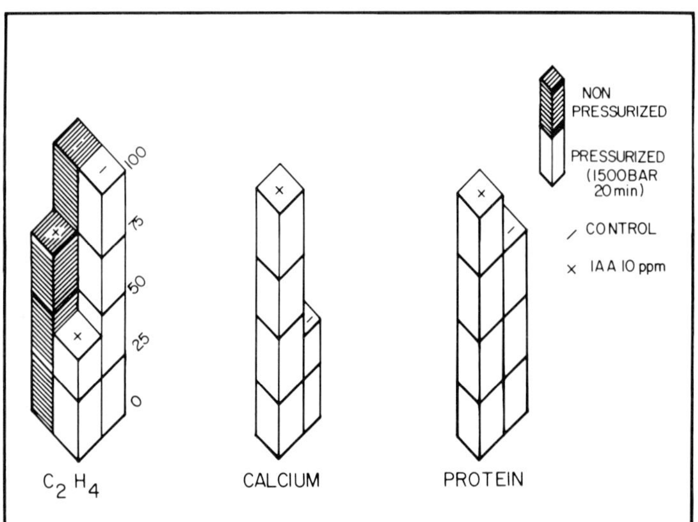

Fig. 2. Ethylene production, Ca^{2+} and protein release from pea foliage membranes as effected by hydrostatic pressure and IAA.

This figure indicates that in the absence of IAA, pressurization had no effect on ethylene production; however, a clear decrease is observed in its presence in both non-pressurized and pressurized treatments, the latter effect being more than halved. The figure moreover shows that the inhibition of ethylene production from the microsomal pellet under pressure with IAA goes hand in hand with accentuated Ca^{2+} release to the supernatant this not being encountered in the absence of IAA; the release of protein followed the same pattern.

These findings are interpreted as an IAA-mediated release of ethylene-receptor and/or of an ethylene-forming enzyme complex. The IAA-mediated Ca^{2+} and protein release to the supernatant suggests that pressurization does indeed, as suggested in the Introduction, induce breakage of Ca^{2+} crosslinks. However the fact that the system still produces ethylene possibly indicates that pressurization under present experimental conditions caused only partial ejection of proteins.

Results of the experiment in which microsomal membranes were enriched with PI are presented in Fig. 3.

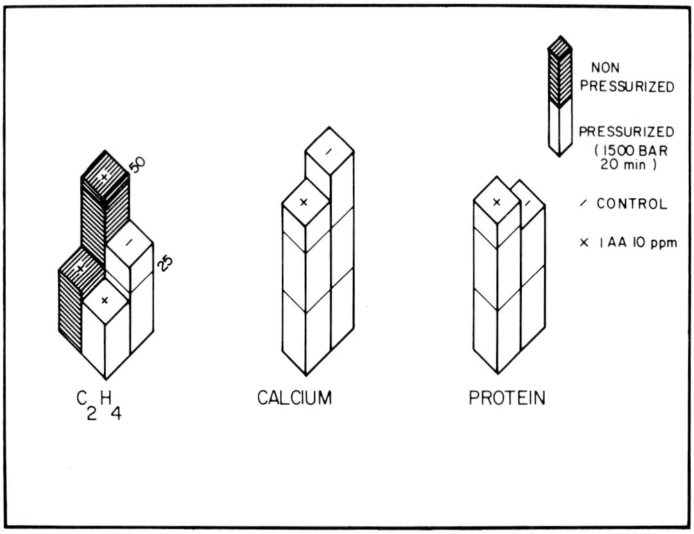

Fig 3: Interaction between auxin and phosphatidylinositol on ethylene production,calcium and protein release as effected by application of hydrostatic pressure on pea foliage microsomes.

This figure indicates that as compared to results obtained where no PI was applied (Fig 2) in general much less ethylene was produced , and less Ca^{2+} and protein released. The overall Ca^{2+} and protein results could be attributable to an enhanced binding of anchored protein by PI which by means of Ca^{2+} crosslinking anchors them more firmly in the PL bilayer-this interpretation possibly supported by the overall lower release of Ca^{2+} by PI treated microsomes (Fig 3) as compared to that of non PI treated ones (Fig 2). In this case IAA as expected is less effective in protein releasing and not at all with Ca^{2+}.

Ethylene as effected by PI (Fig 3) also indicates an overall and significant decrease in production where the IAA inhibiting effect is enhanced by hydrostatic pressure, a trend not observed in non PI treated tissues. These results are apparently problematic since as suggested by the Ca^{2+} and protein results, less receptor/enzyme is released and concomitantly more ethylene would be expected to be formed. At least two explanations may be given to account for this paradox. The first is that the PI enrichment by means of pronounced anchoring while indeed preventing ejection, causes deeper insertion of receptor into the membrane bilayer: thus while not considerably altering its' absolute content nevertheless decrease its' degree of exposure to effectors.

An alternative explanation, not necessarily exclusive of the former is that PI enhances IAA mediated signal transduction (12, 21) which effects ethylene regulation by mechanisms other than the presently discussed one. In this case auxin may stimulate a "Ca^{2+} gating" mechanism (12) where PI hydrolysis causes opening of cell surface Ca^{2+} channels and allows Ca^{2+} ions to pass through the cell membrane. The initial step of this process may be the presently described mechanism which possibly accommodates both interpretations.

Acknowledgements: The authors are deeply grateful to Prof. M. Shinitzsky and his staff, Department of Membrane Research, Weizmann Institute, Rehoboth, Israel, for their aid and advice in application of the hydrostatic pressure procedure.

REFERENCES

1. Berridge MJ Irvine RF (1984) Nature 312:315

2. Bradford M (1976) Analyt Biochem 72:248

3. Cleland RE (1979) In: Skoog F (ed) Plant Growth Substances 1979. Springer Verlag, Berlin p. 71

4. Deckman M Haimovitz R Shinitzky M (1985) Biochim Biophys Acta 821:334

5. Droback BK Ferguson IB (1985) Biochem Biophys Res Comm 130: 1241

6. Heremans K (1982) Ann Rev Biophys Bioeng 11:1

7. Hughes JP Elsholtz HP Freisin HG (1985) In: Posner BJ (ed) Polypeptide hormone receptors. Marcel Dekker, New York-Basel, p 157

8. Leshem YY (1987) In: Stumpf PK Mudd JB Ness UD (eds) Plant lipids: biochemistry, structure and function. Plenum, New York (in press)

9. Leshem YY (1987) Physiol Plant (in press)

10. Leshem YY Sridhara S Thompson JE (1984) Plant Phys 75:329

11. Lorand L Weissman LB Epel DL Bruner-Lorand J (1976) Proc Nat Acad Sci USA 73: 4479

12. Mitchell RH Kirk CJ (1982) Trends Pharm Sci 4: p 140

13. Morre´ DJ (1986) In: Trewavas AJ (ed) Molecular and cellular aspects of calcium in plant development. Plenum, New York, p 293

14. Muller CP Shinitzky M (1981) Exp Cell Res 136:53

15. Onishi ST (1978) Analyt Biochem 85:165

16. Shinitzky M (1984) In: Shinitzky M (ed) Physiology of membrane fluidity. CRC Press, Boca Raton, Florida, Vol I, p 196

17. Thompson JE (personal communication)

18. Weber G Drickamer HG (1983) Quart Rev Biophys 16:89

19. Van Paridon PA De Kruijff B Ouerkerk R Wirtz A (1986) Biochim Biophys Acta 877:216

20. Zbell B Morre´ DJ (1986) In: Newman DW Wilson KG (eds) Models in plant physiology, biochemistry and technology. CRC Press, Boca Raton, Florida (in press)

21. Zbell B (1985) 12th Int Conf Pl Growth Sub Abstr 10.03, p 62

STIMULATION OF IN VITRO H^+ TRANSPORT IN ZUCCHINI MICROSOMES BY THE ETHER LIPID PLATELET ACTIVATING FACTOR AND A SOLUBLE PROTEIN

Günther F. E. Scherer and Georg Martiny-Baron

Botanisches Institut der Universität Bonn, Venusbergweg 22, D-5300 Bonn 1, FRG

SUMMARY

Platelet activating factor (=PAF) stimulates H^+ transport in microsomes from zucchini hypocotyl (1) and ATP hydrolysis. The apparent membrane permeability for protons in membrane vesicles is not changed by PAF and in ATP hydrolysis rather v_{max} than the K_M is changed by PAF. PAF stimulation of H^+ transport is dependent on the presence of a protein factor which can be removed from the membranes. The properties of PAF-dependent protein factor in DEAE-Sephacel chromatography are very similar to a protein kinase. PAF stimulation of ATP hydrolysis and H^+ transport is found in tonoplast and plasma membranes. A possible role for the PAF-dependent factor analogous to animal protein kinase C in plant hormone action is postulated.

INTRODUCTION

Platelet activating factor (=PAF) is a phospholipid with an ether linkage at the C1 position of glycerol, 1-O-alkyl-2-acetyl-sn-glycero-3-phosphocholine, which has been isolated from animal serum and tissue because of its hormone-like properties (2). In animal cells it stimulates platelet aggregation (3), increase of internal Ca^{2+} (4), phosphatidylinositol turnover (5), and protein phosphorylation by protein kinase C (6) at concentrations in the nanomolar range. In plants PAF has been found to stimulate proton transport in vitro at micromolar concentrations (1). A further characterization of the mechanism of this stimulatory action on proton transport in vitro is presented here. PAF needs a soluble protein, PAF-dependent factor, to exert its effect on membrane vesicles from zucchini hypocotyl. The properties of this PAF-dependent factor suggest it to be a protein kinase which could be similar to a hormone-sensitive protein kinase C in the animal tissue.

MATERIALS AND METHODS

Membranes. Light microsomes from zucchini were prepared as described (1,7) and used for most experiments. When membranes were separated from supernatant they were diluted with homogenization buffer and centrifuged at 25 000 rpm with an SW 27 rotor (Beckman) for 45 min. For some experiments heavy microsomes were used (=fraction A3 in ref. 7) in order to avoid contamination by acid phosphatase. Glycerol gradient centrifugation was done as described (7). For the marker enzyme studies (Fig. 10) membranes had been pelleted with an SW 27 rotor for 30 min at 25 000 rpm prior to sucrose step gradient centrifugation. The membrane-containing fractions were layered onto a step

NATO ASI Series, Vol. H10
Plant Hormone Receptors. Edited by D. Klämbt

gradient consisting of layers of 0.5 ml 1.6 M sucrose, 1.5 ml 0.85 M, 1.5 ml 0.7 M, 1.5 ml 0.59 M, 1.5 ml 0.48 M, and 1.5 ml 0.24 M sucrose in 4 % ethanolamine (v/v), 2 mM EDTA, 2 mM DTE, 0.4 M ß-glycerol-phosphate (Sigma, grade III) titrated with acetic acid to pH 7.5. The gradients were spun for 2 h at 25 000 rpm in an SW 27 rotor (slim buckets, Beckman). Fractions at the interphases were collected with a Pasteur pipette.

ATPase assay. ATPase assays were performed with fractions chromatographed on small prepacked Sephadex G25 columns (Pharmacia) in 10mM MES/arginine pH 6.5 containing 0.3 M sucrose (7, 9). Tests contained 25 µl membranes in 100 µl total volume of 50 mM MES/BTP (final assay pH 6.0 or 6.5 or 7.0), 3 mM $MgSO_4$, 3 mM ATP (BTP salt) and 0.1 mM sodium molybdate and 1 mM sodium azide. Anions were added as BTP salts titrated to assay pH or as salts as indicated. Orthovanadate was added at a final concentration of 200 µM (titrated with MES). PAF was added from an ethanolic stock solution (4 mg/ml) usually at 10µg/ml. Assays were incubated at 30 °C for 20 min and stopped by the addition of 0.2 ml phosphate reagent (8) and kept on ice until reading the absorption at 750 nm.

Marker enzymes. Enzyme testes were conducted as described previously (7-10).

H^+ transport assays. Assays were done as described previously (1). Protein fractions for tests had been chromatographed on small prepacked Sephadex G25 columns. All tests with PAF or other lipids were done in the presence of 10 mM phosphate/BTP pH 6.5 if not otherwise indicated. Uncoupling by hexokinase and glucose has been described (9).

DEAE Sephacel column chromatography. Soluble proteins were extracted from 50 g hypocotyls in 60 ml of 20 mM Tris/HCl pH 7.5, 5 mM EGTA, 2 mM EDTA, 3 mM DTE and 0.25 M sucrose (11, 12). After homogenization the pH was adjusted and the homogenate was centrifuged for 1 h in a Ti 60 rotor at 45 000 rpm. The supernatant (90 ml) was applied to a DEAE Sephacel column with a bed volume of 35 ml and the column washed with at least 10 volumes of buffer without sucrose. In some experiments the column was washed in a batch procedure. A 0-0.5 M NaCl salt gradient of a total volume of 150 ml was applied. Fractions of 3 ml were collected at a flow rate of 50 ml/h. Undiluted fractions were assayed for protein, acid phosphatase, and protein kinase. For tests of PAF-dependent stimulation of H^+ transport fractions were equilibrated in 10 mM MES/arginine pH 6.5 in 0.3 M sucrose on small Sephadex G 25 columns. To parallel transport assays 0.7 ml equilibrated column fraction and either 10 µg/ml PAF or ethanol was added.

Protein kinase. To 40 µl 20 mM Tris/HCl pH 7.5 containing 6 mM MgCl and 6 mg/ml histone 5 µl enzyme was added. The reaction was started by the addition of 5 µl 10 nM γ-^{32}P-ATP. Assays were terminated by pipetting aliquots on phosphocellulose filters and washing them with cold phosphoric acid.

RESULTS

Stimulation of H^+ transport by PAF initially was measured using light microsomes from zucchini hypocotyl (1) which were contaminated by cytosolic components. An example of this reaction is given in Fig. 1. A strong stimulation of the ΔpH in the presence of phosphate as a permeant ion was found. Initially, in the presence of chloride only small effects by PAF were obtained (1), presumably, because stimulation of H^+ transport in the presence of chloride can be seen only at low PAF concentrations (Fig. 2).

A comparison of similar lipids suggested a strong specificity for PAF of transport stimulation (1) which is further supported here by comparing other phospholipids with PAF (table 1). Only the anionic PS and PG stimulated H^+ transport to some extent.

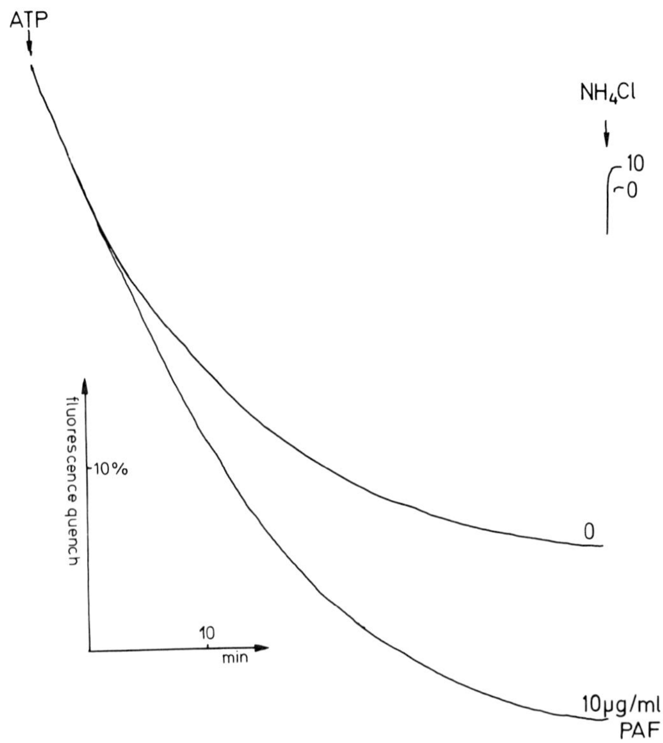

Fig. 1. Stimulation of H^+ transport by PAF in light microsomes containing cytosolic protein (fraction A2, ref. 7) in the presence of 10 mM BTB/phosphate. For PAF stimulation always the steady state ΔpH was measured using the formula $\Delta F = Q\% \, (Q\%-100)^{-1} \, 100$ (7).

Table 1. Stimulation of H^+ transport in light microsomes containing cytosolic protein by various phospholipids. Phospholipids were added at 10 µg/ml each and transport assays were conducted at pH 6.5 in 10 mM phosphate/ BTP. ΔF units were determined and controls set as 100%.

Platelet activating factor	193.5 ± 9.8
Phosphatidylserin	113.7 ± 5.9
Phosphatidylinositol	109.9 ± 4,4
Phosphatidylcholin	107.9 ± 10.3
Phosphatidylethanolamine	103.0 ± 4,0
Phosphatidylglycerol	115.1 ± 1.7

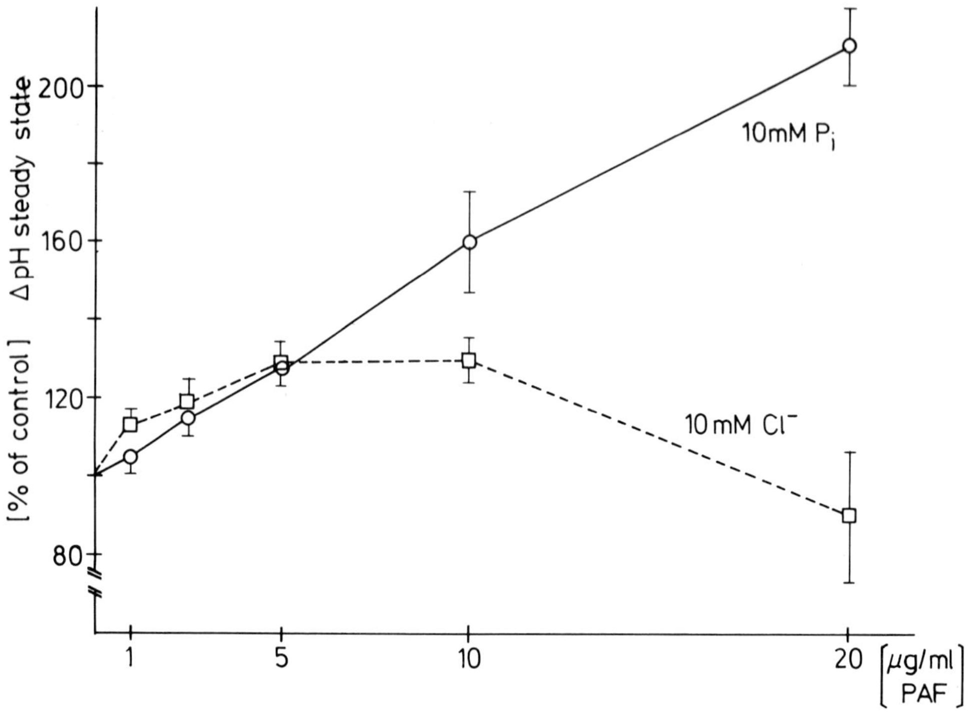

Fig. 2. Comparison of the effect of 10 mM BTP/chloride and 10 mM BTP/phosphate on the stimulatory action of PAF in light microsomes containing cytosolic protein.

We also tested whether PAF stimulates ATP hydrolysis (Fig. 3). When the ATP concentration was varied the effect of PAF on the apparent K_M of ATP was small whereas the v_{max} was increased. This was not surprising since routine test conditions for either H^+ transport or ATP hydrolyses always applied substrate saturation conditions for ATP.

The stimulation of ATP hydrolysis by PAF suggested that stimulation of H^+ transport also was a consequence of stimulation of H^+ transport rather than a mechanism affecting membrane permeability. This assumption was verified by testing the apparent membrane permeability for protons with and without PAF (Fig. 4). Apparent passive proton diffusion (outward) was not changed by PAF independently whether chloride or phosphate was used as an anion. These results make a detergent-like effect of PAF on membranes unlikely as the explanation of transport stimulation. Attempts to measure the membrane potential in the presence of PAF failed because in our routine technique for this purpose extensive bleaching of oxonol VI occurred.

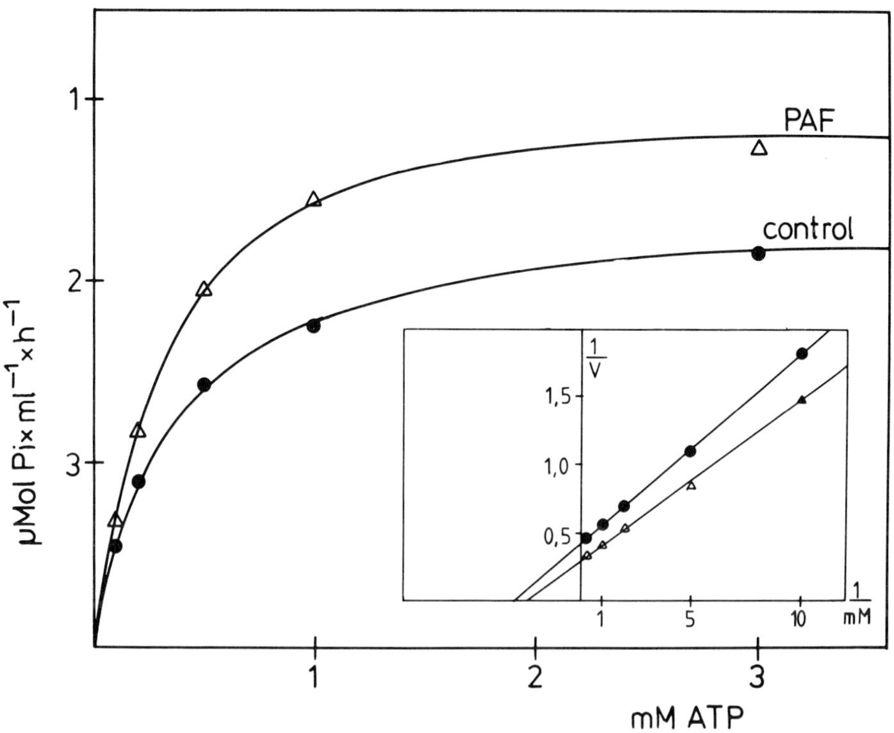

Fig. 3. Effect of PAF on ATP hydrolysis activity at various ATP concentrations. For the tests heavy microsomes (fraction A3, ref. 7) were used.

When the efficiency of various anions was tested to support the stimulatory action of PAF on H^+ transport no clear pattern emerged (Fig. 5). Perhaps the effects of anions reflect complex interactions of the properties of the soluble PAF-dependent factor (see below) or of anion-dependent H^+ transport since no obvious influence of anions on PAF-stimulated ATP hydrolysis was noticed (data not shown): When the PAF-dependent factor was boiled prior of the transport test PAF could not stimulate H^+ transport. This indicated that the factor needed was a protein.

Since PAF is a phospholipid and the transport assay contains ATP we suspected that the PAF-dependent soluble factor could be a protein kinase similar to the animal protein kinase C. The latter is dependent on phosphatidylserine or diglyceride both of which lipids are also to some degree active in transport stimulation (table 1 and ref. 1). Therefore, we tried to purify the PAF-dependent factor by Sephacel column chromatography which is commonly used as a first purification step for protein kinase C (11). Fig. 7 shows that indeed the PAF-

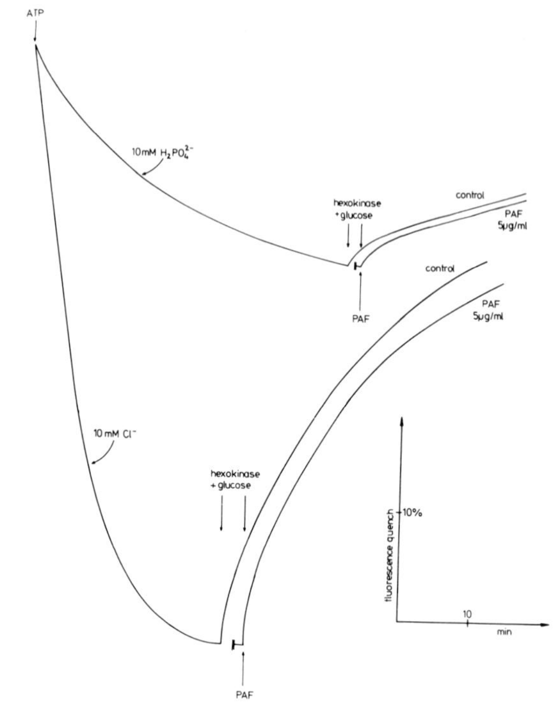

Fig. 4. Effect of PAF on passive membrane permeability. In parallel assays light microsomes containing cytosolic protein were allowed to develop a steady state pH. Glucose and hexokinase were added to hydrolyse ATP and then ethanol or 10 µg/ml PAF were added and the apparent proton diffusion (outward) was recorded. No detectable effect of PAF was found.

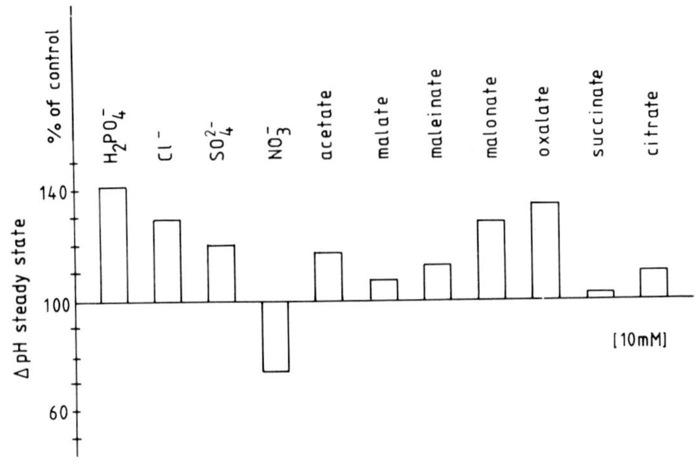

Fig. 5. Effect of anions on the stimulation of H^+ transport in light microsomes. Anions were added at 10 mM as BTP buffered salts and light microsomes (Fraction A2, ref. 7) were tested with 5 µg PAF/ml and ethanol controls were set as 100 %.

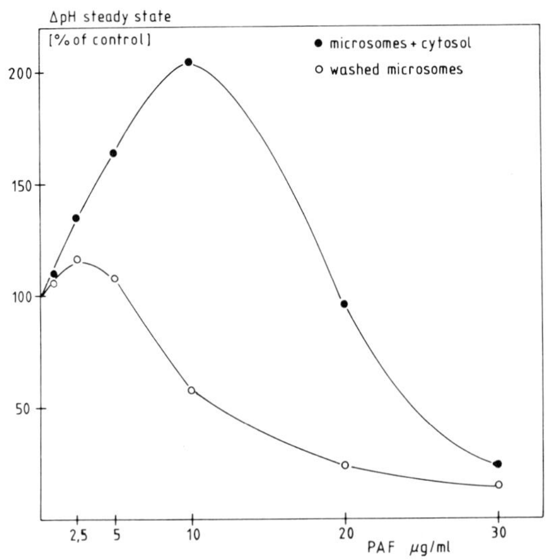

Fig. 6. Effect of PAF on microsomes pelleted and resuspended (o—o) and on microsomes containing cytosolic protein (●—●) in the presence of 10 mM BTP/phosphate.

dependent factor was eluted at about 0.3 M NaCl similarly as animal protein kinase C and similarly as a phospholipid-dependent protein kinase in plants (12, K. Palme, personal communication). The bulk of the protein either did not bind to the column or is eluted with high salt concentrations. The peak of the PAF-dependent activity was always found shifted to somewhat higher salt concentrations with respect to the main protein peak in the middle of the gradient. Sometimes a protein peak eluting with low salt (0.1 M NaCl) was found which did not contain the PAF-dependent factor activity. Interestingly, a decrease in PAF-dependent factor activity was always found coinciding with a peak of acid phosphatase activity. Three peaks of protein kinase were found and the second peak coincided with PAF-dependent factor activity. This protein kinase is phospholipid-dependent and stimulatable by PAF (K. Palme, personal communication). We take the coincidence of both activities and the coincidence of the acid phosphatase peak and the decrease in PAF-dependent factor activity as good evidence that PAF-dependent factor and protein kinase are identical. Experiments are in progress to further support this assumption.

The subcellular localization of H^+-ATPase is tonoplast and plasmalemma (7, 13). When we tested the PAF-dependent transport stimulation in the tonoplast- and plasma membrane-enriched fraction (7) it was found in both (table 2). In order to corroborate this finding we centrifuged the post-mitochondrial supernatant on an isopynic sucrose step gradient. Stimulation of ATPase activity was found in the uppermost and in the two bottom fractions. Addition of soluble proteins did not increase stimulation of ATPase activity by PAF (data not shown). The comparison with marker enzymes showed that the upper fractions contained the highest vanadate inhibition of pH 6.0 ATPase activity. Markers for ER and mitochondria were found at the expected densities and did not coincide with PAF

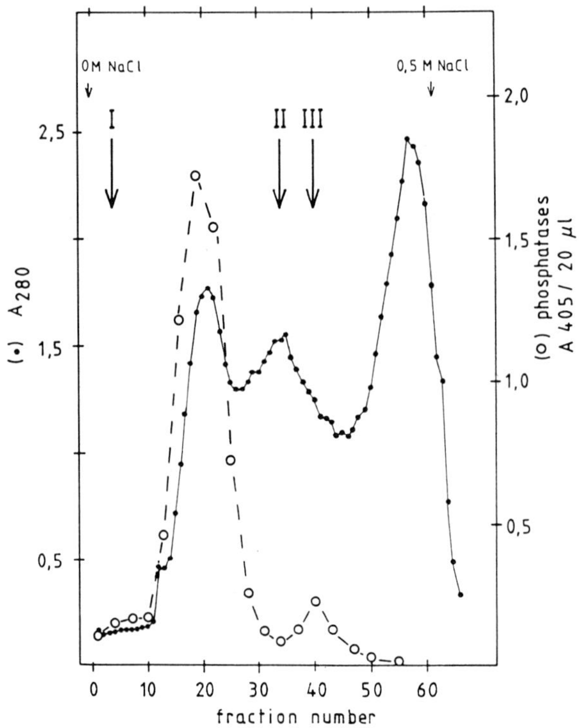

Fig. 7. DEAE Sephacel column chromatography of PAF-dependent factor. Soluble protein was bound to a DEAE Sephacel column (12) and eluted with a linear NaCl gradient. I, II, and III indicate the position of protein kinase activity. Peaks II and III coincided with the transport stimulation by PAF and PAF-dependent factor.

Table 2. Stimulation of H^+ transport in the tonoplast enriched fraction B1 and in the plasma membrane-enriched fraction B5, 6 (ref. 7) measured as steady state ΔpH in the presence of 10 mM BTP/phosphate pH 6.5 and 10 µg/ml PAF. Controls without PAF were set at 100 %.

Tonplast-enriched fraction	147.9 ± 11.2
Plasma membrane enriched fraction	143.6 ± 9.8

stimulation of ATPase activity. The data show conclusively that tonoplast H^+-ATPase and plasma membrane ATPase are the targets of the PAF-dependent factor.

When a two-dimensional thin layer chromatogram of phospholipids of zucchini microsomes was made, always two unidentified phospholipids appeared on the plate. Both are minor phospholipids containing 1-2% of the total lipid phosphorous (Fig. 8). X could be either phosphatidylmethanol, an often observed extraction artifact in plant lipid extracts, or N-acyl-phosphatidylethanolamine both of which would be expected to have a high R_f value in both solvent systems. "Überbleibsel", however, did not exhibit the polarity as expected for lysolipids since LPE and LPC were identified by standards. When PAF was added to the lipid extract to our surprise the "Überbleibsel" spot was increased so that both lipids most probably are similar phospholipids. Recent experiments show that this new plant phospholipid has biological activity.

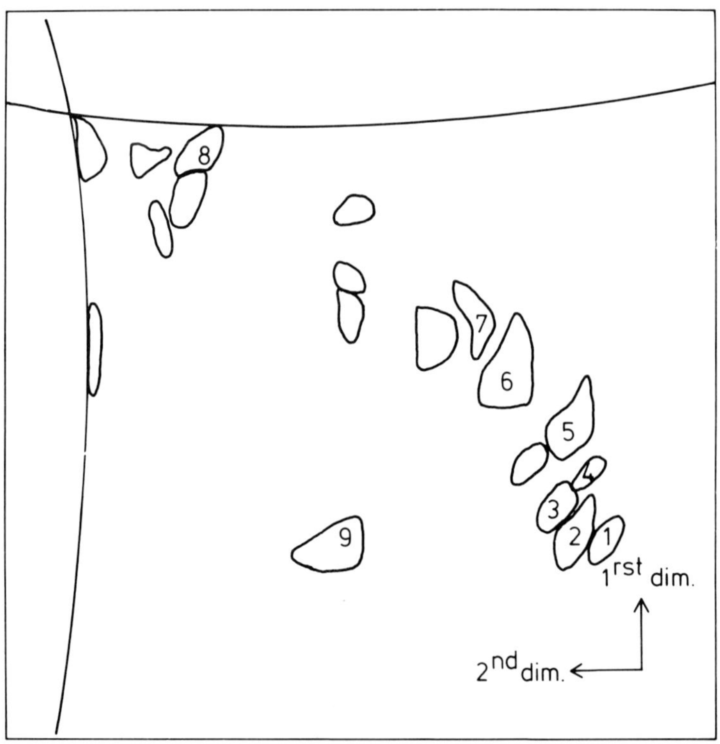

Fig. 8. Two-dimensional thin layer chromatography of membrane lipids from zucchini microsomes. Extraction and chromatography of lipids has been described (25). Phospholipids were identified as phosphorous-containing spots with a phosphate spray and other lipids with iodine vapour. PAF (Sigma chemical company) corresponded exactly to spot X_1 in its position on the plate increasing it in size when added to a lipid extract. (1) Lysophosphatidylcholine. (2) Phosphatidylinositol. (3) Lysophosphatidylethanolamine. (4) "Überbleibsel". (5) Phosphatidylcholine. (6) Phosphatidylethanolamine. (7) Phophatidylglycerol. (8) X. (9) Phosphatidic acid.

DISCUSSION

A functional role for the unusual phospholipid PAF in plants is not known as it is for animal cells (2) so that its effects described here could be purely "pharmacological" since roughly 1000 fold higher concentrations of PAF are needed in the plant system to show effects as compared to animal cells. Nevertheless, with twodimensional thin layer chromatography we found a phospholipid which is at least rather similar to PAF. So the possibility exists that we found a new set of regulatory molecules interacting in the plant membrane: a PAF-dependent factor similar to protein kinase C, a PAF-like phospholipid, and target proteins or subunits associated with the two H^+-ATPases.

PAF exerts its effect on H^+ transport not as such but rather interacts with a protein factor (Fig. 6, 7). This PAF-dependent factor has strong similarities to the hormone-dependent animal protein kinase C (Fig. 8). PAF-dependent factor activity depends on lipids, PAF being the most effective one (table 1 and ref. 1) but PS, PG and diglyceride also stimulate H^+ transport slightly. Physiological concentrations of Ca^{2+} abolish the action of PAF (1) which we interprete as stimulation of PAF-dependent factor by Ca^{2+} so that further stimulation by PAF is no longer possible. PAF-dependent factor is mostly soluble as reconstitution experiments with microsomes show (Fig. 6). Possibly, PAF- dependent factor adheres to plasma membrane more strongly as ATPase activity stimulation in the plasma membrane-enriched fractions shows. In reconstitution experiments similarly as with microsomes but with a plasma membrane-enriched fraction (B5, ref. 7) PAF-dependent factor could not be removed from the membranes (data not shown).

A link to hormonal control of PAF-dependent factor is suggested by the similarities to protein kinase C and by the nature of the target proteins, the H^+-ATPases or proteins associated with them. Protein kinase C is involved in signal transduction in the animal plasma membrane (14). We present here a hypothetical model (Fig. 9) how known hormone receptors could interact with H^+-ATPase via PAF-dependent factor. Other regulatory influences such as by fusicoccin might be accomplished in a more direct manner since cell growth and proton secretion are at least not always linked (15, 16). Certainly, a protein kinase could be responsible for a multiplicity or regulatory events so that the receptors inserted in Fig. 9 could be receptors other than for auxin only.

In our model PAF-dependent factor is depicted as a protein kinase which can be stimulated by lipids or by calcium ions. Increase in cytosolic Ca^{2+} or the formation of stimulatory lipids in turn could be brought about by various events. Ca^{2+} is stored in the ER (10, 18) and might be released by interaction of the auxin binding sites on the ER membranes (19, 20) with a Ca^{2+} gate possibly mediated by inositol-1,3,5-trisphosphate in a similar manner as in animal cells (21). The precursor for IP_3,phosphatidylinositol-1,4-bisphosphate, is present in plant cells (22, 23).

If the PAF-dependent factor is to exert its effect via lipid activation the respective lipids should originate in the tonoplast and/or plasma membrane. Various possibilities for lipid activation exist as exemplified in Fig. 9 which are still hypothetical for plants. A model with a phospholipase C would be similar as for protein kinase C in animal cells (14) although we rather assume that differences exist between plant and animal cells in this respect. Phospholipase D could be another candidate for the production of activating lipids. It has been shown to be associated with plant ER mainly (24, 25, Scherer, unpublished) and would produce PA which in turn is a precursor to PS and PI. Another possible pathway would be the formation of PAF-like lipids via transacetylation.

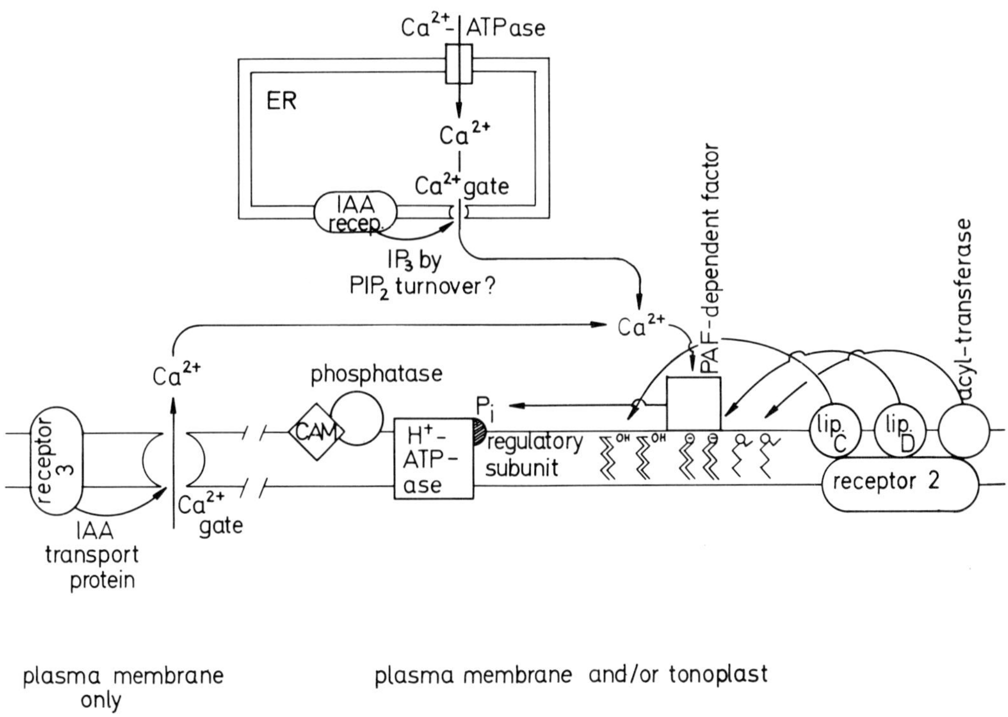

Fig. 9. Model of the action of PAF-dependent factor and interaction with receptors for other signal molecules. Details in the text.

H^+-ATPase in plant plasma membrane presumably consists of a single 100 000 Dalton polypeptide (26) so that either this polypeptide or a regulatory polypeptide associated with it would be the target of the PAF-dependent factor activity whereas the tonoplast is a multisubunit enzyme (26) one of which could be modified by PAF-dependent factor. A calmodulin-dependent phosphatase (calcineurin-like?) probably is present in microsomes from carrot cells the activity of which is counteracted by auxin (27). Initial experiments in our laboratory showed very little effect of calmodulin on H^+ transport in vitro but perhaps microsomes in our preparation were contaminated by endogenous calmodulin that would tend to mask the effect of exogenous calmodulin. In any event, then increases of cytosolic Ca^{2+} would also regulate dephosphorylation so that differential sensitivity to Ca^{2+} for phosphorylation and dephosphorylation would be required.

A third possibility is the hypothetical coupling of IAA transport and Ca^{2+} entry (28) which might result in an overall counterflux of IAA and Ca^{2+} (29). Hence, increased IAA transport would lead to increase in cytosolic Ca^{2+} which could result in a complex regulatory pattern (our model and ref. 28).

In our model all three receptors could act in a parallel fashion whereas in a given tissue not all three receptors might be present or other growth regulators could modulate up or down the same system. The necessity of coupling various enzyme reactions in membranes might also explain why stimulation by IAA of H^+ transport (27, 30) or of ATPase activity (8, 31) has been difficult to measure.

ACKNOWLEDGEMENT

This work has been supported by the Deutsche Forschungsgemeinschaft. Miss K. Kappes provided excellent technical assistance. The protein kinase experiments were done in the laboratory of Dr. Klaus Palme (MPI Züchtungsforschung, Köln, FRG) to whom we are grateful for his advice with these experiments. G. M.-B. received support by the Graduiertenförderung.

LITERATURE

(1) Scherer, G.F.E. (1985) Biochem. Biophys. Res. Comm. 133, 1160-1167
(2) Demopoulos, C.A., Pinckard, R.N., Hanahan, D.J. (1979) J. Biol. Chem. 254, 9355-9358
(3) Vargaftig, B.B., Benveniste, J. (1983) Trends in Pharm. Sci. 4, 341-343
(4) Conrad, G.W., Rink, T.J. (1986) J. Cell Biol. 103, 439-450
(5) Fisher, R.A., Shukla, S.D., Debuysere, M.S., Hanahan, D.J., Olson, M.S. (1984) J. Biol. Chem. 259, 8685-8688
(6) Sugatani, J., Hanahan, D.J. (1986) Arch. Biochem. Biophys. 246, 855-864
(7) Scherer, G.F.E., Fischer, G. (1985) Protoplasma 129, 109-119

(8) Scherer, G.F.E. (1981) Planta, 151, 434-438
(9) Scherer, G.F.E., Martiny-Baron, G. (1985) Plant Science 41, 161-168
(10) Buckhout, T.J. (1982) Planta 159, 84-90
(11) Takai et al. (1979) J. Biol. Chem. 254, 3692-3695
(12) Schäfer, B., et al. (1985) FEBS Lett. 187, 25-28
(13) Scherer, G.F.E. (1984) Planta 160, 348-356
(14) Nishizuka, Y. (1984) Trends in Biochem. Sci. 9, 163-166
(15) Hertel, R. (1979) Plant Organelles. Methodological Surveys (B) Biochemistry vol. 9 pp. 173-183. E. Reid, ed. Halsted Press, New York Chichester Toronto
(16) Kutschera, U., Schopfer, P. (1985) Planta 163, 483-493
(17) Kutschera, U., Schopfer, P. (1985) Planta 163, 494-499
(18) Buckhout, T.J. (1984) Plant Physiol. 76, 962-967
(19) Buckhout, T.J., Young, K.A., Low, P.S., Morré, D.J. (1981) Plant Physiol. 68, 512-515
(20) Ray, P.M., Dohrmann, U., Hertel, R. (1977) Plant Physiol. 59, 537-546
(21) Muallem, S., Schoeffield,. M., Pandol, S., Sachs, G. (1985) Proc. Natl. Acad. Sci. USA 82, 4433-4437
(22) Boss, W.F., Massel, M.O. (1985) Biochem. Biophys. Res. Comm. 132, 1018-1023
(23) Strasser, H., Hoffmann, C., Grisebach, H., Matern, U., (1986) Z. Naturforsch. 41 c, 717-724
(24) Yoshida, S. (1979) Plant Physiol. 64, 241-246
(25) Scherer, G.F.E., Morré, D.J. (1978) Plant Physiol. 62, 933-937
(26) Sze, H. (1985) Annu. Rev. Plant Physiol. 36, 175-208
(27) Zbell, B. (1983) Ph. D. thesis, Berlin, Free University
(28) Hertel, R. (1983) Z. Pflanzenphysiol. 112, 53-67
(29) De la Fuente, R.K. (1984) Plant Physiol. 76, 342-346
(30) Gabathuler, R., Cleland, R.E. (1985) Plant Physiol. 79, 1080-1085
(31) Scherer, G.F.E. (1984) Planta 161, 394-397

IMMUNOCYTOLOGICAL LOCALIZATION OF A WHEAT EMBRYO CYTOKININ BINDING PROTEIN AND ITS HOMOLOGY WITH PROTEINS IN OTHER CEREALS

A. Chris Brinegar and J. E. Fox
ARCO Plant Cell Research Institute
6560 Trinity Court
Dublin, Calif. 94568, USA

INTRODUCTION

A protein (CBF-1) isolated from wheat embryos was shown several years ago to bind in a non-covalent manner and with relatively high affinity N^6-substituted purines which have cytokinin activity in standard assays (7,8). Although the subunit composition of the protein and the kinetics of the binding have been the subject of some controversy in the literature (6,11,12), recent studies in our laboratory (1) demonstrate that the native protein consists of three identical subunits, each with a molecular mass near 54,000 and confirm our earlier finding (6) that one molecule of the cytokinin, N^6-benzylaminopurine (bzlADE) is bound per molecule of the native protein. CBF-1 appears in the developing wheat embryo about two weeks after anthesis and increases rapidly during the next 14 day period reaching a level of approximately 47 µg per embryo, or about 9% of the soluble protein in mature wheat embryos (3). A number of cDNA clones have been isolated for the CBF-1 gene (2). An amino acid sequence for the binding domain and for much of the protein has been obtained and will be reported elsewhere. In this paper we present evidence which demonstrates that CBF-1 is highly tissue specific, being localized to the scutellum, the coleoptile and the coleorhiza and that within the cells of these tissues, the cytokinin-binding protein is packaged in membrane bound protein bodies. In addition we demonstrate that cytokinin-binding moieties which are immunologically related to CBF-1 occur in certain other cereals, confirming preliminary reports on this subject (9,10)

MATERIALS AND METHODS

Seeds.

Wheat (*Triticum durum*, cv. Mexicali), rye (*Secale cereale*, cv. Smolice), triticale (*Triticosecale*, cv. Siskiyou), barley (*Hordeum vulgare*, cv. Himalaya), oats (*Avena sativa*, cv. Kanota), rice (*Oryza sativa*, cv. M101), maize (*Zea mays*, cv. Sweet Temptation), pea (*Pisum sativum*, cv. Alaska), soybean (*Glycine max*, cv. Prize) were used in this study.

NATO ASI Series, Vol. H10
Plant Hormone Receptors. Edited by D. Klämbt

Antibody Preparation.

Polyclonal antibodies specific for CBF-1 were isolated from rabbit serum by Protein A-Sepharose and CBF-1-Sepharose chromatography (3), made 5 mg/ml and 0.02% in bovine serum albumin and sodium azide, respectively, and stored at 4 C. Preimmune antibodies were isolated by Protein A-Sepharose chromatography and treated similarly.

Immunocytological Localization.

Wheat seeds (25 days post-anthesis) were fixed in cacodylate buffered 5% glutaraldehyde and postfixed in cacodylate buffered 2% osmium tetroxide (pH 7.2). The material was dehydrated through an ethanol series and embedded in Spurr's resin. For light microscopy, thick sections (0.5-1.0 μm) were transferred to a drop of water in the frosted rings of antibody slides and dried overnight at 50 C. For electron microscopy, ultrathin sections (60-90 nm) were mounted on parlodion coated 200 mesh gold grids. Both slides and grids were treated with CBF-1 specific IgG or preimmune IgG (20 μg/ml) according to the procedure of Craig and Goodchild (4,5) then incubated with 20 nm colloidal gold conjugated to goat anti-rabbit IgG (Janssen Life Sciences, Beerse, Belgium).

Benzyladenine Binding Assay, Western Blotting and Immunodiffusion.

For benzyladenine (bzlADE) binding assays, embryos (including scutella) of cereals or cotyledons of legumes were excised (25-40 mg) and homogenized in 1 ml of 500 mM KCl, 1 mM dithiothreitol, 50 mM Tris-HCl (pH 8.5). The 10,000*g* supernatants were diluted to 2 ml in the same buffer. [3H]-bzlADE binding activities of 0.5 ml aliquots were measured by equilibrium dialysis according to Erion and Fox (6), except that the dialysis buffer was 0.2 M ammonium bicarbonate (pH 8.0).

For SDS-PAGE the tissue was homogenized in 1 ml of 1% SDS, 10 mM dithiothreitol, 1 mM Na_2EDTA, 50 mM Tris-HCl (pH8.0) at 90 C. After running on 10% SDS-acrylamide gels, the samples were stained with Coomassie Blue R-250 or transferred to nitrocellulose. Western blotting was performed with affinity purified rabbit anti-CBF-1 IgG (2 μg/ml) and goat anti-rabbit IgG conjugated to horseradish peroxidase as described previously (3).

Ouchterlony immunodiffusion was performed in 1% agarose buffered with 0.2 M NaCl, 0.1 M Tricine (pH 8.2) using anti-CBF-1 IgG (0.2 mg/ml) and the KCl-Tris extracts (above) adjusted to 1.4 mg/ml total protein.

RESULTS AND DISCUSSION

Immunological Comparison of Wheat Embryo CBF-1 with other Cereal and Legume Proteins.

Cross-reactivities between CBF-1 specific antibodies and proteins in extracts of several cereal embryos and legume cotyledons were measured qualitatively by Western blotting of SDS-acrylamide gels and by Ouchterlony immunodiffusion. CBF-1 in the wheat embryo extract was detected on the Western blot (Fig. 1) as a single band at 54 kD. Strongly cross-reactive polypeptides with similar molecular weights were also found in rye and triticale embryos. A much weaker reaction (barely visible in the photograph) was observed at a slightly lower molecular weight in barley. In extracts of rice and maize embryos, a very strong reaction occurred with polypeptides near 50 kD. No antigenically similar polypeptides were detected in oat embryos or pea and soybean cotyledons. Minor lower molecular weight polypeptides appearing on the blot are assumed to be degradation products since wheat CBF-1 is known to undergo similar degradation (1). All of the proteins detected with the CBF-1 antibodies also appeared as bands on the Coomassie Blue stained gel.

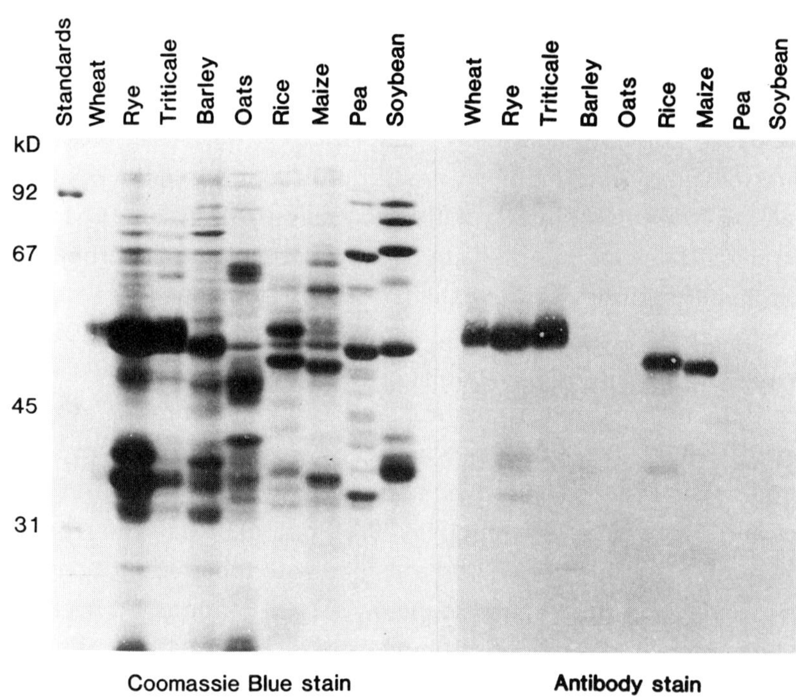

Figure 1. SDS-acrylamide gel of cereal embryo and legume cotyledon extracts: Coomassie Blue stain (left) and Western blot of a duplicate gel probed with anti-CBF-1 IgG (right).

Since the polypeptides on the Western blot had been denatured in SDS prior to exposure to the antibodies, cross-reactivities of the undenatured extracts were also tested by immunodiffusion against anti-CBF-1 IgG (Fig. 2). Strong identity reactions were observed with the wheat, rye, triticale, and barley embryo extracts. No precipitin lines were detected with the oat, rice, maize, pea, or soybean samples, although a weak reaction with oats has been observed with other CBF-1 antisera (10).

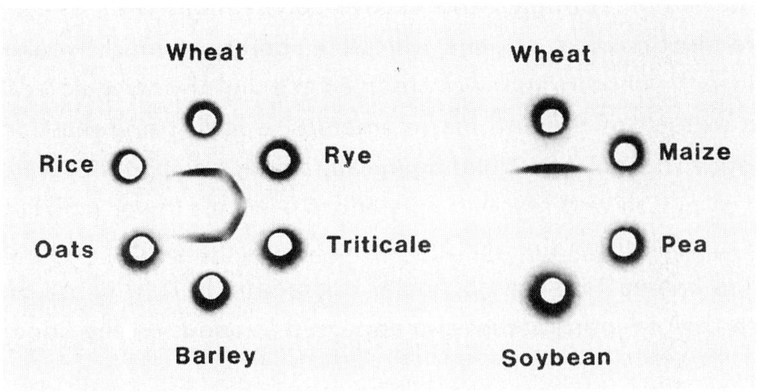

Figure 2. Ouchterlony immunodiffusion (Coomassie Blue stained) of cereal embryo and legume cotyledon extracts against CBF-1 IgG (in center wells)

Benzyladenine Binding Activity in Cereal Embryos and Legume Cotyledons.

The bzlADE binding activities of the various extracts are shown in Table I. When expressed either in terms of dry weight or total protein, the relative binding activities are: oats>wheat≃barley>triticale>rye≃rice>maize≃pea≃soybean. The activity of triticale was intermediate between wheat and rye, probably because triticale is a hybrid of tetraploid wheat (*T. durum*) and tetraploid rye (colchicine treated *S. cereale*). The bzlADE binding activities of maize, pea, and soybean were extremely low and probably reflect non-specific binding.

The data from Western blotting and cytokinin-binding assays were in agreement for wheat, rye, triticale, barley, and rice (cross-reactivity, medium to high BA binding) and for pea and soybean (no cross-reactivity, low BA binding), but not for rice and maize (cross-reactivity, low BA binding) or oats (no cross-reactivity, high BA binding). The Ouchterlony data agreed with the binding assay for wheat, rye, triticale, barley, maize, pea, and soybean, but not for oats and rice. Thus there exists in corn a protein which does not bind cytokinins but which has at least some of the antigenic determinants found in CBF-1, while in oats a protein is found which binds cytokinins strongly but has immunogenic properties considerably different than CBF-1.

Table I.
Benzyladenine Binding Activity in Cereal Embryos and Legume Cotyledons

Embryo/Cotyledon	Activity Relative to Wheat	
	per unit dry wt	per unit protein
Wheat (*Triticum durum*)	100	100
Rye (*Secale cereale*)	32	39
Triticale (*Triticosecale*)	73	85
Barley (*Hordeum vulgare*)	98	93
Oats (*Avena sativa*)	165	199
Rice (*Oryza sativa*)	31	48
Maize (*Zea mays*)	5	7
Pea (*Pisum sativum*)	4	3
Soybean (*Glycine max*)	7	4

Tissue and Subcellular Localization of CBF-1 in Wheat Embryos.

The colloidal gold immunocytological staining technique was used to localize CBF-1 at both the tissue and subcellular levels. By light microscopy at low magnification the embryonic tissues containing the gold particles have a pink coloration which photographed poorly in black and white so these regions are shown here diagrammatically (Fig. 3, left). Tissues which stained (*i.e.*, contained CBF-1) were the scutellum, coleoptile, coleorhiza, and epiblast. There was no detectable staining in the tissues of the embryonic axis (shoot apex, nodal region, and radicle) or the endosperm.

A magnification of the area at the lower scutellum-radicle juncture (Fig. 3, right) shows the contrast in labelling. The scutellar cells contain substantial numbers of dark gold particles, whereas the radicle cells are obviously unlabelled. Similar magnifications of coleorhiza-radicle, coleoptile-shoot apex, and scutellum-endosperm sections (not shown) exhibit the same sharp contrast between the labeled and unlabeled sections indicating a high degree of tissue specificity for the cytokinin-binding protein.

Electron micrographs of scutellar cells treated with preimmune (control) IgG followed by the gold labelled second antibody showed only sparse background labelling (Fig. 4, left). Treatment with anti-CBF-1 IgG, however, gave very specific and extensive labelling of the protein matrix in protein bodies (Fig. 4, right). From these results it is clear that CBF-1 is localized in protein bodies found only in cells of those tissues (scutellum, coleoptile, coleorhiza and epiblast) which surround the embryonic axis and isolate it from the endosperm.

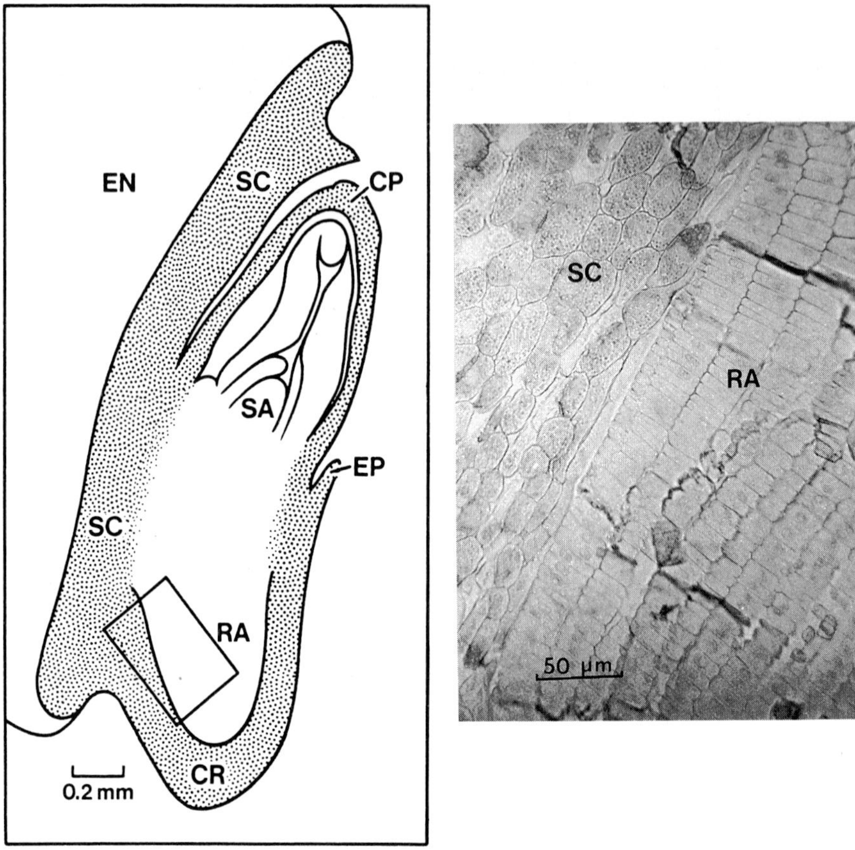

Figure 3. Longitudinal sections of 25-day-old wheat embryos treated with anti-CBF-1 IgG and colloidal gold labelled second antibody. Left: Diagram of the whole embryo including some endosperm. Shaded areas in the scutellum, coleoptile, coleorhiza, and epiblast indicate the gold labeling and the distribution of CBF-1. Right: Micrograph of the scutellum-radicle area (boxed in region of the diagram). The scutellar cells contain dark, gold labelled spheres. EN, endosperm; SC, scutellum; CP, coleoptile; CR, coleorhiza; SA, shoot apex; RA, radicle; EP, epiblast.

Perhaps the localization of CBF-1 (as well as the fact of its great abundance in specific tissues) gives us a clue to the biological function of this cytokinin-binding moiety. One could speculate that this protein functions by sequestering cytokinins, thus preventing their movement in any great amounts from the endosperm, a tissue where they occur abundantly during certain stages of seed development, into the embryo where it might be supposed they would cause unwanted cell division or induce premature germination.

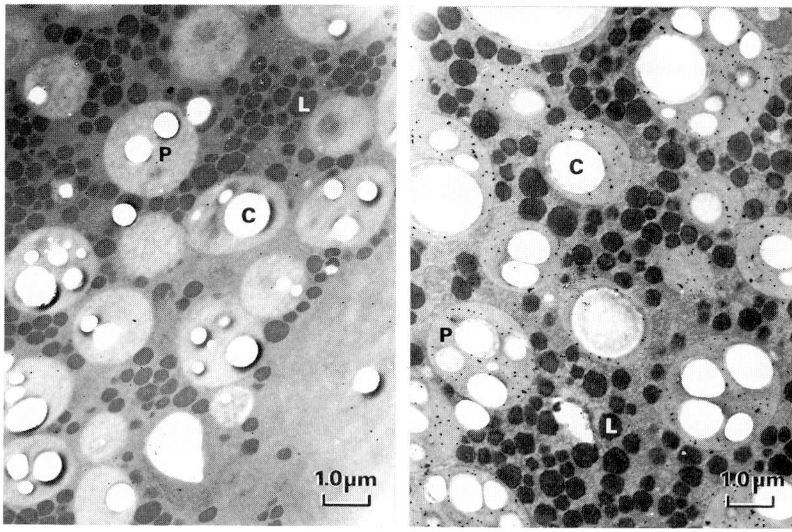

Figure 4. Transmission electron micrographs of scutellar cells treated with preimmune IgG (left) or anti-CBF-1 IgG (right) followed by colloidal gold labelled second antibody. Gold particles appear as black specks. P, protein body; C, crystalloid inclusion (or hole left by inclusion); L, lipid body.

Acknowledgements

The authors would like to thank Margery Marsden and Kathy Kaneko for their help with the light and electron microscopy, Anne Stevens and Joanne Pratt for help with the antibody studies,and Dale Cetlinski for the illustration.

REFERENCES

1. Brinegar, A.C., Fox, J.E., 1985. Resolution of the subunit composition of a cytokinin-binding protein from wheat embryos. Biol. Plant. 27:100-104.
2. Brinegar, A.C., Fox, J.E., 1985. The developmental expression and molecular cloning of the gene for a wheat embryo cytokinin-binding protein. In: "Plant Genetics", UCLA Symposium on Molecular and Cellular Biology, New Series Vol 35. Freeling M. (ed.) Alan R. Liss Inc. New York. pp 147-155.
3. Brinegar, A.C., Stevens, A., Fox, J.E., 1985. Biosynthesis and degradation of a wheat embryo cytokinin-binding protein during embryogenesis and germination. Plant Physiol. 79:706-710.
4. Craig, S., Goodchild, D.J., 1982. Post-embedding immunolabelling. Some effects of tissue preparation on the antigenicity of plant proteins. Eur. J. Cell Biol. 28:251-256.
5. Craig, S., Goodchild, D.J., 1984. Periodate-acid treatment of sections permits on-grid immunogold localization of pea seed vicilin in ER and Golgi. Protoplasma 122:35-44.
6. Erion, J.L., Fox, J.E., 1981. Purification and properties of a protein which binds cytokinin-active 6-substituted purines. Plant Physiol. 67:156-162.
7. Fox, J.E., Erion, J.L., 1975. A cytokinin-binding protein from higher plant ribosomes. Biochem Biophys Res. Comm. 64:694-700.
8. Fox,J.E., Erion, J.L., 1977. Cytokinin-binding proteins in higher plants. In: Plant Growth Regulation, Proc. 9th Int. Conf. on Plant Growth Substances, Pilet, P. (ed.) Springer-Verlag, Berlin. pp 139-146.

9. Fox, J. E., Gregerson, E., 1982. Variation in a cytokinin binding protein among several cereal crop plants. In: Plant Growth Substances 1982, Wareing,P.F. (ed) Academic Press, London, pp. 207-214.

10. Keim, P., Erion, J.L., Fox, J.E., 1981. The current status of cytokinin-binding moieties. In: Metabolism and Molecular Activities of Cytokinins. J. Guern, J., Péaud-Lenoël, C. (eds)., Springer-Verlag, Berlin, pp. 179-190.

11. Moore, F.H. III, 1979. A cytokinin-binding protein from wheat germ. Isolation by affinity chromatography and properties. Plant Physiol. 64:594-599.

12. Polya, G., Davis, A.W. 1978. Properties of a high affinity cytokinin-binding protein from wheat germ. Planta 139:139-148.

STUDIES ON CYTOKININ BINDING PROTEINS

Huang Hai, Lu Jia-ling, Jian Zhi-ying and Tang Yu-wei
Shanghai Institute of Plant Physiology, Academia Sinica
Shanghai, China 200032

It is generally believed that receptors in cell which bind the hormones as a result of this interaction, initiated a series of biochemical reactions leading to the physiological responses (Rubery 1981). In the studies of cytokinin (CTK) receptors, Berridge(1970) first described reversible binding of kinetin(KT) and 6-benzyladenine (6BA) to 83 S plant ribosomes. CTK-binding proteins were found later in ribosomes of wheat germs and tobacco callus(Fox 1975), in moss protonema and in fractions of tobacco callus (Sussman 1978) and carrot callus(Kobayashi 1981). In our recent researches we found that there were CTK-binding proteins in cotton ovary, young wheat seedlings and <u>in vitro</u> chloroplasts of wheat leaves. The property and possible physiological function of CTK-binding proteins are discussed in this paper.

Material and Methods

1. Isolation and measurement of CTK-binding proteins in cotton ovary

Flowers of cotton (Gossypium hirsutum) were collected at 8-10 o'clock AM each day, and ovaries were obtaines immediately after flower being picked and preserved in -37° C.

CTK-binding fractions were prepared according to Kasamo(1976). Generally, 40 g fresh weight of ovary was homogenized in a blender with 240 ml isolation buffer(20 mM Tris-HCl,pH 7.5,containing 250 mM sucrose and 3 mM EDTA-Na_2). The homogenate was filtered through 4 layers of gauze, then centrifuged at 1.000 xg for 15 min. Pellets were suspended with 120 ml isolation buffer and centrifuged again as above.The pellets were suspended with 16ml binding reaction buffer (80 mM Tris maleate, pH 6.5, containing 0.5 M sucrose, 100 mM KCl and 1.5 mM $MgCl_2$) as fraction P1, and the supernatant was centrifuged at 10.000 xg for 30 min and the

NATO ASI Series, Vol. H10
Plant Hormone Receptors. Edited by D. Klämbt

subsequent pellets were suspended with 16 ml binding reaction buffer as fraction P10.

Binding reaction was measured according to the method of centrifugation assay. Each centrifugal tube contained 1 ml binding reaction buffer (without sucrose) and 1 uCi ^{3}H-6BA. These tubes were divided into two parts: tube A had no cold 6BA and tube B contained $4x10^{-4}$ M cold 6BA. To each tube 0.1 ml cell fraction was added followed by incubation at 4° C for 30 min, then centrifugation at 20.000 xg for 30 min. The pellets were suspended with 0.5 ml water and added to 7 ml Bray's solution. Radioactivity in pellets was estimated by a liquid scintillation counter. Radioactive amount in tube A minus that in tube B was considered as specific binding of 6BA of these fractions.

2. Isolation and purification of CTK-binding proteins in young wheat seedlings.

Gel for affinity chromatography of CTK-binding proteins was prepared according to Romanko (1982) with Epoxy-activated Sepharose 6B. Wheat (Triticum aestivum L.) seedlings of 3-d-old were homogenized in a blender with extraction buffer (50 mM phosphate buffer, pH 7.7, containing 0.1 mM EDTA-Na_2 and 70 mM NaCl). The homogenate was filtered through 4 layers of gauze and centrifuged at 20.000 xg for 20 min. From supernatant a fraction was precipitated by ammonium sulfate (40-80 % saturation). The precipitate was suspended with chromatographic buffer (50 mM phosphate buffer, pH 6.9, containing 0.1 mM EDTA-Na_2 and 5 mM NaCl), and loaded on the top of a 6BA-Sepharose 6B column (1.8 x 6.0cm) which was equilibrated with chromatographic buffer previously. The column was washed by chromatographic buffer, then eluted by this buffer containing 0.1 M NaCl. Binding activity of the fraction eluted to 6BA was measured by equilibrium dialysis (Fox 1977).

3. Measurement of CTK-binding proteins in chloroplasts of wheat leaves.

Wheat leaves of 5-d-old were homogenized in a blender with a extraction buffer(25 mM Tris acetated,pH 7.6,containing 60 mM sodium acetate, 5mM magnesium acetate, 0.4 M sucrose and 5 mM ß-mer captoethanol). The homogenate was passed through 4 layers of gauze, then centrifuged at 1.500 xg for 5 min. Chloroplasts were purified according to Miflin (1974) with some modifications. The pellets were suspended by extraction buffer and loaded onto the top of a sucrose gradient (60 % 3 ml, 40 % 8 ml, 20 % 5 ml, w/v), and centrifuged at 400 xg for 2 min then 2.700 xg for 10 min. The zone of chloroplast was removed into 3 volume of buffer (as extraction buffer without sucrose) followed by centrifugation at 1.500 xg for 5 min. The pellets were suspended with 5 ml buffer, and again centrifuged at 1.500 xg for 5 min. The pellets were suspended with 5 ml buffer and treated by an ultrasonic generator at 300 mA for 30 sec, then divided into 400 ul aliquots. In ice bath, 200 ul 6BA or water was added and samples were incubated at 0° C for 1 h, then 400 ul 2 uCi ^{3}H-BA/ml buffer was added. After incubation at 4° C for 12h, the samples were centrifuged at 12.000 xg for 15 min. The supernatant was decanted, and walls of centrifugal tubes were carefully dried by filter paper. In each tube 200 ul 6 N H_2SO_4 was added and the precipitate was suspended at room temperature overnight, then 1 ml water was added, followed by centrifugation at 12.000 xg for 15 min. The supernatant of 400 ul plus 7 ml scintillation solution (Triton X-100/toluene, 1/2, v/v, containing 0.5 % PPO and 0.01 % POPOP) was taken into darkness for 1 day and estimated by a liquid scintillation counter.

Results and Discussion

1. CTK-binding proteins in cotton ovary

Table 1: Binding of ^{3}H-6BA to P1 and P10 fractions of cotton ovary

Fraction	Specific binding (cpm)	Specific binding/ Total binding (%)
P1	7,951	22.6
P10	6,045	17.9

From cotton ovary homogenate two fractions were collected through centrifugation at 1,000 xg(P1) and 10,000 xg(P10). Both fractions could bind 6BA(Table 1). Some cold CTK such as 6BA, KT and zeatin could substitute for ^{3}H-6BA while auxin (IAA), abscisic acid (ABA) adenine and cyclic adenosine monophosphate almost had not this ability (Table 2). This means that P1 could bind CTK specifically.

Table 2: Binding of ^{3}H-6BA to P1 fractions of cotton ovary. Competition with different cold compounds.

Substance added	Specific binding
6BA	6,564
zeatin	4,510
KT	5,592
IAA	0
ABA	0
cAMP	644
adenine	0

The binding activity of ^{3}H-6BA to P1 decreased by P1 preincubated with trypsin at 30°C for 30 min (Fig.1). This suggested that CTK-

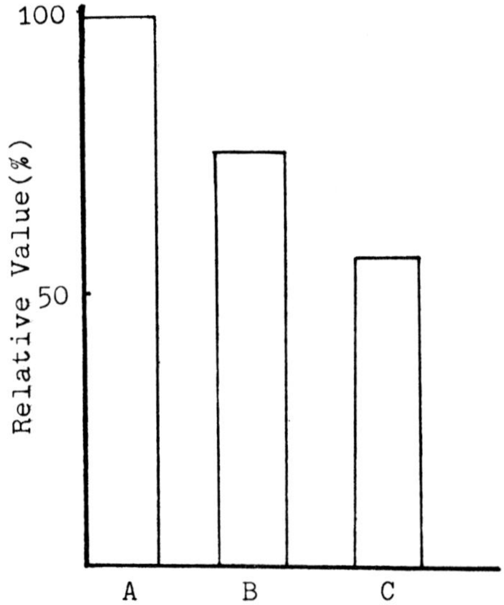

Fig. 1: Effects of trypsin on binding of ^{3}H-6BA to P1. A: P1. B: P1 + trypsin (mg/ml). C: P1 + trypsin (4mg/ml).

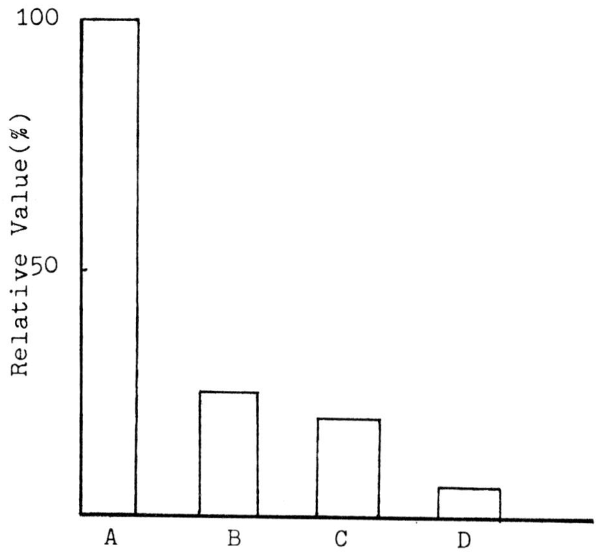

Fig. 2: Effects of dithiothreitol, cysteine and reduced glutathion on binding activity of P1 to ^{3}H-6BA. A: P1. B: P1+dithiothreitol (3 mM). C: P1 + cysteine (3 mM). D: P1 + reduced glutathion (3mM).

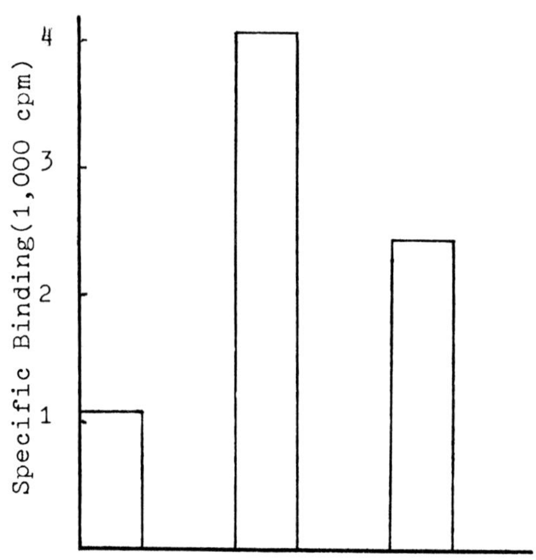

Fig. 3: Binding of ^{3}H-6BA to P1 in different pH mediums.

binding sites in P1 be proteins (Fig.1). Dithiothreitol, cysteine and reduced glutathion depressed the binding activity of ^{3}H-6BA to P1 (Fig.2). These three compounds are reduced agents of disulfide groups in CTK-binding proteins which deal with the binding function of the proteins. The binding activity of ^{3}H-6BA to P1 was higher at pH 6.5 (Fig.3). Treatment with Triton X-100 or SDS decreased binding of ^{3}H-6BA to P1 (Fig.4). It is possible that CTK-binding proteins in P1 are located on some membranes. From cotton ovary homogenate a fraction precipitated by ammonium sulfate (20-50 % saturation) might contain an inactivated factor(s), which could inactivate the CTK-binding proteins by preincubating P1 with this fraction (Fig.5). The factor(s) was thermolabile (Fig.6).

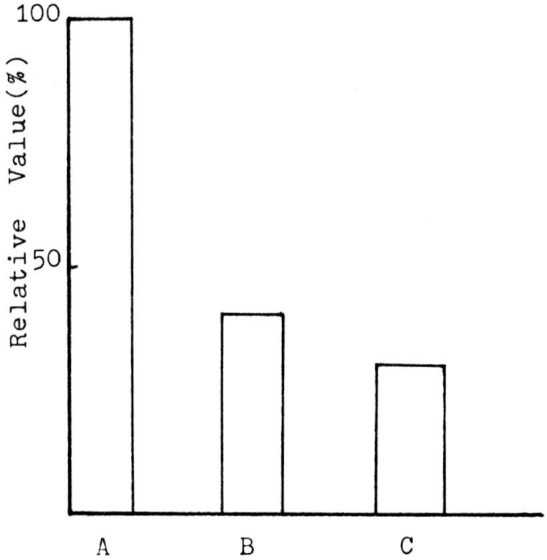

Fig. 4: Effects of Triton and SDS on binding of ^{3}H-6BA to P1.
A: P1. B: P1 + Triton X-100(1%). C: P1 + SDS (0.5 %).

2. Isolation and purification of CTK-binding proteins from young wheat seedlings.

For exploring the properties of CTK-binding proteins we purified the proteins by 6BA Sepharose 6B affinity chromatography from young wheat seedlings. A fraction eluted with 0.1 M NaCl(Fig.7) had the ability to bind ^{3}H-6BA measured by means of equlibrium dialysis (Table 3). The fraction contained several kinds of proteins

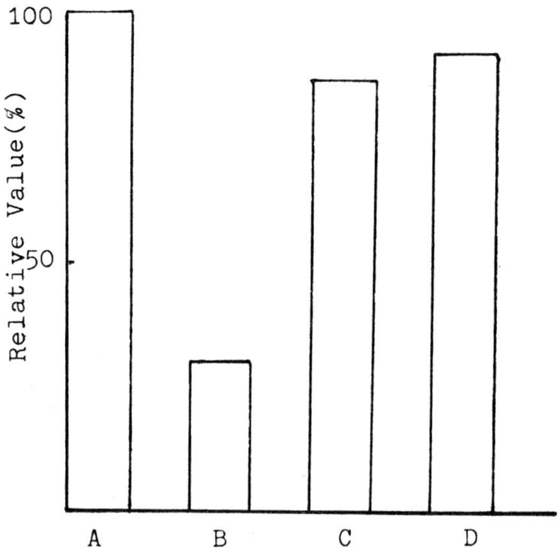

Fig.5: Effects of different fractions precipitated by ammonium sulfate from cotton ovary homogenate on binding of ^{3}H-6BA to P1. A: P1. B: P1 + 138 ug protein precipitated by ammonium sulfate (20-50 % saturation). C: P1 + 155 ug protein precipitated by ammonium sulfate (50-80 % saturation). D: P1 + 86ug protein from supernatant after precipitation of ammonium sulfate (80 % saturation).

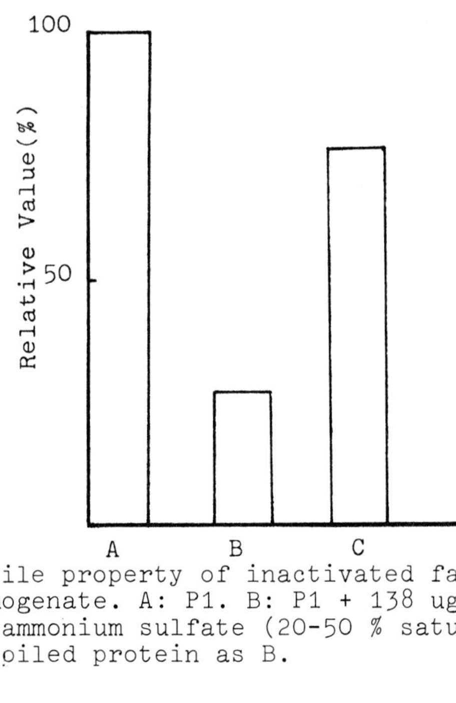

Fig.6: Thermolabile property of inactivated factor(s) from cotton ovary homogenate. A: P1. B: P1 + 138 ug protein precipitated by ammonium sulfate (20-50 % saturation). C: P1 + boiled protein as B.

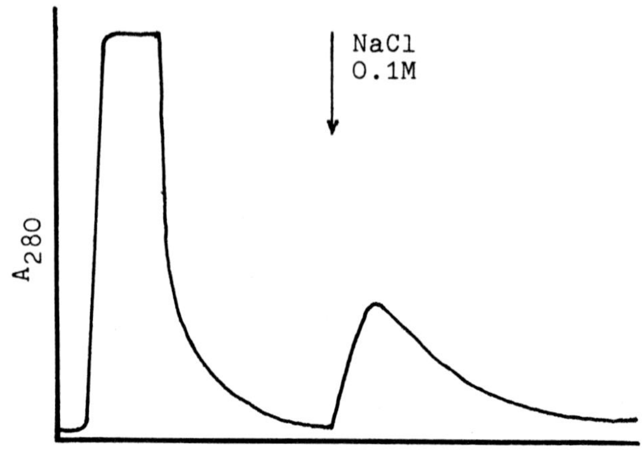

Fig. 7: 6BA Sepharose 6B affinity chromatography for a fraction from young wheat seedling homogenate

Table 3: Binding of ^{3}H-6BA to a fraction eluted from 6BA Sepharose 6B by means of equilibrium dialysis.

Fraction	Binding of ^{3}H-6BA (cpm/mg protein)
Eluted by buffer	32
Eluted by buffer containing 0.1 M NaCl	293

shown by polyacrylamide electrophoreses (Fig.8).We compared the protein patterns from 6BA Sepharose 6B gel with that from a blank Sepharose 6B (preparation as 6BA Sepharose 6B without 6BA), and found that two of the protein bands existing in the fraction eluted from 6BA Sepharose 6B had missed in blank Sepharose 6B (Fig. 8).It was implied that these two protein bands might be CTK-binding proteins. Their molecular weight is 220,000 and 27,000 daltons respectively (Fig. 9).

3. CTK-binding proteins in chloroplasts of wheat leaves

Chloroplasts play a very important role in plant growth and development. They are not only the places for carbon metabolism,but the places for nitrogen metabolism as well.Recently it was reported that 6BA could increase the rate of photophosphorylation in *in vitro* chloroplasts (Huang 1984) which interested us to explore whether or not CTK could be bound to chloroplasts to regulate

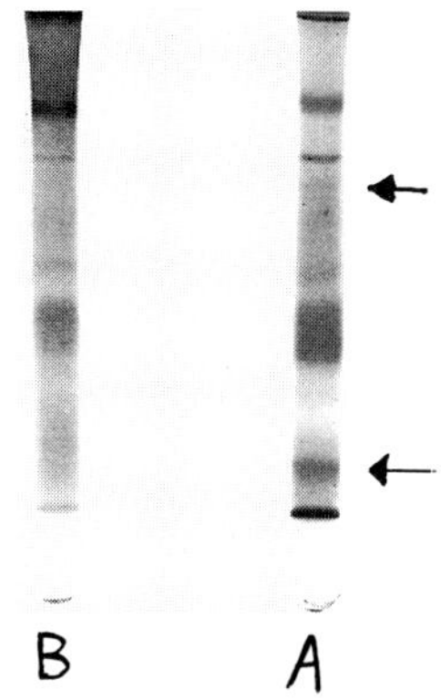

Fig. 8: Electrophoresis of proteins of a fraction from wheat seedlings after affinity chromatography. A: eluted from 6BA Sepharose 6B. B: eluted from blank Sepharose 6B.

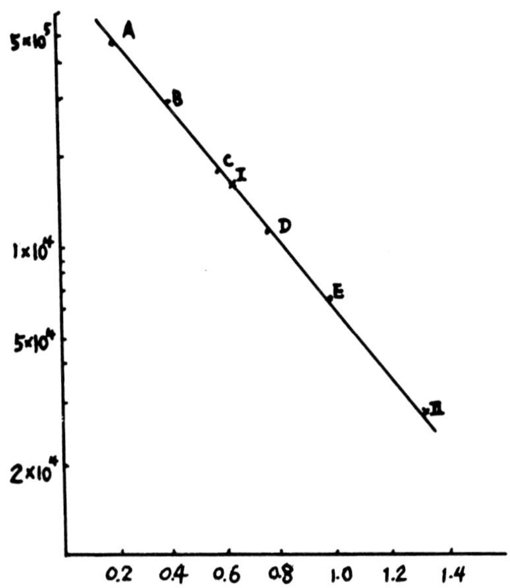

Fig. 9: Measurement of molecular weight of two protein bands from a 6BA Sepharose 6B chromatographic fraction by polyacrylamide electrophoresis. A: Thyroglobulin (669,000 d). B: Ferritin (440,000 d). C: Catalase (232,000 d). D: Lactase Dehydrogenase (140,000 d). E: Bovine Serum Albumin (67,000 d).

photosynthesis or other metabolic processes.There was a distinct obstacle to learn 6BA binding to chloroplasts because of the pigments which have serious quench effects in scintillation solution, and methods of equilibrium dialysis and centrifugation assay which were usually used for assaying hormone-binding protein were not available for chloroplasts. Based on centrifugation assay, we took an additional step by using H_2SO_4 to extract ^{3}H-6BA bound to the pellets. We found that ^{3}H-6BA bound to chloroplast fractions could fall off by treatment of H_2SO_4. Radioactivity in H_2SO_4 could indirectly reflect the amount of ^{3}H-6BA bound to chloroplasts. The radioactivity decreased with cold 6BA concentrations from 1 x 10^{-4} M to 8 x $10^{\ 4}$ M in reaction mediums (Fig. 10).

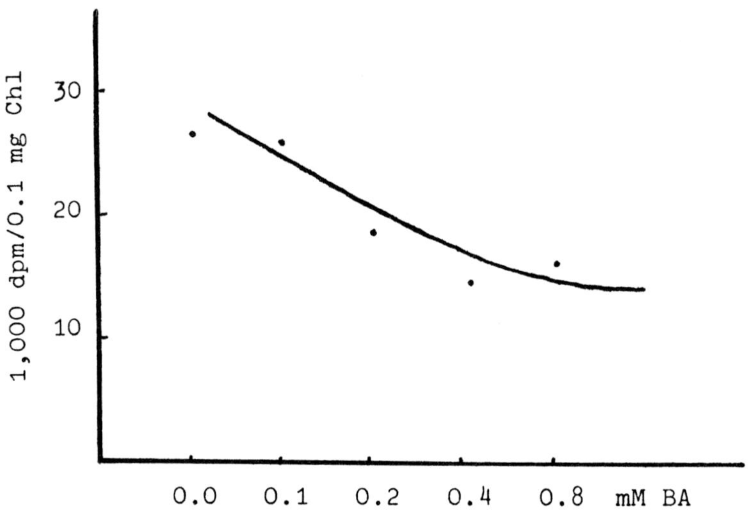

Fig. 10: Binding of ^{3}H-6BA to chloroplast fractions.

Since the chloroplasts had been broken by lower osmotic pressure and treated by ultrasonic before binding reaction, there was not any intact envelope membrane with chloroplast and the binding of ^{3}H-6BA might take place on surface of thylakoids.

When the broken chloroplasts were boiled for 5 min, ^{3}H-6BA-binding activity disappeared (Table 4).It is suggested that CTK-binding sites are proteins. They can bind CTK when they have a certain natural structure. If the proteins are denatured by boiling, CTK-binding ability will be lost. Binding activity changed with reaction mediums in different pH and it was higher at pH 7.6 (Table 5). CTK-binding proteins were specific to bind CTK. When reaction medium contained different natural plant

Table 4: Thermolabile property of CTK-binding proteins in chloroplast fractions

	Water (dpm/0.1 mg	0.4 mM 6mM 6BA Chl)	%
Control	21,353	16,048	-24.8
Boiled for 5 min	22,857	23,188	1.4

Table 5: Binding of ^{3}H-6BA to chloroplast fraction in different pH.

PH	Water (dpm/0.1 mg	0.4 mM 6BA Chl	%
7.6	21,353	16,048	-24.8
7.2	21,048	18,841	-10.5
6.5	22,462	22,274	- 0.8

hormones, only zeatin, a kind of CTK, could substitute for ^{3}H-6BA (Table 6). If the reaction medium contained same concentration of cold 6BA and ABA, the substitution of cold 6BA for ^{3}H-6BA declined (Table 7). There are extensive antagonistic effects of ABA and CTK on plant growth and development (Longo 1981). ABA action against CTK in plant might be caused by ABA altering the properties of CTK-binding proteins.

Compared with the researches on animal hormone receptors, researches on plant hormone receptors began later. Hormone-binding receptors in animals usually act at the level of transcription and translation (Berridge 1970). So at the beginning of CTK-receptor investigation some researchers only paid attention to cell components which involved the transcription and translation such as ribosome. Although CTK was found to be bound to biomembrane (Sussman 1978; Gardner 1978) later, it is difficult to make out the location of binding sites with non-purified cell fractions. Our work with purified chloroplasts has shown that CTK could be bound to chloroplast fractions. It is suggested that besides the regulation of CTK on transcription and trans-

Table 6: Effects of different plant hormones on binding of ^{3}H-6BA to chloroplast fractions

	Water	0.4 mM 6BA	0.4 mM zeatin	0.4 mM IAA	0.4 mM GA_3
dpm/ 0.1 mg Chl	28,191	16,389	21,469	28,676	25,739
%	____	-41.9	-23.8	1.7	-8.7

Table 7: Affects of ABA on binding of ^{3}H-6BA to chloroplast fractions

	Water	0.4 mM 6BA	0.4 mM ABA	0.4 mM 6BA 0.4 mM ABA
dpm/ 0.1 mg Chl	31,676	19,030	30,289	28,253
%	____	-39.9	-4.4	-10.8

lation, if possible, the receptors after binding CTK might regulate the membrane permeability directly. How CTK is bound to membrane in chloroplasts then effects metabolic processes is an interesting problem which should be considered.

(1) Berridge,M.V., Ralph,R.K., Letham,D.S., Biochem.J. 119 (1970) 75-84

(2) Cleland,W.W.: Biochem. 3 (1964) 480-482

(3) Davis.B.J.: Ann.N.Y.Acad.Sci. 121 (1964) 404.408

(4) Fox,J.E., Erion,J.L.: Biochim.Biophys.Res.Commun. 64 (975) 694-700

(5) Fox,J.E., Erion,J.L. In "Plant Growth Regulation", Pilet,P.E. (Ed.) pp 139-146. Berlin-Heidelbaerg-New York: Springer

(6) Fox,J.E., Gregerson,E., (1982) In "Plant Growth Substances" Wareing,P.F.Ed.) pp 207-214. Academic, London, UK

(7) Gardner,G., Sussman,M.R., Kende,H., Plant Physiol.Suppl. 56 (1975) 28

(8) Gardner,G., Sussman,M.R., Kende,H.: Planta 143 (1978) 67-73

(9) Hertel,R., Thomson,K.S., Russo,V.E.A.: Planta 107 (1972) 325-340

(10) Huang,Z.H., Wei,J.M.: Acta Phytophysiologia Sinica 10 (1984) 161-167

(11) Kasamo,K., Yamaki,T.: Plant Cell Physiol. 17 (1976) 149-164

(12) Kobayashi,K., Zbell,S., Reinert,J.: Protoplasma 106 (1981) 145-155

(13) Longo,G.P., Stopelli,B., Rossi,G., Longo,C.P.: Physiol. Plant. 53 (1981) 82-86

(14) Rubery,P.H.: Ann.Rev.Plant Physiol. 32 (1981) 569-596

(15) Sussman,M.R., Kende,H.: Planta 140 (1978) 251-259

THE GIBBERELLIN RECEPTOR

Lalit M. Srivastava
Department of Biological Sciences
Simon Fraser University
Burnaby, B. C. Canada, V5A 1S6

CONTENTS

Introduction

As a class of plant hormones, gibberellins (GAs) are involved in regulating or controlling several different morphogenetic and biochemical responses. These responses include synthesis and/or release of α-amylase and other hydrolases in aleurone tissue of cereal grains; elongation growth of epicotyls, hypocotyls, leaf sheaths; induction of cambial activity and phloem tissue formation in trees; flower formation; antheridia formation in fern gametophytes, etc. (see review by Jones, 1973). The regulation of such diverse activities has prompted the suggestion that gibberellins act at the chromatin level and bring about differential gene activation via receptor molecules (Kende and Gardner, 1976; Varner and Ho, 1976; Stoddart and Venis, 1980). Since seemingly minor changes in the gibberellin structure can result in profound changes in biological activity

NATO ASI Series, Vol. H10
Plant Hormone Receptors. Edited by D. Klämbt

(Fig. 1), and since proteins are the preferred macromolecules for structural specificity, it is further assumed that the receptor molecules are proteins which on the one hand recognize the

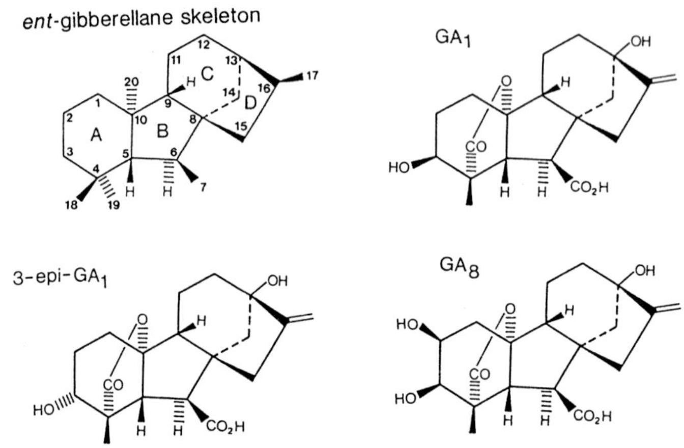

Figure 1. Ent-gibberellane skeleton and structure of GA_1, 3-epi-GA_1 and GA_8. GA_1 is highly active in legumes and cereals, but 3-epi-GA_1 with a 3α-OH or GA_8 with a 2β-OH are both biologically inactive.

gibberellin molecule and on the other the appropriate acceptor site in chromatin to bring about gene activation. The analogy with steroid hormones in animals is often given (for steroid action, see Schrader et al., 1981; Greene and Press, 1986; Scheidereit et al., 1986).

In support of the above scheme for gibberellin action, there has accumulated some very elegant evidence indicating that gibberellin treatment leads to new transcriptional activity and production of new proteins. Most of this work has been done on aleurone tissue of cereal grains and has been summarized by Jacobsen (1983) and in some more recent papers (Jacobsen and Beach, 1985; Mundy et al., 1985; Hammerton and Ho, 1986; Zwar and Hooley, 1986). However, the earlier part of the scheme, that is, interaction of gibberellin with chromatin is still a profound mystery.

In this paper, I would like to present some of the work in our laboratory which suggests that there are indeed protein molecules which recognize gibberellins and which fulfill the criteria associated with hormone receptors. Some possible models for mechanism of gibberellin action and some data on purification of protein from cucumber are presented. I must emphasize that this is not a review article and that I will be selective in my treatment of the published literature.

Criteria for hormone-receptor binding

Criteria for hormone-receptor binding are well known and specifically for plant hormones have been discussed by Kende and Gardner (1976) and Stoddart and Venis (1980). These criteria are reviewed briefly to indicate the sense in which we use them: 1. If receptor molecules exist, then *ipso facto* they must have a defined number and consequently there must be saturability of GA uptake. 2. The receptor binding to hormone is assumed to be noncovalent or at least transient and hence exchangeable with nonradioactive for radioactive ligand or *vice-versa*. 3. The receptor molecules must show a high affinity for biologically active hormone or analogs. The high affinity is a consequence of a stereospecific fit of the hormone in the active site of the receptor. Hence, proportionally speaking, for a hormone with high biological activity, a lesser number of hormone molecules should at any time be required to fill half the available number of binding sites than for a less biologically active ligand. Thus, the numerical value of equilibrium dissociation constant or K_d[1] is low, (e.g., in the range of 10^{-8} M to 10^{-10} M, or <) when the specificity or affinity of the receptor for the ligand is high. 4. Finally the data and kinetics of binding must generally conform to biological activity as known from bioassays, endogenous concentrations of hormones, and known metabolic pathways. In other words binding must be biologically meaningful.

---

[1] K_d is the ratio of k_d/k_a where k_a is the rate of formation of the receptor-ligand complex and k_d is the rate of dissociation of that complex. This use of K_d has much the same meaning as K_m in enzyme kinetics, which refers to the affinity of an enzyme for its substrate.

Even if all these criteria are satisfied the final proof that one is dealing with receptor protein can only be obtained by getting an unambiguous and known gibberellin-induced response in a receptor mutant by administration of the hormone-receptor complex.

[^{3}H]GA-binding *in vivo*

Until a few years ago none of these criteria were satisfied for gibberellins. For instance, several attempts by Kende and his associates to demonstrate an association of radioactivity to macromolecular fractions and saturability of uptake in dwarf pea epicotyls or barley aleurone layers fed [^{3}H]GA_1 or [^{3}H]GA_5 up to 10^{-4} M were unsuccessful (Kende, 1967; Ginzburg and Kende, 1968; Musgrave and Kende, 1970; Musgrave *et al*., 1969, 1972). Instead, in barley aleurone tissue, there was a large accumulation of the biologically inactive [^{3}H]GA_5 methyl ester and this accumulation was directly proportional to GA metabolism (Musgrave *et al*., 1972). Silk *et al*. (1977) also observed no saturation of GA uptake by lettuce hypocotyl segments kept in [^{3}H]GA_1 up to 10^{-2} M concentration. Our results using barley aleurone tissue were similar. These data are summarized in Figure 2 which shows that there is no saturation of GA uptake by these tissues either with time or with increasing concentration.

Figure 2. A diagrammatic plot showing no saturation of [^{3}H]GA_1 uptake by aleurone tissue at 25°C with time or increasing concentration of [^{3}H]GA_1.

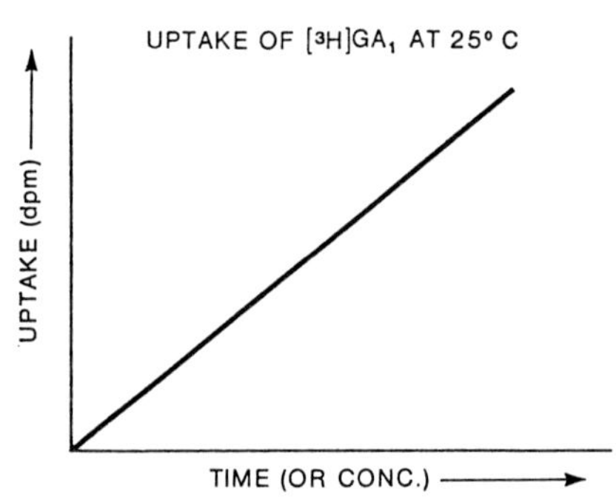

In a pioneering study, Stoddart _et al_. (1974) fed [^{3}H]GA$_1$ to dwarf pea epicotyls for 12 h, homogenized the tissue at 2°C, and passed the 20,000_g_ supernatant through G-200 Sephadex column. They showed that [^{3}H]GA$_1$ was bound to two macromolecular fractions, a high molecular weight (HMW) and an intermediate molecular weight (IMW) fraction, and that the binding was competed for _in vivo_ by biologically active keto GA$_1$ (see Keith and Srivastava, 1981) but not by biologically inactive 3-epi-GA$_1$. While the authors could demonstrate that the radioactivity bound to HMW and IMW fractions was authentic [^{3}H]GA$_1$, and not a metabolite, the bound activity could not be exchanged _in vitro_. Moreover, as in the previous study by Musgrave _et al_. (1972), a large component of radioactivity taken up was present as GA metabolites, especially the biologically inactive GA$_8$.

We argued that GA uptake by plant tissues was essentially nonsaturable at room temperature because plants had evolved to regulate the endogenous levels of active gibberellin not by controlling uptake but by metabolism. Excess GA taken up was conjugated or converted to inactive forms and perhaps sequestered in vacuoles or some other compartment. Accordingly, if uptake experiments were done at a temperature when metabolism was slowed down or stopped, a saturation of GA uptake could be observed (Keith _et al_., 1980). We chose barley aleurone tissue for these experiments for 3 reasons: 1. the tissue was relatively homogeneous being composed of the same kind of cells; 2. the tissue was only a few cell layers thick thus minimizing problems with GA transport; 3. aleurone tissue, among all the other bioassay systems, was supposed to have the least capacity for GA-metabolism (Crozier _et al_., 1970; Reeve and Crozier, 1974).

Figure 3 shows the uptake of [^{3}H]GA$_1$ by barley aleurone layers as a function of temperature. Saturation is obtained at 1° and 1.5°C between 24-48 h; by contrast, at 3° or 4°C saturation does not occur. If the layers are incubated at 0°C and then the temperature raised to 4°C, there is an immediate increase in uptake (see Keith _et al_., 1980). A measurement of GA-metabolites

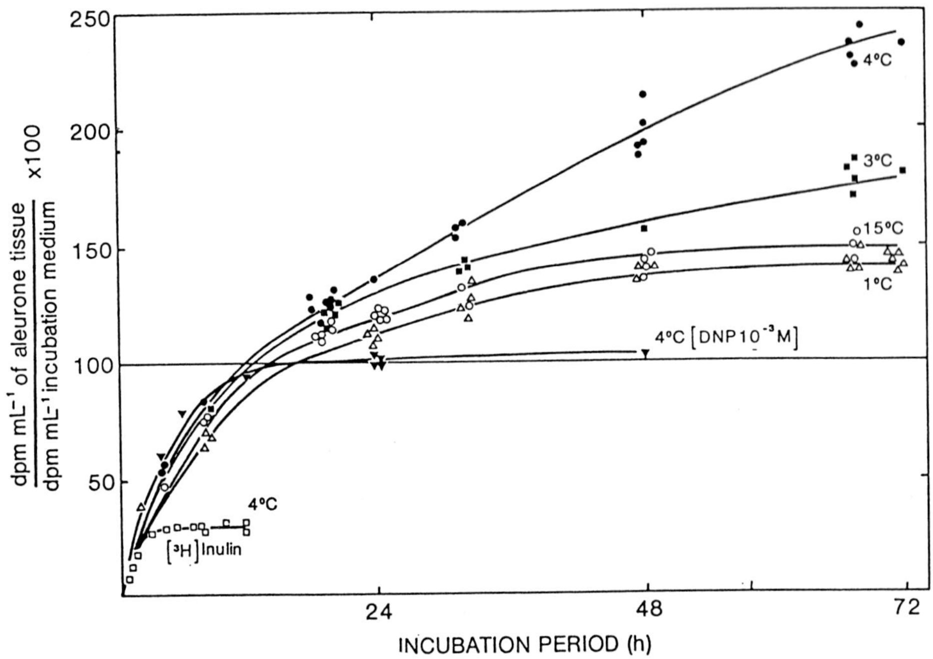

Figure 3. The effect of temperature on the uptake of $[^3H]GA_1$ into barley aleurone layers. Concentration of $[^3H]GA_1 = 4.4 \times 10^{-9}$ M (41.5 Ci mmol^{-1}). Uptake of $[^3H]$inulin into free space, and uptake of $[^3H]GA_1$ (4.4×10^{-9} M) in the presence of DNP (1 mM), both at 4°C, are also shown. Level of $[^3H]GA_1$ activity in incubation medium = 100% = 415×10^3 dpm mL^{-1}. Level of $[^3H]$inulin activity in incubation medium = 100% = 60×10^3 dpm mL^{-1} (adapted from Keith et al., 1980).

produced in aleurone cells at various incubation temperatures shows that at 1.5°C and lower temperatures the radioactivity is mostly in the form of $[^3H]GA_1$, whereas at higher temperatures there is considerable and increasing conversion of $[^3H]GA_1$ into polar metabolites (Fig. 4).

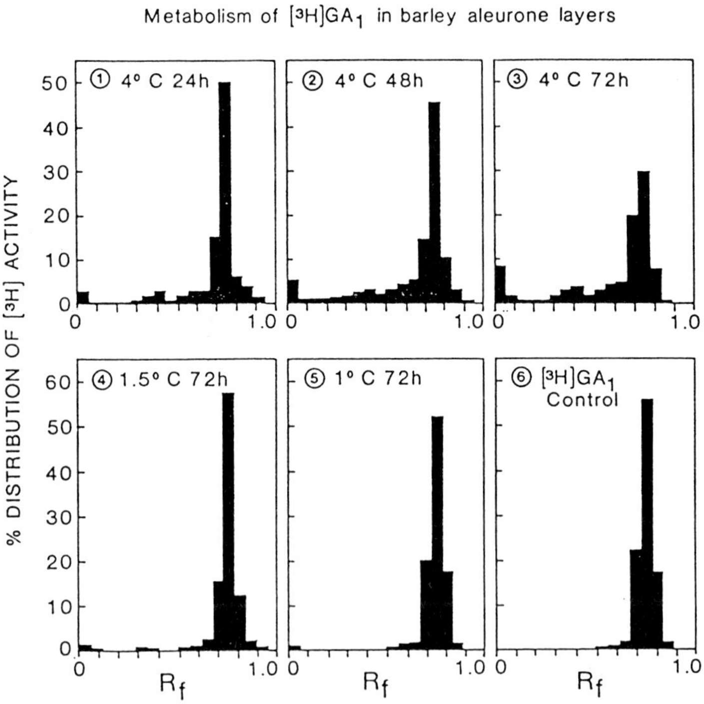

Figure 4. TLC on Chrom-AR of 80% (v/v) ethanolic extracts of barley aleurone layers incubated in $[^3H]GA_1$ at 4°C, 1.5°C and 1°C. Rf of values of metabolites according to Nadeau et al. (1972): Ampho GA_1, 0.09; GA_8-glucoside, 0.41; GA_1-glucoside, 0.52; GA_8, 0.74. Actual cpm x 10^{-3} per 30 aleurone layers at 50% counting efficiency: (1) 29.2, (2) 45.4, (3) 54.6, (4) 26.7, (5) 25.8 (adapted from Keith et al., 1980).

At equilibrium, the internal concentration of $[^3H]GA_1$ in aleurone cells was higher than in the ambient medium (Fig. 3), which suggests binding to a cellular component. Accordingly, we incubated aleurone tissue in a known concentration of $[^3H]GA_1$ for 72 h at 0.5°C with and without a 50 fold excess of nonradioactive GA_1 or GA_8. As the data in Table 1 show, some of the bound radioactivity was competed for by nonradioactive GA_1 but not by the biologically inactive GA_8.

These experiments were repeated with cut slices of dwarf pea epicotyls using $[^3H]GA_1$ and cucumber hypocotyls using $[^3H]GA_4$ with similar results (Keith and Srivastava, 1980, Keith et al., 1981).

Table 1. Effects of nonradioactive GA_1 or GA_8 on uptake of $[^3H]GA_1$ by barley aleurone tissue.

30 aleurone layers per sample (7 samples/series A, B, or C) were incubated in $[^3H]GA_1$ (A), or $[^3H]GA_1$ plus 50 fold nonradioactive GA_1 (B), or $[^3H]GA_1$ plus [50 fold nonradioactive GA_8 (C) for 72 h at 0.5°C. At the end of incubation the aleurone layers were blotted in a standard way and the radioactivity take up was determined by scintillation counting. Specific activity of $[^3H]GA_1$ was 1.09 Ci $mmol^{-1}$. Adapted from Keith *et al.* (1980).

$[^3H]GA_1$ in Incubation Medium ($dpm\ mL^{-1}$)	Aleurone Tissue ($dpm\ mL^{-1}$) A	B	C
160,100 (66.7 nM)	222,379 ±4044	200,445 ±3874	219,177 ± 3426

$[^3H]GA_4$-binding *in vitro*

Our next task was to develop an *in vitro* system where the criteria of saturability and exchangeability could be demonstrated and in addition sufficient experiments could be run in a short time to get meaningful kinetic data on binding. For this work we chose to work with cucumber. The 100,000*g* cucumber cytosol or desalted and resuspended $(NH_4)_2SO_4$ precipitate of the 100,000*g* cytosol were assayed for binding activity (for extraction and assay procedures, see Keith *et al.*, 1981, 1982; Yalpani *et al.*, 1986).

The assays involved incubating the cytosol with $[^3H]GA_4$ with and without a 100 fold excess of nonradioactive GA_4 to determine the amount of specific binding. In earlier assays, aliquots were loaded on G-100 columns and radioactivity bound to protein fractions determined. In these assays it became clear that we were dealing with a soluble macromolecule. Assays done with 20,000*g*, 40,000*g*, or 100,000*g* cytosol showed little difference in the amount of $[^3H]GA_4$ specifically bound, and addition of detergents to extraction buffer did not enhance the level of specific binding. It was also shown that the binding was occurring to a protein as heat and protease treatments destroyed specific binding, but DNase, RNase, and lipases did not. There was a pH optimum of 7.5 and there did not appear to be any requirement for cationic cofactors because specific binding was unaffected by the presence or absence of EDTA (Keith et al., 1981).

The problem with Sephadex filtration was that it was slow and the binding of $[^3H]GA_4$ to protein, being noncovalent, was disrupted as soon as the mixture was loaded onto the column. Accordingly, a modified filter paper assay was developed. The method relies on the protein-binding capacity of DEAE-cellulose filter discs. An 100-200 µL aliquot of cytosol or protein preparation, which has been previously incubated with $[^3H]GA_4$ for 1-2 h on ice, is loaded onto the filter paper and then washed under suction with buffer to remove unbound radioactivity. After the wash, the filter papers are allowed to run dry for 2-5 seconds and then transferred to scintillation vials containing methanol and scintillant and counted. From the time the aliquot is taken to the time the filters are transferred to methanol it takes between 90-120 seconds and up to ten replicates can be processed at the same time (for details, see Keith et al., 1982; Yalpani et al., 1986).

Figure 5 shows the saturability of $[^3H]GA_4$ bound to macromolecules in cucumber cytosol. Saturation is reached between 30-60 minutes and that value remains more or less

Figure 5. Total binding with time of 100,000g cucumber cytosol incubated in $[^3H]GA_4$ (125 nM) and the effect of addition of unlabeled GA_4 (1000 fold) supplied after 3 h (indicated by arrow) on that binding (from Keith et al., 1981).

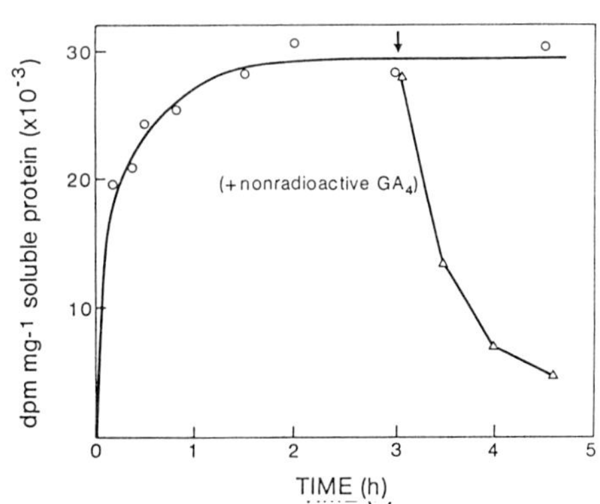

constant for 4 to 5 h. If an excess of nonradioactive GA_4 is added after saturation is reached, there is an immediate decline in the amount of radioactivity bound to protein which confirms that a good proportion of radioactive ligand molecules is exchangeably bound. The half-life of dissociation under these conditions in different experiments is about 6 to 10 minutes.

These experiments are done at 0-2°C when we have previously shown that metabolism of GAs is almost nonexistent and that the radioactivity bound to the protein fraction cochromatographs with authentic $[^3H]GA_1$ (Keith et al., 1980). Recently, this was reconfirmed by using GLC (data not shown).

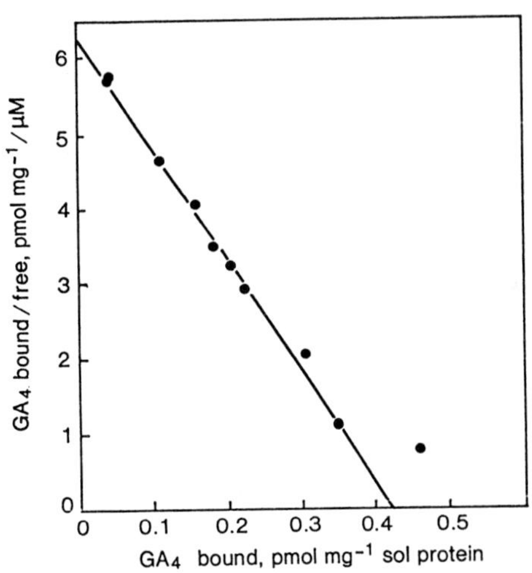

Figure 6. A Scatchard plot of $[^3H]GA_4$ binding to cucumber cytosol. Triplicate aliquots of cytosol were incubated with $[^3H]GA_4$ (concentration range tested, 6-600 nM) with (series B) and without (series A) a 100 fold excess of unlabeled GA_4. $[^3H]GA_4$ binding was determined by the DEAE-cellulose filter assay. Specific binding was calculated by subtracting series B from series A. The straight line corresponds to K_d = 70 nM and n = 0.4 pmol mg^{-1} soluble protein (after Keith et al., 1982).

The kinetics of binding of $[^3H]GA_4$ to cucumber cytosol are plotted in Figure 6. A single class of binding sites is indicated and the number of binding sites is about 0.4 pmol mg^{-1} soluble protein. The K_d or the concentration at which half the binding sites are occupied was calculated to be about 70 nM.

To see whether the biologically active GAs compete for the same binding site, the cucumber cytosol was incubated with a range of $[^3H]GA_4$ concentrations alone and in the presence of nonradioactive but biologically active GA_7 and GA_4 and biologically inactive GA_{26}. A double reciprocal plot showed that GA_7 and GA_4 competed for the same site whereas GA_{26} showed no competition (Fig. 7).

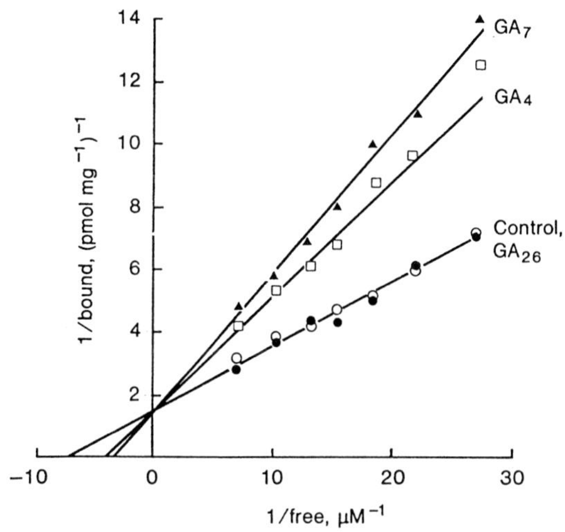

Figure 7. Double-reciprocal plots of $[^3H]GA_4$ binding to cucumber cytosol in the presence of various GAs. Samples (0.5 ml) of cytosol were incubated with $[^3H]GA_4$ at a range of concentration in the presence or absence of a suitable concentration of each GA. After 1 h incubtion, binding was assayed by DEAE-cellulose filtration. GAs tested as competitors of binding were: o, control, no addition; Δ, GA_7, 0.12 μM; , GA_4, 0.12 μM; •, GA_{26}, 430 μM (from Keith et al., 1982).

From several experiments using a range of gibberellins, GA derivatives, and nonGAs it can be concluded that biologically active GAs compete for these binding sites in proportion to their activity in the cucumber hypocotyl bioassay, whereas inactive GAs such as GA_8, GA_{26}, methyl esters of GA_3, GA_4, GA_7,and nonGAs such as IAA, (±) ABA, or kinetin do no compete (Keith et al., 1981, 1982; Yalpani and Srivastava, 1985). (The judgement that a ligand is competing or not is to some extent subjective and is based on a certain maximal concentration of the ligand. For our purposes if a ligand does not show competition at 100 to 1000 fold concentration of the 'natural' ligand we assume that it is noncompetitive.)

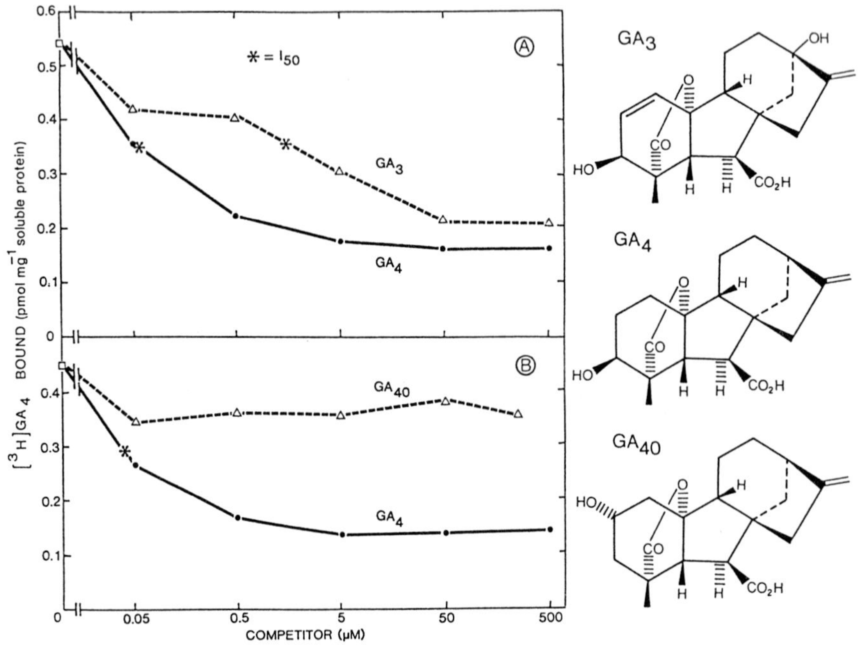

Figure 8. Displacement of $[^3H]GA_4$ binding from cucumber cytosol by GA_3 (A) and GA_{40} (B) compared to GA_4. A 60% $(NH_4)_2SO_4$ pellet of cucumber cytosol was desalted and incubated with 50 nM $[^3H]GA_4$ in the absence or presence of the competing ligands at the concentrations shown. Binding of $[^3H]GA_4$ to the GA binding protein was measured using the DEAE-filter paper assay. I_{50} represents the concentration of the competitor which displaces 50% of the exchangeable, specific binding of $[^3H]GA_4$ from the binding protein (after Yalpani and Srivastava, 1985).

In another series of experiments the so-called I_{50} value for various GAs and GA derivatives was determined (Yalpani and Srivastava, 1985). The I_{50} value is that concentration of a competing analog which displaces 50% of the $[^3H]GA_4$ specifically bound to cucumber cytosol or protein preparation. In these experiments the $[^3H]GA_4$ concentration was kept constant at 50 nM and the concentration of the competing ligand was varied from 50 nM to 250 μM, in some cases even 500 μM. As the plots in Figure 8 show the I_{50} value for GA_4 was 50 nM, and that for GA_3 was about 2.5 to 5.0 μM; GA_{40} even at 250 μM was unable to displace 50% of $[^3H]GA_4$ specifically bound.

Table 2. Relative *in vitro* affinities of various ligands to $[^3H]GA_4$ binding protein in cucumber cytosol and their *in vivo* biological activity in cucumber hypocotyl bioassay.

Concentration of $[^3H]GA_4$ in the incubation medium was 50 nM and that of competing ligands ranged from 50 nM to 0.5 mM, except for GAs with superscript[c]. Relative affinity was measured using I_{50} values determined for each ligand using the DEAE-filter paper assay.

I_{50} Competitor	Competitor
50 nM	GA_4(+++)[a], DiMeGA_4(+++)[b], GA_7(++++)
0.5 μM	GA_2(++)
5.0 μM	GA_1(++), GA_3(++)
50 μM	GA_5(+), GA_9(+++), GA_{30}[c](+)
0.5 mM	3-epi-GA_4(0)[d]
>0.5 mM	GA_3ME(0), GA_4ME(0), GA_6[c](+), GA_7ME(0), GA_8(0), GA_{13}(0), GA_{14}(0), GA_{15}(++), GA_{16}[c](+), GA_{20}(0), GA_{22}(0), GA_{27}(0), GA_{36}[c](+++), GA_{40}[c](++), ABA(0), IAA(0), Kinetin(0)

[a]Relative activity in the cucumber hypocotyl bioassay (++++, very high; +++, high; ++, moderate; +, low; 0, very low or inactive, after Crozier and Durley, 1983, except for those w superscript[d] or[b]). [b]Relative *in vivo* activity after Hoad *et al*.,(1981). [c]Competitor assayed *in vitro* at concentratio ranging from 50 nM to 0.25 mM. [d]N. Yalpani, L.M. Srivastava (unpublished data). (From Yalpani and Srivastava, 1985).

Values for some other GAs and derivatives are shown in 2. While generally confirming the published information fr bioassays using cucumber hypocotyls (Crozier, 1981; Crozier Durley, 1983), the data refine some information and point me important exceptions. Thus: 1. GA_4, GA_7 and 2,2-dimethyl have the same affinity for the $[^3H]GA_4$ binding protein. In the bioassays 2,2-dimethyl GA_4 and GA_7 are reported to have an activity equal to or slightly higher than that of GA_4. We suspect that that is probably due to their high permeability in plant tissue compared to that of GA_4 (see also Hoad *et al.*, 1981). 2. The important exceptions are GA_9 and GA_{36} (Fig. 9), both of which are reported to have high activity in the cucumber bioassay (Crozier and Durley, 1983), but show no affinity for the $[^3H]GA_4$ binding protein. This is especially significant because GA_{36} is reported to be the immediate precursor of GA_4 in cell free extracts of cucurbits (Hedden, 1983). Also, the 2α-hydroxylated GA_{40} showed no binding affinity in our assay (Fig. 8), though it is reported to have some activity in the cucumber bioassay (Sponsel *et al.*, 1977).

Figure 9. Structure of GA_9 and GA_{36}. Both show high biological activity in cucumber bioassay but show little affinity for the GA-binding protein *in vitro*.

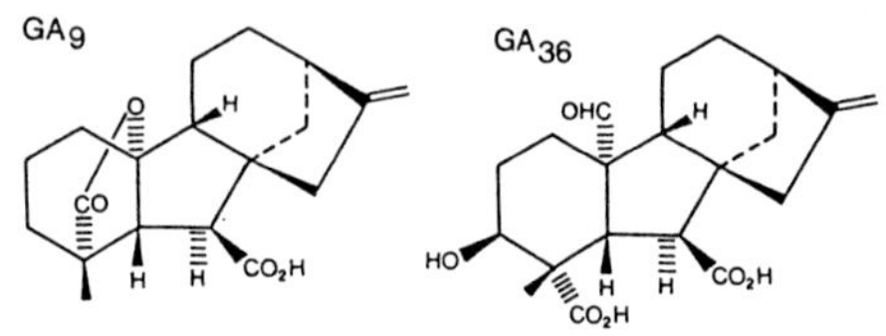

Although our work to date has concentrated on cucumber, we have done $[^3H]GA_4$ binding assays using 100,000*g* cytosol and desalted, resuspended ammonium sulfate precipitate of cytosol from dwarf and tall pea. Specific and exchangeable binding is btained in each case. The kinetic data indicate a K_d = 130 nM d n = 0.66 pmol·mg^{-1} soluble protein for dwarf pea and a K_d = nM and n = 0.43 pmol mg^{-1} soluble protein for tall pea (Liu d Srivastava, 1986, and unpublished results).

relation with dose-response curves and endogenous gibberellin entrations

Appendix I gives some other saturation data and correlates t inetics of binding with published information on dose-response curves in cucumber bioassay and estimated endogenous gibberellin concentrations in vegetative tissues. It can be seen that the saturating concentration for cucumber cytosol is within the concentration range for biological response and further that the saturation *in vivo* occurs at about 1 to 2 orders of magnitude less than the exogenous concentration at which the biological response is saturated. It can also be calculated that the saturating concentration of GA_4 in target tissue of cucumber is 46.5 $\mu g \cdot Kg^{-1}$ fresh tissue as opposed to estimated endogenous content in vegetative tissues of about 1 to 10 $\mu g \cdot Kg^{-1}$ fresh tissue (Stoddart, 1983).

Binding to a receptor vs. an enzyme

On the basis of the above data we think that we are dealing with a gibberellin receptor. Nonetheless, we are aware that we may be witnessing specific binding of $[^3H]GA_4$ to a metabolizing enzyme, such as a 2β-hydroxylase or an enzyme in the synthetic pathway from GA_{12} aldehyde to biologically active C-19 GAs. While the latter possibility cannot be excluded with the data currently available, we doubt it for several reasons: 1. The 2β-hydroxylases and the enzyme that converts the C-20 GA_{36} to C-19 GA_4 have strict cofactor requirements such as Fe^{2+} or Fe^{3+}, NADPH or ascorbate, and 2-oxoglutaric acid (Hedden, 1983; Smith and MacMillan, 1986). These cofactors are not necessary, in fact inhibit specific binding of $[^3H]GA_1$ to 100,000*g* cowpea cytosol fractions (Dr. Brian Keith, personal communication). 2. Our *in vitro* competition assay shows no affinity of the protein for the C-20 GA_{36} which would be expected if we were dealing with an enzyme for which this GA was a substrate. 3. The binding protein shows no affinity for the 2β-hydroxylated GA_8 (we recognize that this GA has not been reported in the cucurbits), though it does show the same high affinity for 2,2-dimethyl GA_4 as for GA_4.

Architecture of the active site in Cucumber and Pea

Since the structures of various GAs and GA derivatives are known, the data from the I_{50} values can be used to deduce some characteristics of the active site of the putative receptor. The data suggest a structural specificity of the binding protein in cucumber for γ-lactonic C-19 GAs with a C_3-hydroxyl and a C_6-carboxyl group. Additional hydroxylations of C_{16} in the D ring and of C_{13} and C_{12} in the C ring progressively impede binding. Changes in the hydroxylation pattern of the A ring either curtail binding affinity or completely eliminate it. The data suggest the presence of strong hydrophobic environments in the active site of the receptor protein corresponding to C_{18} and the 1α-, 2α- and β-, and 3α-positions, and strong polar or ionic interaction in the vicinity of 3β-OH, the γ-lactone ring, and the C_6-carboxyl (see also Serebryakov _et al_., 1984).

Halogenated and alkyl derivatives of GAs have often been used to deduce structure-activity relationships using bioassays (see e.g., Hoad, 1983). While these data are difficult to interpret because of possible removal of substituent groups as well as permeability problems in the living tissue, nonetheless, some of these data support the architecture of the active site as postulated above. Thus, 1β-iodo GA_1 and GA_4 were much less potent than the parent GA_1 and GA_4 in several bioassays (Keith _et al_., 1979; Hoad, 1983); 3,13-difluoro GA_3 was much less active than GA_3 in the barley aleurone assay (Jones, 1976); and while 2β-methyl and 2,2-dimethyl derivatives were equally or more potent than the parent GA, increasing the alkyl chain length resulted in lessened activity (Hoad _et al_., 1981). Thus, not only the polarity but also the size of the substituent atom or group are important in biological activity. These data reinforce the notion of a precise stereospecific fit of active GAs with a receptor protein and the importance of weak interactions, operative over 2-4 Å, in that fit.

Based on dose-response relationships, GA_1 and GA_3 appear to be considerably more active in pea epicotyl and barley aleurone assays than GA_4 and GA_7, while the reverse seems to be true in cucumber hypocotyl bioassay (Katsumi _et al._, 1965; Keith _et al._, 1979; Crozier, 1981). Our competition data using [^{3}H]GA_4 or [^{3}H]GA_1 and protein preparations from the 100,000_g_ cucumber or dwarf pea cytosol indicate that the cucumber protein has about 50 to 100 times more affinity for GA_4 and GA_7 than for GA_1 or GA_3 (Yalpani and Srivastava, 1985). By contrast, the protein preparation from the dwarf pea cytosol shows 50 to 100 times more affinity for GA_1 and GA_3 than for GA_4 or GA_7 (unpublished data). Since our data are obtained under conditions of little or no metabolism, the likely explanation is that these differences between legume and cucurbit families in respect to GA response represent differences in the amino acid composition of the receptor protein in the vicinity occupied by the 13α-OH of the gibberellin structure. A proof of this assumption must await the purification and sequencing of these proteins.

Receptor in nontarget cells

There have been very few experiments dealing with specific binding of [^{3}H]GAs to fractions from nontarget regions and no data are available for computation of K_d or number of binding sites at saturation. However, both in cucumber and in dwarf pea specific [^{3}H]GA binding was greater in the GA-responsive apical region than in the nonresponsive basal region on a fresh weight basis, but not on a mg^{-1} soluble protein basis (Stoddart _et al._, 1974; Keith and Srivastava, 1980; Keith _et al._, 1981, 1982). These data suggest that the receptor molecules continue to be present in the nontarget regions. However, in the absence of precise data about 'n' or K_d not much can be made out of these data.

Possible mechanism of gibberellin-receptor action

The picture that has emerged then is that there are soluble proteins in target, and possibly nontarget, regions of cucumber and pea which recognize and bind active gibberellins with high affinity. What does this tell us about the mode of action of GA and how does it relate to the excellent data that are now emerging with regard to transcription of α-amylase genes in the barley aleurone. In that tissue, gibberellin action leads to a *de novo* synthesis of at least 4 different types of α-amylases and other hydrolases including protease and α-glucosidases, and an increase in amount, presumably by new synthesis, of some polyadenylated mRNAs (see Jacobsen, 1983; Jacobsen and Beach, 1985; Hammerton and Ho, 1986). The α-amylases belong to two groups of isoenzymes which are coded for by different structural genes on different chromosomes and which differ from each other in several ways including isoelectric point. The low pI (pH 4.6) group of isoenzymes appears first, reaches a plateau at about 16 h, stays high and then declines at about 48 h after GA_3 application. The high pI group appears later, increases sharply to 12 to 16 h and then declines (see Jacobsen, 1983, 1986; Ho, 1986). The mRNAs for these groups of isoenzymes follow similar rise and declines (Chandler *et al*., 1986). The low pI group enzymes apparently respond to about 1/50 of the GA_3 concentration necessary to induce the high pI isoenzymes (see Jacobsen, 1983).

The synthesis of several new enzymes following GA_3 application suggests that the gibberellin-receptor complex must bind to an acceptor site on chromatin which releases a sequence of functionally-related genes for transcription (see also Jacobsen and Beach, 1985). Such cascading of related genes is common for several steroid receptors (Schrader *et al*., 1981; Greene and Press, 1986). Furthermore, it is possible that we are dealing with several acceptor sites of higher and lower affinities for the hormone-receptor complex (Fig. 11a). Thus, the low pI group of isoenzymes may require a much lower molar concentration of gibberellin-receptor complex for binding to its acceptor site (from the data of Chandler *et al*., 1986, the mRNA production of this group of isoenzymes is saturated at 10^{-9} M GA_3), whereas the high pI group may require much higher

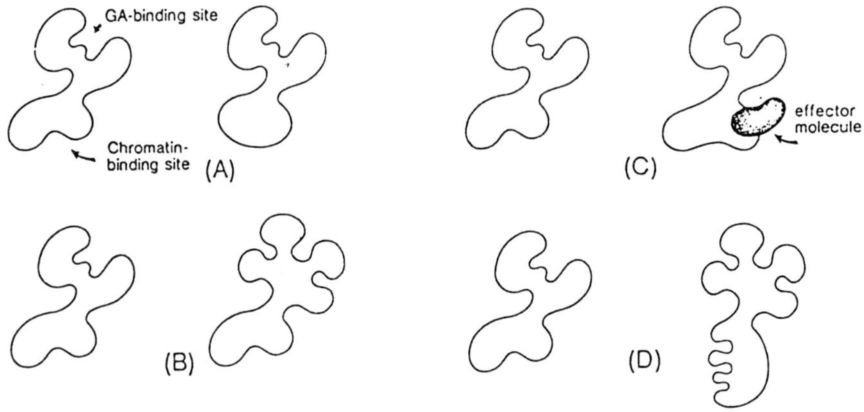

Figure 11. Models of gibberellin receptor which may explain differential gene activation in the same or different tissues. The receptor molecule is assumed to have at least two binding sites or domains, a gibberellin-binding site, and a chromatin-binding site, or a site that recognizes an 'acceptor site' on chromatin (the acceptor site may be a nonhistone protein or a sequence of nucleotides upstream or downstream from the promoter sequence for the structural gene(s). It is likely that there is at least one more allosteric site which recognizes an effector or signal molecule. The presence (or absence) of the effector molecule may alter the conformation of the chromatin-binding site and thus may allow recognition of new acceptor sites on chromatin. In (A) the GA-binding sites are the same but the chromatin-binding sites are different with differing affinities for different acceptor sites on the chromatin. In (B) the GA-binding sites are different but the chromatin-binding sites are the same. The work on cucumber and pea to date does not support this model. Model (C) is similar to (A) but the affinity of the chromatin-binding site is altered by the binding of an effector molecule. Model (D) shows receptor molecules with different GA-binding sites as well as different chromatin-binding sites. (Note: It must be emphasized that these models represent only four of a multitude of possibilities).

concentrations of the gibberellin-receptor complex.[2] Another possibility is that there may be receptor molecules with different binding affinities for a gibberellin (Fig. 11b). Whilst this possibility cannot be ignored, the limited data at hand suggest only one class of binding site for gibberellin receptor in cucumber and possibly in tall and dwarf pea.

If we transfer our attention to different gibberellin responses in the same plant, e.g., stem elongation and flowering (Michniewicz and Lang, 1962),the conclusion is inescapable that the gibberellin-receptor complex must be able to recognize different sets of gene sequences at different times in a plant's life cycle. How is this accomplished? There are at least three possibilities: 1. Reference has already been made to the possibility that the chromatin-binding site has different affinities for different acceptor sites in the same tissue (Fig. 11a). This may well extend to different tissues anddifferent organs in the life cycle of the plant. 2. A variation on this theme is that there are 'effector' or 'signal' molecules characteristic of an organ or tissue which bind to the receptor molecule and bring about a conformational change in the chromatin binding site, thus allowing the hormone-receptor complex to bind to a new acceptor site (Fig. 11c). 3. Still another possibility is that the receptor molecules in different tissues of the same plant are different both in respect to the architecture of the GA-binding site and the chromatin-binding site (Fig. 11d).

At present there is no way to choose between these alternatives, though intuitively the first and second alternatives (Figs 11a,c) are appealing because they offer the greater economy of design and also explain the transition from a target to a nontarget region by the loss of the effector molecule

---

[2]The possibility of acceptor sites with different affinities for the hormone-receptor complex may also explain the observation that gibberellin-induced responses often occur over 3-4 decades (see Kende and Gardner, 1976; Stoddart, 1983). An explanation of this phenomenon must also include the possibility that exogenous GA concentration may not reflect the concentration in the target cells because a good deal of it may be metabolised and/or not reach the target zone.

or a change in the affinity of the hormone-receptor complex for the acceptor site. Certain observations in the literature should be noted, however, if for no other reason than to promote greater investigation. Thus, in *Myosotis*, GA_3 caused only stem growth, but GA_7 caused both stem growth and flowering (Michniewicz and Lang, 1962); in *Chrysanthemum* and *Hydrophyllum*, GA_5 was the most active in promoting flowering, whereas GA_7 was the most active in promoting stem growth (Zeevart, 1983); and in Pinaceae, GA_4 and GA_7 were the most active in promoting cone formation (Pharis and Kuo, 1977), but endogenous GA_1 content rose sharply in *Picea* shoots during active growth (Dunberg, 1976). Thus, we cannot ignore possibility number 3 either (Fig. 11d).

Much more work needs to be done on a more careful characterization of the gibberellins and their required concentrations for different responses in the same plant and in different plants, and we need to look at many other tissues besides barley aleurone for specific proteins that are synthesized as a result of GA action. More *in vitro* assays are needed from different target tissues to see whether we are dealing with different or same receptors and these receptors need to be isolated, purified and sequenced.

Purification of the receptor

In the last few months our major effort has gone into purification of receptor protein from cucumber. The 100,000*g* cytosol is cut with 60% $(NH_4)_2SO_4$, the precipitate washed once with 60% $(NH_4)_2SO_4$, passed through a G-25 or G-50 Sephadex column and lyophilized. In a lyophilized state the powder can be stored at -20°C almost indefinitely if repeated thawing and freezing are avoided. The frozen powder is resuspended in buffer, and passed through DE32 ion-exchange resin. The fraction eluting between 0.2 to 0.28 M KCl shows the maximum specific binding activity. These fractions are pooled from different runs and passed through a hydroxylapatite column. The protein binds to the column and is eluted with 0.4 M phosphate buffer. The fractions from the hydroxylapatite column are again pooled, concentrated and desalted by washing through Amicon filter of 30,000 Dalton cut off and passed through a gel filtration column. The purification protocol to date has yielded some improvement in binding activity (Table 3).

Table 3. Purification of the cucumber cytosol and improvement in binding.

	$[^3H]GA_4$ specifically bound (pmol•mg^{-1} soluble protein)	Yield of protein (mg•100 g^{-1} fresh tissue	Total pmol $[^3H]GA_4$ specifically bound (pmol•100 g^{-1} fresh tissue
100,000*g* cytosol	0.12	99	11.88
desalted $(NH_4)_2SO_4$ precipitate	0.28	53	14.84
ion-exchange column (DE32)	0.44	14	6.34

We are trying to devise an affinity column for large scale routine separation of the receptor protein. The C-16 methylene group has been derivatized to a keto group for linkage to an NH_2 group of a spacer arm of an affinity gel (Affigel 102 from BioRad).

Acknowledgment

I am grateful to my colleagues Nasser Yalpani and Zin Liu for critically reading the manuscript and offering many valuable suggestions. The research reported here has been supported by the Natural Sciences & Engineering Research Council, Canada, Grant #A2905.

References

Chandler, P.M., Z. Ariffin, and L. Huiet. 1986. Hormone regulation of specific mRNA levels in barley aleurone. J. Cellular Biochem. Supplement 10B: 7 (Abstract).

Crozier, A. 1981. Aspects of the metabolism and physiology of gibberellins. In Advances in Botanical Research, vol. 9, ed. H.W. Woodhouse, Academic Press, N.Y. pp. 33-149.

Crozier, A. and R.C. Durley. 1983. Modern methods of analysis of gibberellins. In The Biochemistry and Physiology of Gibberellins, Vol. 1, ed. A. Crozier, Praeger, New York, pp. 485-560.

Crozier, A., C.C. Kuo, R.C. Durley and R.P. Pharis. 1970. The biological activities of 26 gibberellins in nine plant bioassays. Can. J. Bot. 48: 867-877.

Dunberg, A. 1976. Changes in gibberellin-like substances and indole-3-acetic acid in Picea abies during the period of shoot elongation. Physiol. Plant. 38: 186-190.

Ginzberg, C. and H. Kende. 1968. Studies on the intracellular localization of radioactive gibberellin. In Biochemistry and Physiology of Plant Growth Substances, ed. F. Wightman and G. Setterfield, Runge Press, Ottawa, pp. 333-340.

Greene, G.L. and M.F. Press. 1986. Structure and dynamics of the estrogen receptor. I. Steroid receptor structure (including monoclonal antibodies and new methods of determination). Structure and dynamics of the estrogen receptor. J. steroid Biochem. 24: 1-7.

Hammerton, R.W. and T.H.D. Ho. 1986. Hormonal regulation of the development of protease and carboxypeptidase activities in barley aleurone layers. Plant Physiol. 80: 692-697.

Hedden, P. 1983. In vitro metabolism of the gibberellins. In The Biochemistry and Physiology of Gibberellins, vol. 1, ed. A. Crozier, Praeger, New York, pp. 99-149.

Hoad, G.V. 1983. Gibberellin bioassays and structure-activity relationships. In The Biochemistry and Physiology of Gibberellins, Vol. 2, ed. A. Crozier, Praeger, New York, pp. 57-94.

Hoad, G.V., B.O. Phinney, V.M. Sponsel and J. MacMillan. 1981. The biological activity of sixteen gibberellin A_4 and gibberellin A_9 derivatives using seven bioassays. Phytochemistry 20: 703-713.

Ho, T.H.D. 1986. Hormonal, genetic and environmental regulation of enzyme synthesis in the aleurone layers of cereal grains. J. Cellular Biochem. Supplement 10B: 3 (Abstract).

Jacobsen, J. 1983. Regulation of protein synthesis in aleurone cells by gibberellic and abscissic acid. In The Physiology and Biochemistry of Gibberellins, Vol. 2, ed. A. Crozier, Praeger, New York, pp. 159-187.

Jacobsen, J.V. 1986. Studies on gibberellin and abscissic acid action in barley aleurone cells. J. Cellular Biochem. Supplement 10B: 4 (Abstract).

Jacobsen, J.V. and L.R. Beach. 1985. Control of transcription of α-amylase and r-RNA genes in barley aleurone protoplasts by gibberellin and abscissic acid. Nature (Lond.) 316: 275-277.

Jones, R.L. 1973. Gibberellins: Their physiological role. Ann. Rev. Plant Physiol. 24: 571-98.

Jones, R.L. and I.D.J. Phillips. 1966. Organs of gibberellin synthesis in light grown sunflower plants. Plant Physiol. 41: 1381-1386.

Jones, T.W.A. 1976. Biological activities of fluorogibberellins and interactions with unsubstituted gibberellins. Phytochemistry 15: 1825-1827.

Katsumi, M., B.O. Phinney and W.K. Purves. 1965. The roles of gibberellin and auxin in cucumber hypocotyl growth. Physiol. Plant. 18: 462-473.

Keith, B. and L.M. Srivastava. 1980. *In vivo* binding of gibberellin A_1 in dwarf pea epicotyls. Plant Physiol. 66: 962-967.

Keith, B. and L.M. Srivastava. 1981. The biological activity of ketogibberellin A_1. Can. J. Bot. 59: 2173-2174.

Keith, B., R. Boal and L.M. Srivastava. 1980. On the uptake, metabolism and retention of [^{3}H]gibberellin A_1 by barley aleurone layers at low temperatures. Plant Physiol. 66: 956-961.

Keith, B., S. Brown and L.M. Srivastava. 1982. *In vitro* binding of gibberellin A_4 to extracts of cucumber measured by using DEAE-cellulose filters. Proc. Natl. Acad. Sci. (USA) 79: 1515-1519.

Keith, B., L.M. Srivastava and N. Murofushi. 1979. The biological activities of some iodinated gibberellins. Agric. Biol. Chem. 43: 141-143.

Keith, B., N.A. Foster, M. Bonettemaker and L.M. Srivastava. 1981. In vitro gibberellin A_4 binding to extracts of cucumber hypocotyls. Plant Physiol. 68: 344-348.

Kende H. 1967. Preparation of radioactive gibberellin A_1 and its metabolism in dwarf peas. Plant Physiol. 42: 1612-1618.

Kende, H. and G. Gardner. 1976. Hormone binding in plants. Ann. Rev. Plant Physiol. 27: 267-290.

Liu, Z.-H and L.M. Srivastava. 1986. In vitro binding of gibberellin A_4 in epicotyls of dwarf pea and tall pea. In Molecular Biology of Plant Growth Control. Proc. UCLA Symposium, Feb. 1986, Lake Tahoe, CA., Alan R. Liss, Inc., N.Y. In press.

Mundy, J., A. Brandt and G.B. Fincher. 1985. Messenger RNAs from the scutellum and aleurone of germinating barley encode (1→3, 1→4)-β-D-glucanase, α-amylase and carboxypeptidase. Plant Physiol. 79: 867-871.

Michniewicz, M. and A. Lang. 1962. Effects of nine different gibberellins on stem elongation and flower formation in cold-requiring and photoperiodic plants grown under non-inductive conditions. Planta 58: 549-563.

Musgrave, A. and H. Kende. 1970. Radioactive gibberellin A_5 and its metabolism in dwarf peas. Plant Physiol. 45: 56-61.

Musgrave, A., S.E. Kays and H. Kende. 1969. In vivo binding of radioactive gibberellins in dwarf pea shoots. Planta 89: 165-177.

Musgrave, A., S.E. Kays and H. Kende. 1972. Uptake and metabolism of radioactive gibberellins by barley aleurone layers. Planta 102: 1-10.

Nadeau, R., L. Rappaport and C.F. Stolp. 1972. Uptake and metabolism of [^{3}H]-gibberellin A_1 by barley aleurone layers: response to abscissic acid. Planta 107: 315-324.

Pharis, R.P. and C.C. Kuo. 1977. Physiology of gibberellins in conifers. Can. J. Forest. Res. 7: 299-325.

Reeve, D.R. and A. Crozier. 1974. An assessment of gibberellin structure-activity relationships. J. Exptl. Bot. 25: 431-445.

Scheidereit, C., P. Krauter, D. von der Ahe, S. Janich, O. Rabenau, A.C.B. Cato, G. Suske, H.M. Westphal and M. Beato. 1986. Mechanism of gene regulation by steroid hormones. J. steroid Biochem. 24: 19-24.

Schrader, W.T., M.E. Birnbaumer, M.R. Hughes, N.L. Weigel, W.W. Grody and B.W. O'Malley. 1981. Studies on the structure and function of the chicken progesterone receptor. In Recent Progress in Hormone Research Vol. 37: 583-633. Academic Press, N.Y.

Serebryakov, E.P., V.N. Agnistikova and L.M. Suslova. 1984. Growth-promoting activity of some selectively modified gibberellins. Phytochemistry 23: 1847-1854.

Silk, W.K., R.L. Jones and J.L. Stoddart. 1977. Growth and gibberellin A_1 metabolism in excised lettuce hypocotyls. Plant Physiol. 59: 211-216.

Smith, V. and J. MacMillan. 1986. The partial purification and characterization of gibberellin 2β-hydroxylases from seeds of *Pisum sativum*. Planta 167: 9-18.

Sponsel, V.M. (nee Frydman), G.V. Hoad and L.J. Beeley. 1977. The biological activities of some new gibberellins (GAs) in six plant bioassays. Planta 135: 143-147.

Stoddart, J.L. 1983. Sites of gibberellin biosynthesis and action. In The Biochemistry and Physiology of Gibberellins, Vol. 2, ed. A. Crozier, Praeger, New York, pp. 1-55.

Stoddart, J.L. and M.A. Venis. 1980. Molecular and subcellular aspects of hormone action. In Hormonal Regulation of Development. 1. Molecular Aspects of Plant Hormones. Encyclopaedia of Plant Physiology, New Series, vol. 9, ed. J. MacMillan, Springer-Verlag, New York, pp, 445-510.

Stoddart, J., W. Breidenbach, R. Nadeau and L. Rappaport. 1974. Selective binding of [^{3}H]gibberellin A_1 by protein fractions from dwarf pea epicotyls. Proc. Natl. Acad. Sci. (USA) 71: 3255-3259.

Yalpani, N. and L.M. Srivastava. 1985. Competition for *in vitro* [^{3}H]gibberellin A_4 binding in cucumber by gibberellins and their derivatives. Plant Physiol. 79: 963-967.

Yalpani, N. and L.M. Srivastava. 1986. Partial purification of a gibberellin binding protein from cucumber hypocotyls. In Molecular Biology of Plant Growth Control. Proc. UCLA Symposium, Feb. 1986, Lake Tahoe, CA., Alan R. Liss, Inc., N.Y. In press.

Yalpani, N., Z.-H. Liu and L.M. Srivastava. 1986. Extraction and assay of gibberellin-binding proteins from cucumber and pea. In this volume.

Varner, J.E. and D.T.-H. Ho. 1976. The role of hormones in the integration of seedling growth. In The Molecular Biology of Hormone Action, ed. J. Papaconstantinou, Academic Press, N.Y., pp. 173-194.

Zeevart, J.A.D. 1983. Gibberellins and flowering. In The Biochemistry and Physiology of Gibberellins, Vol. 2, ed. A. Crozier, Praeger, New York, pp. 333-374.

Zwar, J.A. and R. Hooley. 1986. Hormonal regulation of α-amylase gene transcription in wild oat (*Avena fatua* L.) aleurone protoplasts. Plant Physiol. 80: 459-463.

Appendix 1

In the following calculations we compare the kinetic data of $[^3H]GA_4$ binding in cucumber: K_d = 70 nM; n = number of binding sites = 0.42 pmol mg^{-1} soluble protein (Keith *et al.*, 1982) against: 1. saturation of total (specific and nonspecific) binding sites in other cucumber preparations, 2. dose-response curve in cucumber bioassay, and 3. estimated endogenous hormone concentrations in vegetative tissues.

For these calculations -

1 Bq = 1 disintegration sec^{-1}

MW of GA_4 = 332

1 mg soluble protein = 1 g fresh tissue = 1 mL solution

3 cucumber seedlings = 1 g fresh tissue

1. Saturation of total (specific and nonspecific) binding in a desalted 60% $(NH_4)_2SO_4$ precipitate from 100,000*g* cytosol from cucumber hypocotyls occurred at 10,900 dpm•mg^{-1} soluble protein (unpublished data from Yalpani).

   Specific activity of $[^3H]GA_4 = 1.6 \times 10^{12}$ Bq•$mmol^{-1}$

   $= 2.66 \times 10^{10}$ dpm•$mmol^{-1}$

   Saturation at 10.9×10^3 dpm•mg^{-1} soluble protein

   $$= \frac{10.9 \times 10^3}{2.66 \times 10^{10}} \text{ mmol } GA_4 \cdot mg^{-1} \text{ soluble protein}$$

   $= 0.41$ nmol $GA_4 \cdot mL^{-1}$

   $= 410$ nM

2. Cucumber hypocotyl bioassay:
Saturating concentration = 140 nM
(Keith *et al.*, 1982)

$$= 140 \times 332\ ng{\bullet}L^{-1}$$
$$= 46.5\ ng{\bullet}g^{-1} \text{ fresh tissue}$$
$$= 15.5\ ng{\bullet}seedling^{-1}$$

The saturating concentration of GA_4 in the cucumber bioassay, as judged from saturation of the growth response, is 100 to 1000 ng GA_4 applied per seedling (e.g., see Katsumi *et al.*, 1965; Hoad *et al.*, 1981). However, metabolism and transport of GA_4 in intact seedlings may well reduce the concentration of active GA in target cells by 1-2 orders of magnitude.

3. Relationship of K_d and n to endogenous hormone concentration:

A. From #2 above, saturating concentration of GA_4 in plant target tissue = 46.5 $ng{\bullet}g^{-1}$ fresh tissue
= 46.5 $\mu g{\bullet}Kg^{-1}$ fresh tissue

Estimated endogenous GA concentrations in vegetative tissues such as young leaves vary from 1-10 $\mu g{\bullet}Kg^{-1}$ fresh tissue (Jones and Phillips, 1966; Stoddart, 1983). However, the endogenous values are likely to be lower than expected because recovery of extracted GA may not be 100% and the concentration of active GA in the target tissue may be higher than the average value for the extracted material.

B. n = 0.42 pmol mg^{-1} soluble protein assumes 1:1 binding of hormone to receptor
0.42 pmol GA_4, equivalent to 0.42 nM GA_4, are bound mg^{-1} soluble protein
For saturation to occur requires 140 nM GA_4

This means that 1 molecule of GA_4 is bound for each 333 ($\frac{140}{0.42}$ = 333) molecules. This low ratio is likely due to 2 reasons: (1) Our estimate of 'n' is lower than actual because of possible degradation of binding sites during extraction and purification. (2) Our estimated K_d is higher than actual because measurements of bound radioactivity are not possible under equilibrium conditions and some dissociation of bound hormone from protein invariably occurs during the assay.

ETHYLENE BINDING SITES

A.R. Smith, D. Robertson, I.O. Sanders, R.A.N. Williams and M.A. Hall
Department of Botany and Microbiology, University College of Wales,
Aberystwyth, Dyfed SY23 3DA, Wales, U.K.

ABSTRACT

An account is given of the purification of the ethylene binding protein from developing cotyeldons of *Phaseolus vulgaris*. A correlation has been demonstrated between ethylene binding and sensitivity to the hormone in epicotyls of *Pisum sativum*. As this tissue also metabolises ethylene, this system could provide evidence for the hypothesis that ethylene metabolites are not acting as receptors but as modulators by interacting with the binding sites along with ethylene to bring about a biochemical response.

INTRODUCTION

Over the past few years several reports have presented evidence that binding sites for ethylene, exist in various tissues and organs of a variety of plant species (1). However, none of these binding sites can as yet be assigned a role as functional receptors. It is also well established that higher plants have the capacity to metabolise ethylene and the possible involvement of ethylene metabolism in the mechanism of action of ethylene has been suggested (2).

It is intended here to describe firstly the purification of the ethylene binding site from developing cotyledons of *Phaseolus vulgaris* and secondly to address the question of the involvement of binding and metabolism in the mechanism of ethylene action in seedlings of *Pisum sativum* and in *Vicia faba*.

MATERIALS AND METHODS

Many of the methods used in the studies presented here have been published elsewhere or are discussed in the chapters by Sanders *et al* and Williams *et al* in this volume. Assay of ethylene binding activity was as described by Bengochea *et al* (3,4), extraction and purification of the binding protein from *Phaseolus vulgaris* (5,6,7), extraction and assay of the ethylene monooxygenase from *Vicia faba* (8) and *in vivo* assay of ethylene oxidation in *Pisum sativum* (9). For the assessment of binding and metabolism in *Pisum sativum* and *Vicia faba* a displacement assay was developed which could distinguish between ethylene metabolites in the tissue and specifically

NATO ASI Series, Vol. H10
Plant Hormone Receptors. Edited by D. Klämbt

bound ethylene. Tissue or extracts were placed in sealed flasks incorporating a minivial containing 1 cm^3 3M NaOH. $^{14}C_2H_4$ was then injected to give concentrations in solution of between 5 x 10^{-9}M and 5 x 10^{-7}M. In an equivalent set of flasks, $^{12}C_2H_4$ was injected in addition to give a concentration of 5 x 10^{-6}M. After incubation the minivials were removed and the radioactivity in CO_2 and ethylene oxide estimated (8).

RESULTS AND DISCUSSION

The presence of a binding site with high affinity and specificity for ethylene has been demonstrated in developing cotyledons of *Phaseolus vulgaris* (3,4). Using marker enzymes and high resolution autoradiography, it has been shown that the binding site is localised on elements of the endomembrane system (10,11). Although the binding site displays some of the expected properties of an ethylene receptor there is no known role for ethylene during the ontogeny of *Phaseolus* cotyledons. Attempts to demonstrate a relationship between the binding site and the control of protein synthesis or protein kinase activity in cell-free preparations have met with no success. Spraying *Phaseolus* plants with a range of concentrations of silver thiosulphate had no effect on ethylene binding activity or any obvious effect on seed development.

Because it seems likely that the binding domain in *Phaseolus* cotyledons will show homologies with binding sites for ethylene both in other parts of the same species and also in tissues of other species which are ethylene sensitive, investigations into the extraction and purification of the binding site from *Phaseolus* have been carried out.

It has emerged that only extraction with detergents does not drastically reduce ethylene binding and maintains the affinity of the binding sites for ethylene (5). When a comparison is made of the characteristics of the membrane-bound ethylene binding sites with those of Triton X-100 solubilised preparations, the affinity and the specificity are not much altered as a result of solubilisation (5,12). Using isokinetic sucrose gradients and gel permeation chromatography, the molecular weight of the detergent/protein complex has been estimated to be between 52,000 and 60,000 daltons.

In addition to previously reported work, the hydrophobic nature of the binding site has been investigated using pH 12 precipitation techniques (Table 1). A crude prelabelled homogenate was centrifuged at 100K x g, the supernatant decanted and the pellet resuspended in 0.1M KCl. This was recentrifuged at 100K x g and over 95% of the radioactivity was pelleted. A substantial increase in specific activity is also obtained using this technique.

TABLE 1 - pH 12 precipitation of a prelabelled 100K x g pellet

Sample	Total DPM	SPECIFIC ACTIVITY ($dpm\ mg^{-1}$ protein)
Crude Homogenate	1.84×10^6	591
pH 12 100K x g pellet (0.1M KCl)	1.15×10^6	3774
pH 12 100K x g supernatant	3.3×10^4	403

Using protein body membrane preparations obtained by sucrose density centrifugation followed by pH 12 treatment, a similar distribution of radioactivity was obtained.

The hydrophobicity of the protein has seriously complicated efforts to purify it since any treatment which tends to dissassociate detergent and protein leads to precipitation of the latter. Nevertheless a significant degree of purification has been obtained using a combination of techniques particularly FPLC. Using 10% SDS discontinuous PAGE, a marked decrease in the number of bands detectable is apparent during purification but as ethylene dissociates from the binding protein in dissociating gels a discrete band cannot be associated with binding activity. Unfortunately it is proving difficult to obtain a discrete band of radioactivity associated with the binding protein using non-dissociating gels.

As was mentioned previously, the binding protein from cotyledons of *Phaseolus* has many of the characteristics of a receptor but is situated in a tissue in which there is no known developmental or biochemical response to ethylene. Present evidence suggests that a binding site exists in abscission zones of *Phaseolus* where the primary ethylene response has been characterised partially (13) but the maximum specific activity of ethylene available is insufficient to enable localisation or accurate measurement of binding site concentration in a zone a few cells thick.

In an effort to resolve this problem a programme to produce antibodies to the binding site was emabarked upon in order to develop an immunoassay for use in probing the transduction of the response. As the detection assay used involved the binding of protein radiolabelled with ethylene to antibodies present in serum, it was considered appropriate to maintain the protein in a solubilised form. However using Ochterlony double diffusion tests and a pelleting assay using donkey, anti-rabbit serum, we were unable to detect antibodies specific to the binding site. There are many reasons which could explain this lack of success and it is now intended to take at least two further approaches. If a sample of native binding protein which can be identified using non-denaturing electrophoresis can be obtained then this could be used directly as the antigen and hence use Western blotting as a screening procedure for the serum. Alternatively, if bands could be identified indirectly on SDS gels by a process of elimination then these could also be used as an antigenic source.

Ethylene Metabolism

Several years ago it was established that higher plants could metabolise ethylene and it was suggested by Beyer (2) that the metabolism of ethylene was linked to the mechanism of action of the growth regulator. Our investigations in this context have centred around the ethylene metabolising systems in cotyledons of _Vicia faba_ and in seedlings of _Pisum sativum_.

The system in _Vicia_ has a high affinity for ethylene and is capable of metabolising physiologically active analogues of ethylene, such as propylene. Because the enzyme requires a source of reducing power in the form of reduced pyridine nucleotide (8) and because heavy oxygen (O^{18}) is incorporated into the primary product of metabolism (Beyer, pers.comm.), it has been proposed that the enzyme is an NADPH-dependent mono-oxygenase catalysing the reaction

$$C_2H_4 + O_2 + NADPH + H^+ \qquad C_2H_4O + NADP^+ + H_2O$$

Until recently, there was no evidence that metabolism and ethylene binding were simultaneously present in the same tissue and it was suggested that each might represent reflections of the same system. Measurement of ethylene binding in _Vicia_ using the methods employed with _Phaseolus_ proved impracticable due to the high rates of metabolism observed in cell-free preparations. Using the displacement assay procedure outlined for seedlings of _Pisum_, we now have evidence that seedlings of _Vicia_ do bind ethylene saturably and with high affinity _in vivo_. It has not proved possible as yet to demonstrate this _in vitro_ as the binding activity in _Vicia_ appears to be much more labile than in _Phaseolus_.

Ethylene Metabolism in *Pisum*

Early work by Beyer (2) demonstrated the presence of two ethylene metabolising systems in seedlings of *Pisum sativum*; one metabolising ethylene to ethylene glycol via the production of ethylene oxide and the other metabolising ethylene directly to CO_2. The metabolism of ethylene in *Pisum* differs from that in *Vicia* in that ethylene is metabolised rapidly in *Vicia* at physiological concentrations and that only one primary product is formed, namely ethylene oxide. It has now been established that the systems metabolising ethylene in peas have relatively low affinities for ethylene thus explaining the low rates of metabolism at physiological ethylene concentrations (Table 2). However, the affinities of the systems in peas for propylene are similar to that in *Vicia faba*.

Table 2 - Properties of the ethylene metabolising systems in *Vicia faba* and *Pisum sativum* (*in vivo*)

	Vicia	*Pisum*	
		OX	TI
K_m ethylene (M)*	4.2×10^{-10}	0.9×10^{-6}	1.6×10^{-6}
V_{max} mol g^{-1} dw h^{-1}	6.4×10^{-10}	2.4×10^{-10}	4.5×10^{-10}
K_i propylene (M)*	5.0×10^{-6}	7.0×10^{-6}	3.7×10^{-7}

*Concentration given for aqueous phase

We have recently embarked on studies to characterise the ethylene metabolising systems in *Pisum in vitro*. It does not appear that the tissue incorporation system in peas is directly analogous to that in *Vicia* as the system in peas is not NADH, NADPH or FADH-dependent and activity is not reduced when assayed at reduced oxygen tensions.

Binding sites in *Pisum*

It is now clear that both binding and metabolism of ethylene occur in epicotyls of *Pisum*. Fig.1 shows the results of Scatchard analysis of ethylene binding by 5-day-old etiolated epicotyl tips of *Pisum*. The analysis indicates the presence of two classes of binding sites for ethylene depending on the length of the association period. After 1h incubation, it appears that one rapidly associating high affinity binding site exists whereas after 20h incubation there is a change in the slope of this line which suggests that another binding site with slightly lower affinity with a slower rate of association may also be present. However, it is not clear whether the lower affinity site is absent initially or whether the technique failed to detect it in the early stages of incubation.

When radioactivity incorporated is plotted against time it can be determined that the rates of association and dissociation of ethylene are rapid with approximately half association or dissociation occurring within 30 min (Fig 2). It is also evident that the capacity to metabolise ethylene is essentially absent initially but develops during incubation.

Fig 3 illustrates the extent of ethylene binding and ethylene metabolising activity in serial sections down 5-day-old etiolated epicotyls of *Pisum*. The quantity of ethylene binding and ethylene metabolism to ethylene oxide is highest in the top 1 cm portion of the epicotyl but ethylene incorporation into the tissue or into CO_2 appear to occur generally throughout the tissue. The structural analogue of ethylene, acetylene, inhibits ethylene binding in epicotyl tips competitively with a K_i of 1.03×10^{-6}M (liq.) compared to a K_D for ethylene of 9.53×10^{-10}M (liq.).

CONCLUSIONS

The fact that at least one high affinity ethylene binding site exists in *Pisum* seedlings renders questionable the belief that ethylene metabolism has any direct role in the mode of action of ethylene. Although the binding site in *Pisum* is less well characterised than that in *Phaseolus*, it does possess the appropriate features of a receptor in terms of affinity and rate constants of association and dissociation. It is of relevance that half maximal responses are obtained in pea epicotyls at ethylene concentrations of 4.5×10^{-10}M and 4.5×10^{-9}M (14) and that effects on growth of epicotyls are observed within 6 min of ethylene application (15). Another interesting correlation which emerges from these studies is that two of the components of the triple response, namely inhibition of hook closure and isodiametric cell expansion, occur in the region where binding activity is highest.

However, the fact remains that ethylene is metabolised in higher plant tissues and in most cases it seems unlikely that the function is to control endogenous ethylene concentrations. An alternative role for ethylene metabolism is now being considered particularly in the light of the observations that ethylene oxide, which has no ethylene-like activity in itself, enhances ethylene action in peas synergistically (2). We propose therefore that ethylene oxide could interact with the binding site along with ethylene to bring about the appropriate biochemical response. Such a hypothesis is a modification of Beyer's original ideas since in this case the ethylene oxidising enzyme would not be acting as a receptor but rather by providing a "modulator". Although this hypothesis explains many of the physiological observations, the validity of such a hypothesis can only be assessed if the components of the system can be isolated, characterised and reconstituted.

REFERENCES

1. Smith AR, Hall MA (1985) In: Roberts JA, Tucker GA (eds) Ethylene and plant development. Butterworths, London, p.101.
2. Beyer EM Jr (1985) In: Roberts JA, Tucker GA (eds) Ethylene and plant development. Butterworths, London, p.125.
3. Bengochea T, Dodds JH, Evans DE, Jerie PH, Niepel B, Shaari AR, Hall MA (1980) Planta (Berl) 148: 397.
4. Bengochea T, Acaster MA, Dodds JH, Evans DE, Jerie PH, Hall MA (1980) Planta (Berl) 148: 407.
5. Thomas CJR, Smith AR, Hall MA (1984) Planta (Berl) 160: 474.
6. Thomas CJR, Smith AR, Hall MA (1985) Planta (Berl) 164: 272.
7. Hall MA (1985) In: Proceedings of the 16th FEBS Congress, Part C. VNU Science Press, The Hague, p. 383.
8. Smith PG, Venis MA, Hall MA (1985) Planta 163: 97.
9. Evans DE, Smith AR, Taylor JE, Hall MA (1984) Plant Growth Regula 2: 187.
10. Evans DE, Bengochea T, Cairns AJ, Dodds JH, Hall MA (1982) Plant Cell Environ 5: 101.
11. Evans DE, Dodds JH, Lloyd PC, ap Gwyn I, Hall MA (1982) Planta (Berl) 154: 48.
12. Howarth CJ, Smith AR, Hall MA (1985) In: Roberts JA, Tucker GA (eds) Ethylene and plant development. Butterworths, London, p.117.
13. Sexton R, Lewis LN, Trewavas AJ, Kell P (1985) In: Roberts JA, Tucker GA (eds) Ethylene and plant development. Butterworths, London, p.173.
14. Burg SP, Burg EA (1967) Plant Physiol. 42: 144.
15. Warner LN, Leopold AC (1971) Biochem. Biophys. Res. 44: 989-994.

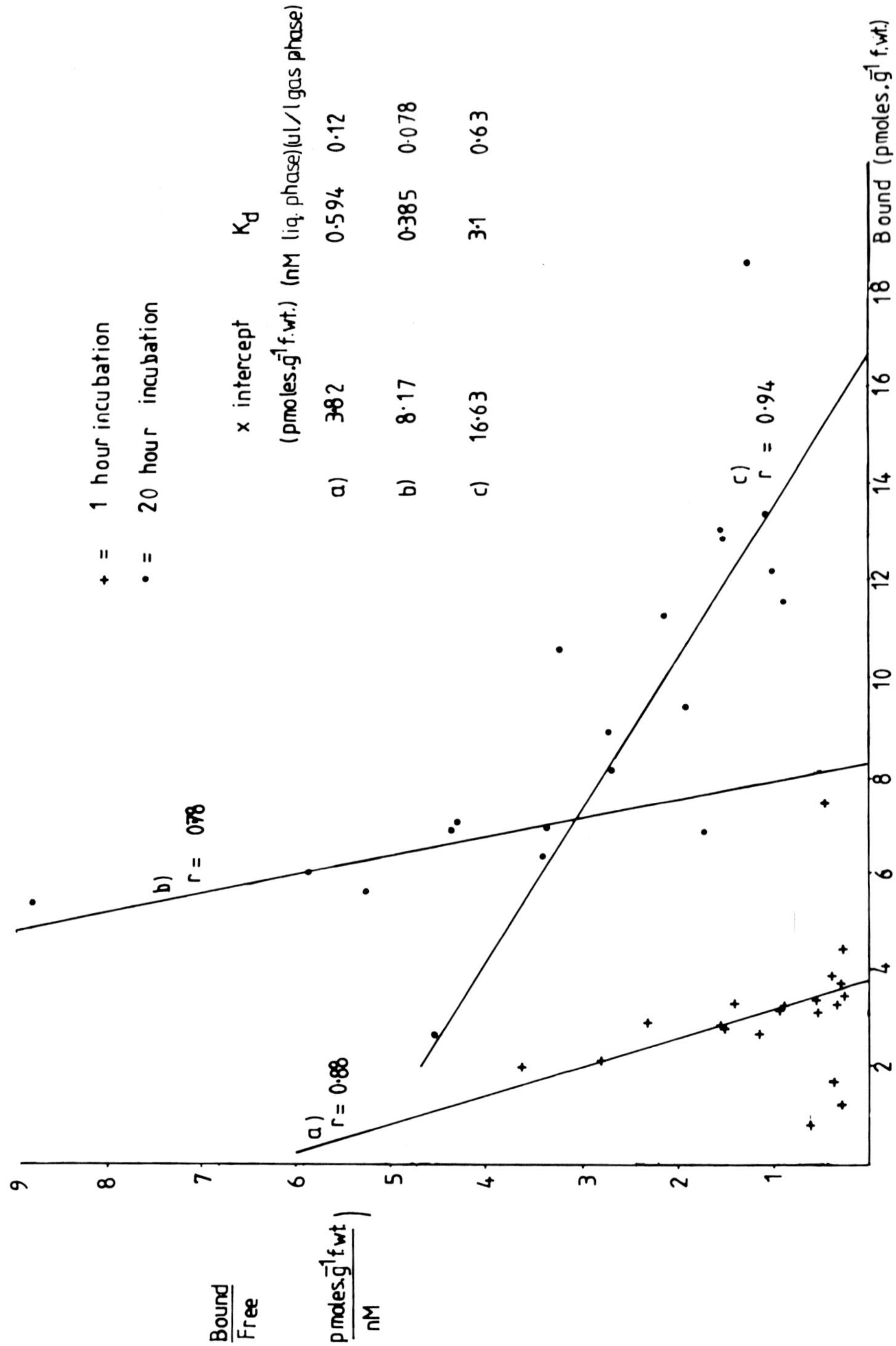

Figure 1 - Scatchard plots of ethylene binding by 5-day-old 1cm etiolated epicotyl tips of *Pisum sativum* L. cv. Alaska *in vivo*.

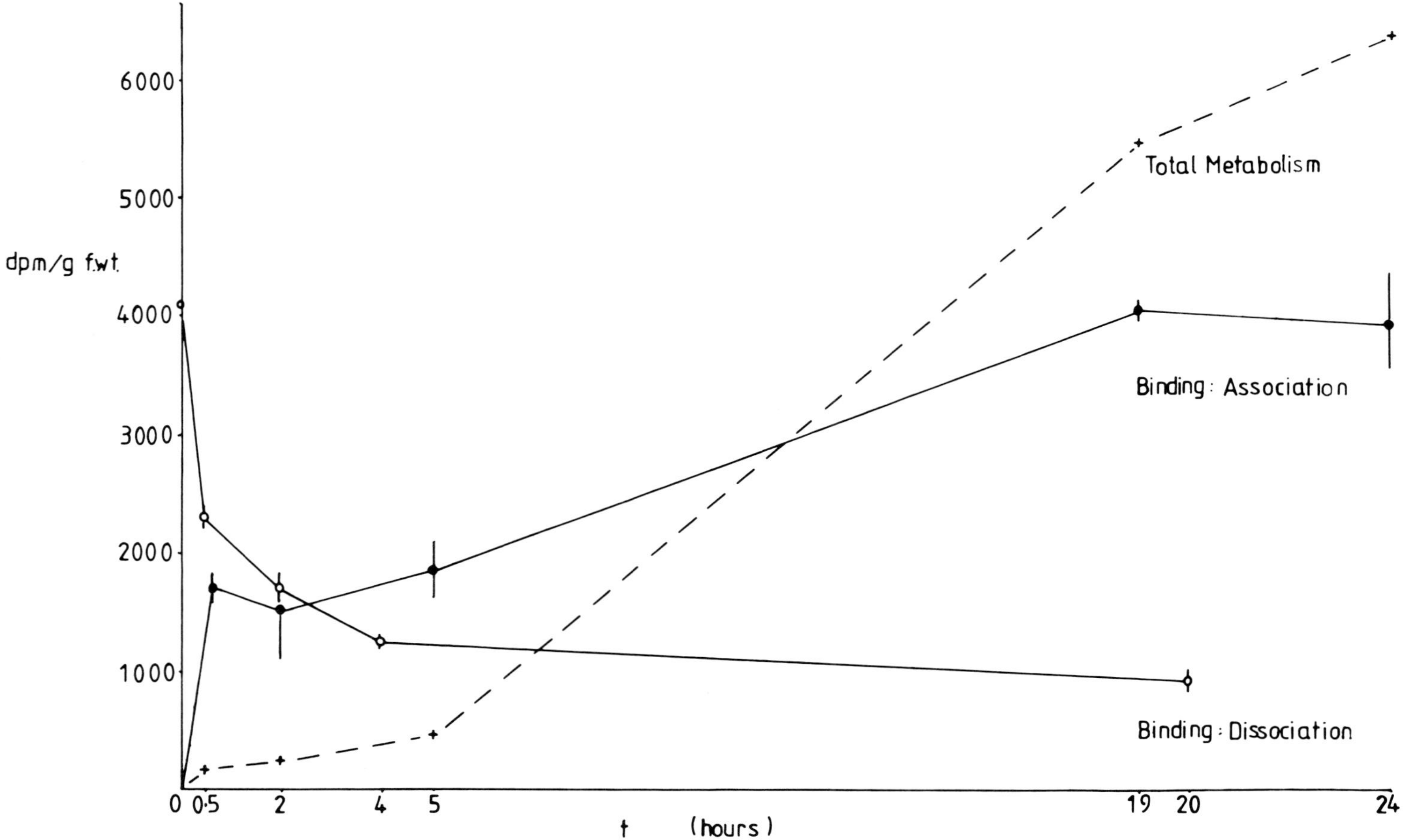

Figure 2 - Association (O) and dissociation (O) curves for ethylene binding and for metabolism (+---+) by 5-day-old 1 cm etiolated epicotyl tips of *Pisum sativum*.

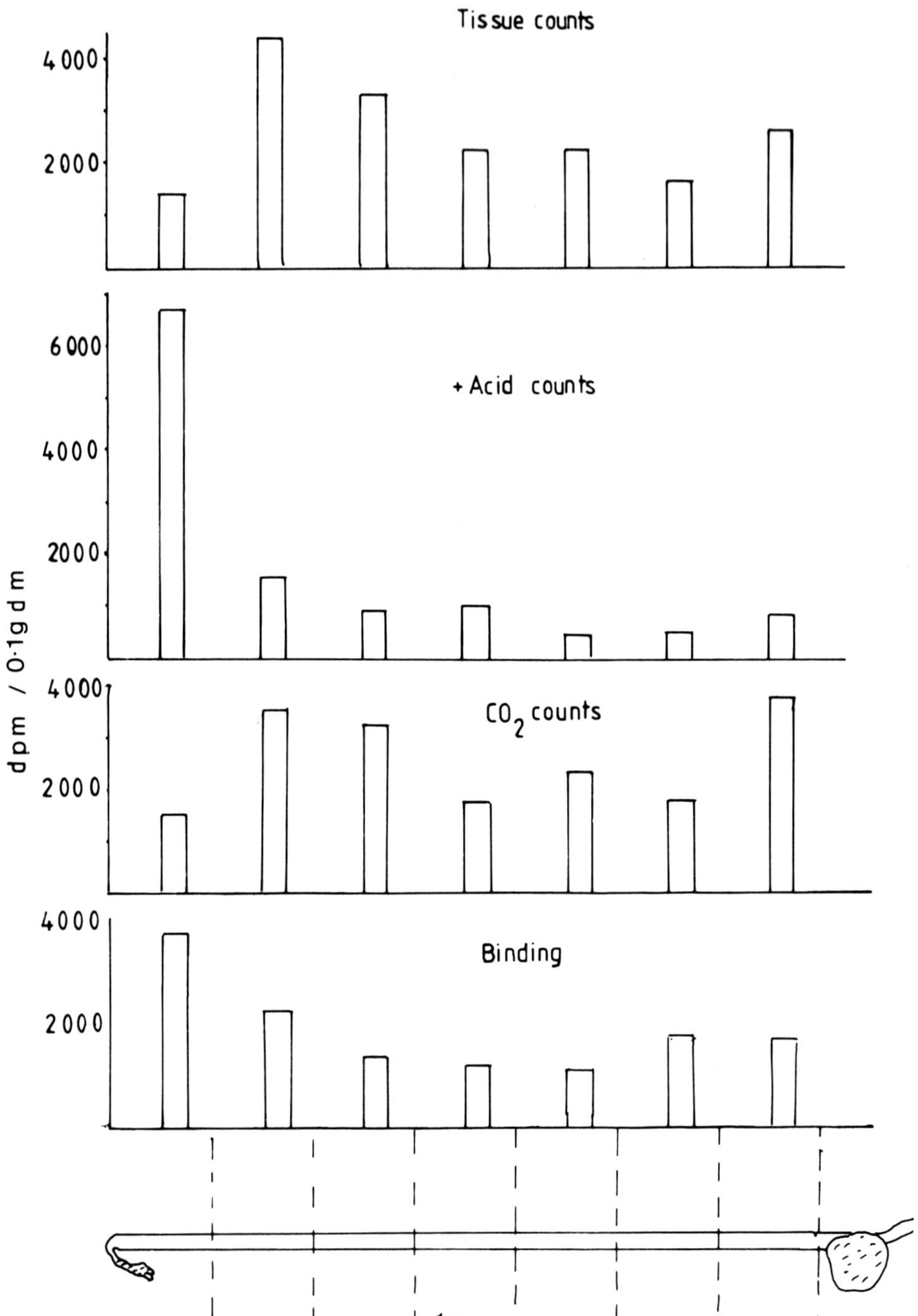

Figure 3 - Binding and metabolism of ethylene by 1cm serial sections of 5-day-old pea epicotyls exposed to ^{14}C-ethylene (7 $\mu l\ l^{-1}$) for 20h.

ETHYLENE BINDING AND EVIDENCE THAT BINDING _IN VIVO_ AND _IN VITRO_ IS TO THE PHYSIOLOGICAL RECEPTOR

Edward C. Sisler and Carmen Wood*
Department of Biochemistry
North Carolina State University
Raleigh, North Carolina 27695-7622 USA

A major question concerning the role of ethylene is, "How does it act?" Does it act as a hormone by binding to a receptor, or does it act as a cofactor in some reaction, or does it act as a product of an enzyme which catalyzes the reaction of ethylene with something else? All of these questions have been considered and attempts have been made to determine which is correct. Abeles et al. (4) and Beyer (6) have conducted experiments to determine if hydrogen is exchanged during ethylene action. The results with deuterium were negative. Provided that the method is sufficiently sensitive, hydrogen exchange seems to be ruled out. Experiments have also been conducted to see if the conversion of ethylene to ethylene oxide and some subsequent action of that metabolite is the mode of action of ethylene (7). Although the correlation of ethylene oxidation and action was fairly good, it has been shown that inactivation of the ethylene-oxidizing system does not prevent the action of ethylene (1). This seems to rule out the oxidation of ethylene as the primary mode of action of ethylene.

Does ethylene bind to a receptor and act through a hormone-receptor complex as other hormones do? Evidence has accumulated that it does and that the hormone receptor can be measured and purified.

MATERIALS AND METHODS

Ethylene binding. When intact plant material was used, binding was essentially as previously described (14). The material was harvested and cut to the desired size and allowed to stand overnight to let wound-generated ethylene subside. The material was placed in desiccators and treated. After about 2 hours the desiccator was opened, and the plant material was shaken in air and placed in a jar containing a scintillation vial, with 0.3 ml of mercury perchlorate as a trapping agent (24). A piece of fiberglass was

---

*R. J. Reynolds Research Apprentice.

NATO ASI Series, Vol. H10
Plant Hormone Receptors. Edited by D. Klämbt

included to increase the surface area. All treatments included duplicate samples exposed to [^{14}C]ethylene in the presence and absence of unlabeled ethylene.

For measurements with extracts, the material was treated essentially as previously described (15). The material was blended with 0.1 M potassium phosphate, pH 6.0, and 2% Triton X-100. Cellulose fiber (0.7 g/ml of fluid) was included. The samples were placed in pans and put in a desiccator to be exposed to ethylene. After about 2 hours the samples were removed, allowed 3 minutes' exposure to the air, and then placed in jars with a vial containing mercury perchlorate. After heating to 60 C for at least 2 hours, and allowing another 8 hours at room temperature to collect the [^{14}C]ethylene, the vials were removed and counted in the presence of liquid scintillation fluid. All treatments included duplicate samples treated in the presence and absence of saturating amounts of ethylene.

Abscission test. This test was run essentially as previously described (2, 13). Explants were cut, placed in agar, and aged before exposure to the various compounds.

Tobacco leaf respiration test. Leaf segments were cut and placed in Warburg vessels. They were exposed to the compounds and each day the flasks were removed and the rate of respiration measured using a Warburg respirometer. Respiration increases after exposure to ethylene (13) and the increase in respiration due to other compounds were also measured.

Isoelectric focusing. Immobilized ampholites were prepared in acrylamide according to the directions supplied by LKB. After prefocusing the gels, the binding preparation was placed in a well near the center. Voltage was applied gradually, keeping the temperature below 10 C until the desired voltage was reached, with the maximum being 3000 V.

For isoelectric focusing in granulated gel, ampholyte (pH 5-7), water, and the extract were adjusted to 3 M in urea. This was evaporated with a fan until the urea concentration was 5 M, the ampholyte was 0.5%, and the granulated gel (washed Sephadex G 25) was 5%. Lysine was used at the cathode end and glutamic acid was used at the anode end. All isoelectric focusing experiments were performed with a flat bed LKB 2117 Multiphore II unit.

Extract. Mung bean seeds were soaked in water overnight at 5 C and blended successively with 5 mg, 2 mg, and 2 mg of Triton X-100 for each g of the original dried seed. One ml of 0.25 M K_2HPO_4 was added for blending. After centrifuging, the extracts were combined and adjusted to 2% Triton X-100. The binding component was salted from solution by adding 250 g of NaCl and 100 g of $(NH_4)_2SO_4$ to each liter of solution. After separation into two phases the material was centrifuged and the lower phase was discarded. The oily upper phase was dialyzed and concentrated in dialysis bags with carboxymethyl cellulose. This preparation was used for isoelectric focusing.

RESULTS AND DISCUSSION

Compounds giving an ethylene response. It has been known for many years that carbon monoxide and other olefins were active, similar to ethylene (8). Also, ethylene was known to induce cyanide-insensitive respiration (21) and did appear to be active as cyanide. How would these compounds, diverse in structure, be able to act to induce the same response? There is an effect in inorganic chemistry, first alluded to by Werner in 1893 (23), now called the *trans* effect. The order of compounds giving a *trans* effect is: $H_2C{=}CH_2$, $C{\equiv}O$, $HC{\equiv}N > PR_3$, $H^- > S{=}C(NH_2)_2$, $CH_3^- > C_6H_5^-$, NO_2^-, I^-, $SCN^- > Br^-$, $Cl^- >$ Amines, NH_3, OH^-, H_2O (9). Not all of these are suited for biological systems. The order of magnitude of ligand substitution from highest to lowest is 10^6. The compounds highest in the series are all π acceptors. They bind to a metal through σ bonds and back-accept electrons into their own vacant π orbitals. Could this be the mechanism by which the compounds act? This possibility has been tested and it was found that these and other acceptor compounds, such as isocyanides, do induce a rise in respiration in tobacco leaves and abscission in beans (Table I), both ethylene responses. Later it was shown that they compete with ethylene for binding. These findings would suggest that ethylene acts by binding, similar to other plant hormones, and it should be possible to measure the sites to which ethylene binds.

Measurement of ethylene binding. Leaves or other plant material exposed to ethylene will bind ethylene. It is important to show that the binding is to a specific binding site. Henry's law states that at a fixed temperature the weight of a gas dissolved in a fixed quantity of liquid is proportional to its partial pressure, i.e. $K = P/N$ (P is the pressure in mm of Hg, N is the mole

Table I. Ethylene-like Response to Some π-Acceptor Compounds

Compound	Structure	Tobacco Leaf Respiration		Bean Abscission Test	
		Concentration[a] µl/l	Activity % of Ethylene	Concentration[a] µl/l	Activity % of Ethylene
Ethylene	$H_2C{=}CH_2$	10	100	1	100
Tetrafluoro-ethylene	$F_2C{=}CF_2$	10	50	1	100
Tetrachloro-ethylene	$Cl_2C{=}CCl_2$	--	--	100	inactive
Carbon monoxide	C≡O	5000	100	500	100
Methyl isocyanide	CH_3-N≡C	100	80	10	100
n-Butyl isocyanide	CH_3-$(CH_2)_3$-N≡C	100	70	5	100
Cyclohexyl isocyanide	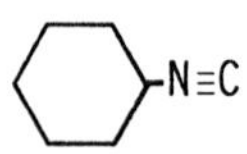	100	66	10	100
Benzyl isocyanide	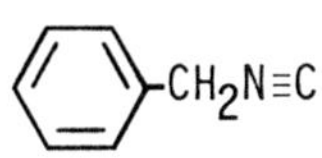	250	60	10	100
Phosphorous trifluoride		50	45	100	100

[a]Concentrations reported are the lowest which gave a maximum response and are for the compound as a gas (13).

fraction in solution, and K is a constant which is equal to 8.6×10^6 for ethylene in water). Dalton's law states that in a mixture of gasses, every gas exerts the same pressure that it alone would if it alone were confined to the same volume. [^{14}C]Ethylene and unlabeled ethylene should behave as separate gasses unless specific binding is involved, i.e. the presence of one should not alter the distribution of the other. Chemically, these gases should behave as if they were the same gas (neglecting any isotope effect) and the presence of one should affect the other. It is known (3) that most ethylene responses saturate at about 10 µl/l. It should be possible to measure the binding site by using a very small amount of [^{14}C]ethylene in the presence and absence of unlabeled ethylene. If [^{14}C]ethylene is allowed to compete with unlabeled ethylene, we can, using a Scatchard plot, determine the K_d (1/2 saturation concentration). This should compare favorably with the 1/2 saturation point (K_m) for ethylene action for the tissue. This has been done for a number of tissues (see Table II). In tobacco leaves the K_d for displacement is 0.27 μl/l and the K_m for action is 0.30 μl/l. The K_d value for displacement for tomato leaves (16) is reported to be 0.3 μl/l, and 0.09 μl/l for carnation leaves (20). A value of 0.1 μl/l is reported for carnation petals, while the K_d value for a Triton X-100 extract of mung bean sprouts is 0.12 μl/l (5). These values are all close to the range of values obtained for ethylene in causing a biological response (3), and the fact that ethylene binding and ethylene action both saturate at very similar concentrations is one line of evidence that binding of ethylene is to a physiological receptor.

Table II. Ethylene-Binding Site in Plant Vegetative Tissue

Tissue	K_d	Apparent Concentration[a] of Receptor moles/kg fresh weight	Reference
Tobacco leaves	0.30	3.5×10^{-9}	(14)
Bean leaves	0.14	2.0×10^{-9}	(12)
Citrus leaves	0.15	5.7×10^{-9}	(12)
Ligustrum leaves	0.31	6.8×10^{-9}	(12)
Mung bean sprouts	--	Identified[b]	(15)
Tomato leaves	0.30	1.9×10^{-9}	(16)
Tomato fruit	0.30	7.0×10^{-11}	(16)
Carnation petals	0.10	6.0×10^{-9}	(20)
Carnation leaves	0.09	2.0×10^{-9}	(20)
Bean cotyledons	0.30	4.8×10^{-9}	(12)
Bean roots	--	Identified[b]	(12)
Pea epicotyl	--	Identified	--
Wheat germ	--	Identified	--
Apple pulp	--	Identified	--
Cucumber leaves	--	Identified	(10)
Mushroom sporophore	--	Not detected	--

[a]Not corrected for the amount occupied by endogenous ethylene.
[b]The concentration was not determined.
[c]For the concentration in seeds, see reference 17.

Effect of compounds other than olefins on binding. Since carbon monoxide, isocyanides, phosphorous trifluoride, and cyanide (13) have been reported to give a response similar to ethylene, it would be of interest to see if they influence ethylene binding. Unfortunately, little more than simply testing of these compounds has been done; however, *in vivo* carbon monoxide has been shown to compete with ethylene for the binding site (14). There is some discrepancy between the 1/2 maximum for action (266 μl/l) vs. that for binding (746 μl/l); however, carbon monoxide does compete. Cyanide, phosphorous trifluoride, *n*-butylisocyanide, and carbon monoxide have been shown to compete for the binding site in non-living tissue (14), and carbon monoxide and cyanide have been shown to compete with ethylene in detergent extracts of plant tissue (15). The fact that these compounds do compete with ethylene for the binding site represents a line of evidence that binding is to a physiological receptor.

Diffusion of ethylene to and from the binding site. Ethylene responses are usually slow, requiring exposure of the tissue to ethylene for several hours. There are some physiological effects, however, which respond to ethylene rapidly. Ethylene affects the growth of pea rapidly (3, 22). Growth is measurably inhibited by ethylene within 20 minutes, and is measurably uninhibited in 20 minutes after removal of ethylene. The diffusion of ethylene to and from the binding site would then be expected to be of a similar time magnitude as the more rapid responses. A number of measurements have been made on time of diffusion to and from the site. Typical data are shown in Figure 1. Most of the ethylene is bound within 20 minutes and unbinds within 20 minutes. There is a small amount which seems to persist for a much longer time. The significance of this aspect is not known. It could represent a different kind of binding site, giving the plant a means of retaining a "memory" of exposure to ethylene.

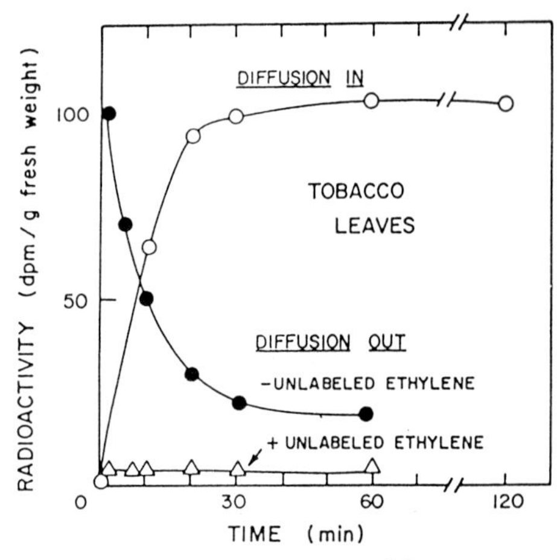

Figure 1. Diffusion of [^{14}C]ethylene into and out of tobacco leaves (11).

In tomato leaflets, much of the ethylene diffuses from the binding site much more rapidly than any other plant tested (16). Binding in bean leaves, *Citrus* leaves, *Ligustrum* leaves (12), carnation leaves, and carnation petals (20) is similar to that in tobacco. The fact that the diffusion rates to and from the binding site are of a similar magnitude to the rapid responses of ethylene provides further evidence that binding is to a physiological receptor.

Compared to intact tissues, diffusion from the ethylene binding site in Triton X-100 extracts of mung bean placed on cellulose is much slower. After 1 hour, at room temperature, most of the ethylene remains bound. Why this longer time is required is not known, but it may be that the physical structure *in vivo* is more favorable.

Compounds which prevent ethylene action. Ethylene action is partially reduced by carbon dioxide (8), and although the effect appears to be competitive for ethylene action, carbon dioxide does not appear to inhibit ethylene's binding competitively; moreover, carbon dioxide is not the type of compound that usually binds to metals to which ethylene binds (9). It is not known how carbon dioxide brings about inhibition of ethylene action.

Silver ion (7) inhibits ethylene action and silver ion does inhibit ethylene binding, both *in vitro* (15) and *in vivo* (10, 20). The kinetics of binding suggest a noncompetitive binding, and since silver is known to bind to sulfhydryl groups, such a group might be present near the binding site. The fact that silver ion inhibits both ethylene action and ethylene binding is evidence that such binding is to a physiological receptor.

Cyclic olefins inhibit ethylene action (18, 19), and 2,5-norbornadiene is the most effective on a concentration basis. Other cyclic olefins, such as cyclopentene and cyclohexene, are also effective. *cis*-2-Butene also causes an inhibition of ethylene action. The kinetics of action of these compounds indicate that they are competing with ethylene for the binding site. Cyclic olefins also compete with ethylene for ethylene binding both *in vivo* and *in vitro*. The competition for binding *in vitro* appears competitive. This would suggest that binding is to the physiological receptor.

Purification of the receptor. Attempts to isolate the receptor from mung bean sprouts and seeds have been made. The receptor appears to be a membrane-bound protein which can be solubilized by detergents. The properties of the receptor from sprouts and of that from seeds seem to differ. Approximately

80% of the component from sprouts binds to CM-Sephadex, while very little of that from seeds goes on CM-Sephadex. Of the portion which binds to CM-Sephadex, the isoelectric point appears to be between pH 6.5 and 5.5, although there is a small amount whose pI is below pH 4.0. Much of the component from seeds appears to have a very low pI (Figs. 2 and 3). Some may be lower than pH 3.0 and thus be inactivated by the low pH as higher voltages are used (Fig. 3). It is possible that the low pH is the result of an aggregate of proteins which have a collectively low pI. The use of 6 M urea to reduce hydrophobic interaction did not substantially improve the purification process (Fig. 4). Even in the presence of this concentration of urea, three bands appear and the major portion may be inactivated due to the low pH. Urea at 8 M seems to inactivate the binding component under electrophoretic conditions, but not in the absence of applied voltage. It may be possible to achieve further purification by isoelectric focusing if a way is found to eliminate a large portion of the inactive protein prior to focusing. Another possibility is to carry out a partial enzymatic digestion to release the binding component; however, without some previous step, success seems unlikely, due to the low pH.

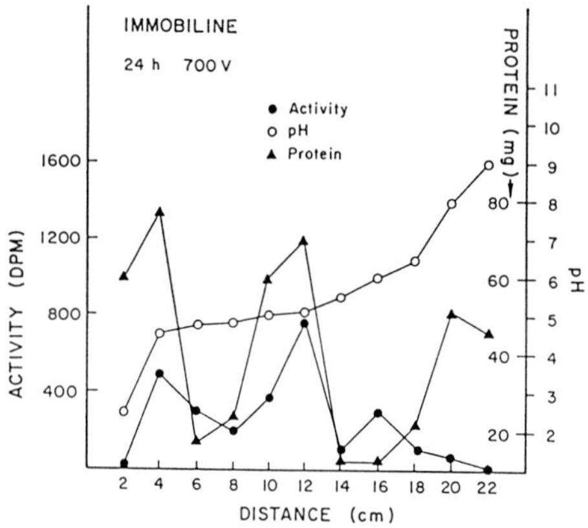

Figure 2. Isoelectric focusing of the ethylene-binding component in mung bean seeds using pH 4-6 immobilized ampholyte and 700 V for 24 h.

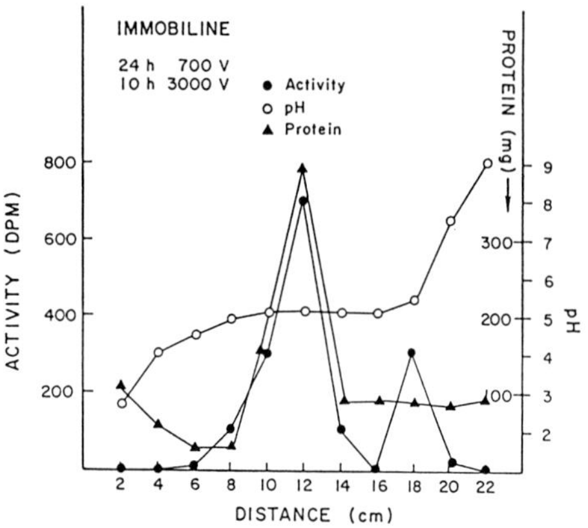

Figure 3. Isoelectric focusing of the ethylene-binding component in mung bean seeds using pH 4-6 immobilized ampholyte and 700 V for 24 h, followed by 3000 V for 10 h.

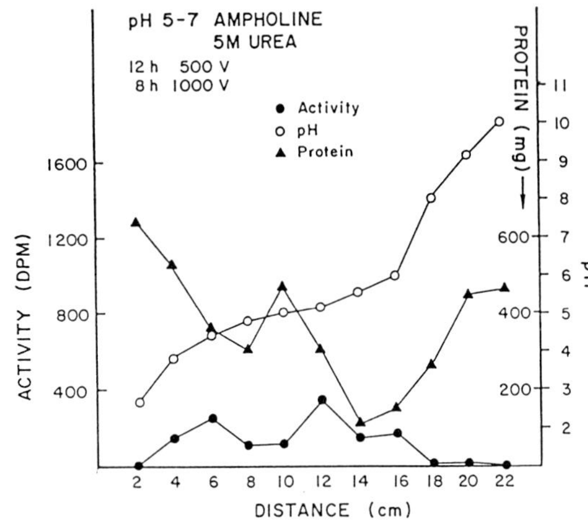

Figure 4. Isoelectric focusing of the ethylene-binding component using pH 5-7 ampholyte containing 5 M urea.

Summary. Several lines of evidence, then, indicate that ethylene is binding to a physiological receptor. The type of compound that elicits a response also competes with ethylene for binding. The 1/2 maximum for activity and binding are very close; the time of diffusion of ethylene to and from the site is similar to the more rapid ethylene responses, and inhibitors of ethylene responses and inhibitors of ethylene action inhibit ethylene binding. Further work is needed to purify the receptor. This should help in identifying the specific mechanism(s) by which ethylene elicits a response.

LITERATURE CITED

1. Abeles, F. B. A comparative study of ethylene oxidation in *Vicia faba* and *Mycobacterium paraffinicum*. J. Plant Growth Regul. 3:85-95. 1984.
2. Abeles, F. B., and R. E. Holm. Enhancement of RNA synthesis, protein synthesis, and abscission by ethylene. Plant Physiol. 41:1337-1342. 1966.
3. Abeles, F. B. Ethylene in Plant Biology. Academic Press, New York. 302 p. 1973.
4. Abeles, F. B., J. M. Ruth, L. E. Forrence, and G. R. Leather. Mechanisms of hormone action. Use of deuterated ethylene to measure isotopic exchange with plant material and the biological effects of deuterated ethylene. Plant Physiol. 49:669-671. 1972.
5. Beggs, M. J., and E. C. Sisler. Binding of ethylene analogs and cyclic olefins to a Triton X-100 extract from plants: Comparison with *in vivo* activities. Plant Growth Regul. 4:13-21. 1986.
6. Beyer, E. M. Jr. Mechanism of ethylene action. Biological activity of deuterated ethylene and evidence against exchange and *cis-trans* isomerization. Plant Physiol. 49:672-675. 1972.

7. Beyer, E. M. Jr. Effect of silver ion, carbon dioxide and oxygen on ethylene action and metabolism. Plant Physiol. 63:169-173. 1979.
8. Burg, S. P., and E. A. Burg. Molecular requirements for the biological activity of ethylene. Plant Physiol. 42:144-152. 1967.
9. Cotton, F. A., and G. Wilkinson. Advanced Inorganic Chemistry. John Wiley and Sons. New York. 1980. 1396 p.
10. Goren, R., A. K. Mattoo, and J. D. Anderson. Ethylene binding during leaf development and senescence and its inhibition by silver nitrate. J. Plant Physiol. 117:243-248. 1984.
11. Goren, R., and E. C. Sisler. Ethylene binding: Some parameters in excised tobacco leaves. Tobacco Sci. 28:110-115. 1984.
12. Goren, R., and E. C. Sisler. Ethylene-binding characteristics in *Phaseolus*, *Citrus*, and *Ligustrum* plants. Plant Growth Regul. 4:43-54. 1986.
13. Sisler, E. C. Ethylene activity of some π-acceptor compounds. Tobacco Sci. 21:43-45. 1977.
14. Sisler, E. C. Measurement of ethylene binding in plant tissue. Plant Physiol. 64:538-542. 1979.
15. Sisler, E. C. Ethylene-binding properties of a Triton X-100 extract of mung bean sprouts. J. Plant Growth Regul. 1:211-218. 1982.
16. Sisler, E. C. Ethylene binding in normal, *rin*, and *nor* mutant tomatoes. J. Plant Growth Regul. 1:219-226. 1982.
17. Sisler, E. C. Distribution and properties of ethylene-binding component from plant tissue. In Y. Fuchs and E. Chalutz, eds., "Ethylene: Biochemical, Physiological and Applied Aspects," pp. 45-54. Martinus Nijhoff/Dr. W. Junk Publishers, The Hague. 1984.
18. Sisler, E. C., and A. Pian. Effect of ethylene and cyclic olefins on tobacco leaves. Tobacco Sci. 17:68-72. 1972.
19. Sisler, E. C., and S. F. Yang. Anti-ethylene effects of *cis*-2-butene and cyclic olefins. Phytochemistry 23:2765-2768. 1984.
20. Sisler, E. C., M. Reid, and S. F. Yang. Effect of antagonists of ethylene action on binding of ethylene in cut carnations. Plant Growth Regul. 4:213-218. 1986.
21. Solomos, T., and G. C. Laties. Similarities between actions of ethylene and cyanide in initiating the climacteric and ripening of avocados. Plant Physiol. 54:506-511. 1974.
22. Warner, H. L. The synthesis and mode of action of ethylene in etiolated pea seedlings. Ph.D. thesis, Purdue University, Lafayette, Indiana. 1970.
23. Werner, A. Beitrag zur Konstitution anorganischer Verbindungen. Zeitschrift fur Anorganische Chemie 3:267-330. 1893.
24. Young, R. E., H. K. Pratt, and J. B. Biale. Manometric determination of low concentrations of ethylene. Anal. Chem. 24:551-555. 1952.

ETHYLENE-INDUCED GROWTH IN AMPHIBIOUS PLANTS

Roger F. Horton
Department of Botany, University of Guelph
Guelph, Ontario, Canada, N1G 2W1

INTRODUCTION

Aquatic plants have long been of interest to physiologists. Many species are major weeds in many parts of the world, and others are commercially important crop plants. Furthermore, aquatic plants have provided us with a number of valuable systems for continuing studies of photosynthesis and ionic relations which are the basis of our understanding of these processes in the entire plant kingdom.

The regulation of growth and development of aquatic and semi-aquatic plants is, however, poorly understood. Tissues under water are subjected to a very different range of environmental factors from those experienced on land. The effects of shifts from one environment to another is of particular interest in amphibious plants where temporally- and spacially-separated submergence may induce major physiological and morphological changes.

In this study, I will examine alterations in ethylene metabolism in leaves and petioles of an amphibious buttercup *Ranunculus sceleratus* L., relate these changes to the normal pattern of growth and discuss how these studies might help us in our consideration of the basis of action of plant regulators at specific binding sites preceeding demonstrable physiological activity.

PETIOLE GROWTH IN *R. SCELERATUS*

R. sceleratus L. has a circumboreal distribution and can be readily grown as a terrestrial plant from achenes collected in the field. In controlled environments it exhibits many of the attributes of a long-day flowering plant (19). Large, uniform populations can be derived in successive generations from a

NATO ASI Series, Vol. H10
Plant Hormone Receptors. Edited by D. Klämbt

single achene. Many submerged aquatic species have proved difficult to grow under controlled environments, and our ability to grow this plant terrestrially on a year-round basis and then subject it to aquatic conditions is a valuable starting point for large-scale physiological studies. Upon submergence, petioles elongate until the blades are again at the water surface when growth ceases or continues at a much slower rate allowing the maintenance of blade separation in a floating rosette. This acceleration of growth has been ascribed to an enhancement of levels of ethylene within the tissues caused by the reduced diffusivity of ethylene in water compared to that in air (16). Submergence-induced growth is exhibited most markedly by petioles of recently-expanded blades. Older leaves, and very young ones, are not so responsive. Alterations in competence to respond to submergence, and the nature of the response, has been studied in detail in other amphibious plants (18). In R. sceleratus, where petiole growth at this stage is due entirely to cell elongation, we do not know whether competence is related to ethylene metabolism, growth potential or the availability of receptors. We have routinely used 6-8 week-old plants in the vegetative state and in these most of the leaves are capable of responding to submergence. Distal portions (25 mm) of the petioles, with attached leaf blades, respond to submergence in a very similar manner to identical tissue on the intact plant (20) - there is no evidence of a major change in the response pattern due to excision and isolation.

THE EFFECT OF REGULATORS

Petioles with attached leaf blades are responsive to treatment in air with ethylene, IAA and gibberellic acid (20). Gibberellic acid is not as active as IAA in promoting growth, and, although gibberellins have been ascribed a critical role in the growth of starwort (16) and deep-water rice (4) stem tissues, it appears likely that shifts in gibberellin levels are not of major importance in R. sceleratus. Auxin is as effective as ethylene in promoting growth (9) when the tissues are exposed to high concentrations (10^{-4}M - 10^{-3}M), although, again, there is no evidence of a significant change in auxin levels following submergence (17).

THE ROLE OF THE LEAF BLADE

If completely isolated petioles, without leaf blades, are submerged (their normal status under a floating leaf) or treated with ethylene, there is no acceleration of growth (20). They remain sensitive to high levels of auxin, and are also sensitive to ethylene if they are bathed in low levels of IAA (10^{-7}M) insufficient of itself to cause significant growth (20). If blades are treated with 1-N-naphthylphthalamic acid or morphactins - phytotropins which reduce the activity of auxins with a specific role at auxin transport sites (7,12) - the petioles become non-responsive to ethylene, but remain responsive to high levels of auxin (9). Petioles treated with silver ions - which inhibit ethylene action - are again not responsive to ethylene but remain responsive to auxin (9). These data suggest that ethylene-promoted growth is dependent on the presence of low levels of auxin, and that these are normally supplied by the leaf blade. The blade also appears to be the primary site of ethylene production (21), and it is this ethylene that is trapped in the blade-petiole air spaces after submergence. There is no evidence of a significant requirement for a blade-buoyancy induced tension for growth; petioles will grow if submerged in an inverted position or if only the blade is submerged.

THE EFFECT OF SUBMERGENCE

The vacuum-extractable ethylene levels rise rapidly after submergence. While care must be taken in the interpretation of data obtained by this procedure (25) it is evident that ethylene levels can rise rapidly enough to sufficiently high levels to explain the growth. Similarly, ethylene levels fall if blades are allowed to float. Although the preponderance of stomata are on the upper surface of the leaf in this species, the degree of stomatal opening is unlikely to critically control the efflux of ethylene (10).

It should be emphasized that in this system we would appear to have a good example of an environmentally-controlled shift in leaves of a naturally-produced growth regulator which exhibits a demonstrable physiological effect. Further, the process of ethylene enhancement and accelerated growth can be rapidly reversed when the system is allowed to dissipate the promoter into the atmosphere.

We have examined some of the shifts in ethylene metabolism that occur during these processes. It has been demonstrated for a number of systems that enhanced ethylene production by plant tissue may be accompanied by a rise in the levels of 1-amino-cyclopropane-1-carboxylic acid (ACC, the likely precursor of ethylene). This rise is clearly measurable over a 24h submerged growth period (21). However, in some shorter term growth experiments, accelerated growth and enhanced ethylene levels can occur without a dramatic rise in ACC levels (Table 1). Thus at least the initial rise in ethylene may be related to the entrapment of the gas, although the enhanced capacity to synthesise ethylene may play a later role in the maintenance of the enhanced growth rate.

Table 1 Levels of C_2H_4, ACC and MACC in *R. sceleratus* petioles during and after submergence

Treatment	C_2H_4	ACC	MACC	Petiole Growth
	---------nmoles.g^{-1}	fw---		(mm)
0h	1.2	0.24	3.07	0
submergence				
3h	10.7	0.42	4.06	2.2
floating				
6h	1.8	0.30	4.13	2.7

Values are means of three replicates

ACC can be metabolised to N-malonyl-ACC, and this was believed to represent an essentially irreversible process (3). Subsequently, there has been some evidence presented to show MACC could be metabolised, albeit slowly, back to ACC in some plants (13). (Formyl-ACC, a synthetic analogue, is readily metabolised

(11,14)). Petioles and blades of R. sceleratus contain MACC, and its level rises in tissue along with that of ACC. However, the higher level is maintained when ethylene levels are reduced after flotation, and there is no clear evidence that MACC represents a significant source of ACC or ethylene during these experimental periods.

ETHYLENE PRODUCTION, ETHYLENE ACTION AND CO_2

Carbon dioxide is known to counteract the physiological effects of ethylene in many systems (1), but does not appear to be critical in the growth of R. sceleratus petioles (8). However, carbon dioxide levels were shown to regulate the production of ethylene by blades of R. sceleratus (8) and this observation was expanded in subsequent investigations with a wide range of plant species (6,8). Carbon dioxide could be acting to promote ethylene biosynthesis, and the possibility remains that it has some direct action at a binding site and that this, in part, is reflected in the measured emanation rates (6). The internal levels of carbon dioxide in water plants are very varied. In our experiments, even in the light, levels often exceed 2000 ul $CO_2.l^{-1}$ air. The possible significance of this should be bourne in mind when both the synthesis and action of ethylene in these systems are considered.

ETHYLENE ABSORPTION

Ethylene in air is active in promoting petiole growth over a wide range of concentrations (21). Similar, apparently log/linear responses have been described for many regulators, and their significance continues to be actively discussed (23,24). At our present stage of understanding they do not preclude the existence and importance of binding sites (24). In an attempt to demonstrate the existence of binding sites in R. sceleratus we have tried to demonstrate the discriminatory absorption of ethylene from an ethylene-acetylene gas mixture presented to tissues in which the capacity to evolve ethylene has been severely reduced by treatment with aminoethyoxyvinyl glycine (AVG, an inhibitor of ACC-synthase) and cobalt ions (an inhibitor of ACC to ethylene conversion). These tissues retain their ability to respond to ethylene (see 22). These experiments - similar to those in which plant tissue was used as the solid phase in a gas chromatography system in attempts to demonstrate tissue ethylene-binding sites (2) - could not establish any differential absorption of the two gases during a period of hydrocarbon-induced growth in the presence of low levels of IAA.

ETHYLENE BINDING AND GROWTH

On the evidence available any consideration of ethylene (and auxin) binding in these tissues must remain largely conjectural. However, if may be useful to emphasise some points about this system and its uses. The growth of petioles in *R. sceleratus* is clearly *functionally*-related to a rapid rise in internal ethylene levels caused by a well-defined environmental shift. The changes in the levels of the regulator correlate well to the levels of the regulator that need to be applied to mimic the effect of submergence. The induced growth is rapid, reversible and repeatable as are the accompanying changes in the associated regulator. Control of the synthesis of the regulator can be separated from the control of its biological activity. The determinations of the internal levels of ethylene in the tissue will require considerable extension - we will need to have data on the concentration of ethylene at various sites in the system, and, perhaps most importantly, on the flux of ethylene in epidermal and subepidermal cells which probably regulate the tissue growth rates. These ethylene levels will then have to be related to concurrent shifts in carbon dioxide and oxygen levels, both of which are known to affect ethylene biosynthesis, ethylene action, and the growth potential of plant tissue.

Our inability to demonstrate differential absorption of ethylene from an ethylene/acetylene mixture may be due to the over-simplicity of the experiment. But the effects of ethylene are clearly rapidly reversed when the regulator is allowed to dissipate. The cells of these tissues exhibit a changing sensitivity or competence to respond to ethylene and this competence may be controlled, in part, by other regulators. In an attempt to classify cells in higher plants in terms of their competence to respond to regulators, Osborne (17) has suggested that petiole cells of R. sceleratus are Type 3 cells, capable of responding to either auxin or ethylene by elongating, and that there may be two separate systems for controlling their growth (5). However, notwithstanding the demonstration of an ethylene-mediated proton-efflux pump in petioles of Nymphoides (15) similar to those described for many auxin-induced growth systems, all the evidence is also consistent with the idea that, although the shift in the ethylene levels may be the functional controller of growth rate, growth is in fact regulated by auxin. Ethylene could thus be considered as the predominant factor (among a number of others) regulating the sensitivity of the tissue to auxin, by acting to modulate the number and/or activity of auxin-binding sites rather than possessing a growth-linked binding site of its own.

Much of the existing knowledge of hormonal binding (and much of our knowledge of hormone-mediated growth) has been derived from studies using young, often dark-grown, seedlings. In these, the lack of clear temporal and spacial distribution of tissue sensitivity to the plethora of regulators, together with their lack of some major structural and physiological attributes, may render them unsuitable as test systems. Using more highly developed plant systems, the investigation of growth (rather than reserve mobilization or senescence) in specific, judiciously-chosen, organs may give us a more direct picture of the normal role of regulators. If, in R. sceleratus, petiole growth is regulated by epidermal cell growth, it should be possible to remove the vascular tissue together with all auxin transport sites to which phytotropins bind (12). This could be done surgically or enzymatically in these tissues, in which, even when grown terrestrially, there exists a preponderance of aerenchyma between the bundles and the epidermis. This may allow an approach to separating the ethylene/auxin/phytotropin interactions at a putative binding site in an elongating epidermis from, possibly related (7), sites in the auxin transport pathway.

REFERENCES

1. Abeles FB (1973) Ethylene in plant biology. Academic Press, London.
2. Abeles FB (1984) Plant Physiol. (Bethesda) 74:525
3. Amrhein N, Breuing F, Eberle J, Skorupka H, Tophof S (1982) In: Wareing PF (ed) Plant growth substances 1982. Academic Press, London, p 249.
4. Bleecker AB, Rose-John S, Kende H (1985) Proc 12th Int Conf Plant Growth Subs, Heidelberg, p 40
5. Cookson C, Osborne DJ (1978) Planta (Berl) 144:39
6. Grodzinski B, Boesel I, Horton RF (1982) J Exp Bot 33:344
7. Hertel R (1983) Z Pflanzenphyiol 112:53
8. Horton RF (1985) In: Roberts JA, Tucker GA (eds) Ethylene and plant development. Butterworths, London, p 37
9. Horton, RF, Samarakoon AB (1982) Aquat Bot 13:97
10. Horton RF, Saville B (1984) Plant Sci Lett 36:131
11. Horton RF, Vanhinsberg N (1986) Plant Physiol Suppl 40:591
12. Jacobs M, Short TW (1985) In: Bopp M (ed) Plant Growth Substances 1985. Springer-Verlag, Berlin, p 218
13. Jiao XZ, Philosoph-Hadas S, Yang SF (1986) Proc 12th Int Conf Growth Subs, Heidelberg, p 33
14. Lurssen K, Konze J (1985) In: Roberts JA, Tucker GA (eds) Ethylene and plant development. Butterworths, London, p 363.
15. Malone M, Ridge I (1983) Planta (Berl) 157:71
16. Musgrave A, Jackson MB, Ling E (1972) Nature (New Biol) 238:93
17. Osborne DJ (1976) In: Pilet PE (ed) Proceedings in life sciences. Plant growth regulation. Springer-Verlag, Berlin, p 161
18. Ridge I (1985) In: Roberts JA, Tucker GA (eds) Ethylene in plant development. Butterworths, London, p 229
19. Samarakoon AB, Horton RF (1981) Can J Bot 59:1386
20. Samarakoon AB, Horton RF (1983) Can J Bot 61:3326
21. Samarakoon AB, Horton RF (1984) Ann Bot 54:263
22. Samarakoon AB, Woodrow L, Horton RF (1985) Aquat Bot 21:33
23. Trevawas A (1981) Plant Cell Environ 4:203
24. Wareing PF (1986) In: Bopp M (ed) Plant growth substances 1985. Spinger-Verlag, Berlin, p 1
25. Yeang HY, Hillman JR (1981) J Exp Bot 32:381

APPENDIX

TECHNIQUES IN PLANT HORMONE RECEPTOR RESEARCH

AFFINITY CHROMATOGRAPHY IS A POWERFUL TOOL TO PREPARE AUXIN RECEPTORS

D. Klämbt, M. Löbler
Botanisches Institut, University of Bonn, Meckenheimer Allee 170
D 53 Bonn 1, FRG

INTRODUCTION

Different approaches have been used to purify auxin binding proteins from a variety of plant species. For the purification of membrane associated auxin binding sites different auxin affinity matrices have been used. Venis (1977) reported the preparation of a protein fraction, using 2,4 D-lysine-Sepharose. Unfortunately the eluted protein fraction did not show any auxin binding activity. A quite similar phenomenon was observed by Tappeser et al. (1981), using 2-hydroxy-3,5-diiodobenzoic acid-Sepharose as affinity matrix and eluting the binding protein with NAA. Elution of the matrix with 0.15 NaCl yielded a reasonable purification of auxin binding activity. This method in combination with immunoaffinity chromatography was successful in isolating and identifying an auxin receptor (Löbler, Klämbt 1985 a,b). Recently Shimomura et al. (1986) described a purification procedure, which is modified for a simple and quick preparation of this receptor.

EXPERIMENTAL PROCEDURES

300 g etiolated maize coleoptiles including the primary leaves were homogenized in 50 mM Tris/citric acid pH 8.0, 0.25 M sucrose, 1 mM EDTA, 0.1 mM $MgCl_2$ and squeezed through 4 layers of gauze. The remainder was reextracted twice. The combined filtrates were centrifuged at 5,000xg for 10 min. After adding $CaCl_2$ to a final concentration of 10 mM to the supernatant and stirring for 10 min the mixture was centrifuged at 10,000xg for 30 min, resulting a Ca^{2+} promoted sedimentation of membrane vesicles (Shimomura et al 1986, Schenkman, Cinti 1978).

For membrane solubilization the frequently used acetone method was replaced by the n-butanol method according to Maddy et al. (1972). The pellet was suspended in 2.5 mM phosphate buffer

NATO ASI Series, Vol. H10
Plant Hormone Receptors. Edited by D. Klämbt

pH 8.0, 0.5 mM EDTA, 1/3 volume of the homogenized fresh weight. After adding the same volume of n-butanol the mixture was few times shaken and kept on ice for 20 min. After centrifugation at 2,000xg for 15-20 min the lower phase was aspirated and dialysed overnight against TM buffer (10 mM Tris HCl pH 7.4, 5 mM $MgCl_2$).

The dialysate was applied to a DEAE Servacel column (30 ml) equilibrated with TM buffer. After washing the column with TM buffer (150 ml) overnight, the receptor containing fraction was front-eluted with 0.3 M NaCl in TM buffer. The eluate was directly applied to a NAA-linked AH-Sepharose 4B column (50 ml), equilibrated with TM buffer. After immediate washing of the column with 0.5 M NaCl in TM buffer overnight the column was eluted with 10 mM NAA, 0.4 M NaCl in TM buffer.

The resulting 10-12 ml eluate was desalted and freed from NAA by application of 2.5 ml to PD 10 column equilibrated and developed with 1/5 TM buffer. The final filtrates were lyophilized and stored at -15° C. For further uses the stored residues were resolved in 1/5 the volume of aqua dest.

CONCLUSION

The described procedure ends up with a highly purified preparation of the membrane associated auxin receptor and few proteins of larger molecular mass. All of them cross-react with anti auxin receptor antibodies. It may be possible however, that these protein bands are artefacts or native covalently linked receptors or higher glycosylated receptor molecules after SDS-gel-electrophoresis.

Reexaminations of NAA elution of DIBA-Sepharose with binding buffer pH 5.5 (Tappeser et al. 1981) led to very similar results. The observed reduction of binding capacity of these eluates may be due to the acid labile auxin receptor (Shimomura et al. 1986).

REFERENCES

Löbler,M. & Klämbt,D. (1985a) J.Biol.Chem. 260, 9848-9853

Löbler,M. & Klämbt,D. (1985b) J.Biol.Chem. 260, 9854-9859

Maddy,A.H., Dunn,M.J. & Kelly,P.G. (1972) Biochim.Biophys.Acta 288, 263-276

Schenkman,J.B. & Cinti,D.L. (1978) in Methods in Enzymology Vol.52, pp 83-99 Academic Press, New York

Shimomura,S., Sotobayashi,T., Futai,M. & Fukui,T. (1986) J.Biochem. 99, 1513-1524

Tappeser,B., Wellnitz,D. & Klämbt,D. (1981) Z.Pflanzenphysiol. 101, 295-302

Venis,M.A. (1977) Nature 266, 268-269

SOLUBLE AUXIN BINDING PROTEINS IN PEA: PREPARATION AND BINDING ASSAYS

K. Hajek, H.-J.Jacobsen and D.Hess
Institut für Genetik, Universität Bonn
D - 5300 Bonn - 1, West Germany

Plant material

Seeds of the pea variety "Dippes Gelbe Viktoria " were washed with tap water and imbibed for 15-20 h. They were germinated in moist vermiculite for 10 d in the dark. Epicotyls were harvested without apical hooks in the cold under low light intensity. They were immediatly used for cytosol preparation or frozen for following investigation.

Chemicals

^{3}H-indole-acetic-acid (IAA, specific activity: 29Ci/mol) was from CEA, Saclay, France, unlabelled IAA was from Sigma. All other chemicals used were of analytical grade.

Preparation of crude cytosol

All procedures are carried out at 4° C. 200-250 g (fresh weight) of epicotyls are homogenized in a Waring Blendor with addition of 50 mM Tris-HCL, pH 7.8 (adjusted in the cold). Using fresh epicotyls, the added buffer volume is 50 ml, for frozen material volume is doubled. Homogenisation in the Waring Blendor was 3 min (with intervals to prevent heating of the fresh material) and 6 min for the frozen tissues. The pap are filtered through three layers of nylon material with defined pore sizes (1oo µm, 60 µm, 10 µm) to remove the nuclei. The filtrate is centrifuged at 145000 x g for 2 h at 4° C. The supernatants are concentrated by ultrafiltration to 10 ml. Remaining low molecular weight compounds are removed by Sephadex G 25 gel filtration (column: width 2.5 cm, height ca 40 cm) with extraction buffer. The protein fractions are pooled and concentrated again by ultrafiltration to 1 - 3 ml. The concentrate is centrifuged at 2000 x g

NATO ASI Series, Vol. H10
Plant Hormone Receptors. Edited by D. Klämbt

and 4° C for 5 minutes to avoid clogging of the steal capillaries of the pump by protein particles.

Preparative chromatofocusing

The cytosol preparation is applied to a pressure resistant glass column (width 12.7 mm, height 109 cm, Latek, Heidelberg, FRG) filled with ion exchange gel PBE 94 (Pharmacia Fine Chemicals, Sweden). The column is equilibrated with 25 mM imidazole- HCL, pH 7.3 and proteins are eluted with 100 mM Na-cacodylate-HCL, pH 5.0. Finally strongly bound proteins are eluted with 1 M NaCL in elution buffer. For a constant flow rate of 1 ml/min, an HPLC-pump is used. Normally, the pressure is not higher than 6-7 bar.A preparative run requires up to 8 h. Therefore all buffers and the gel were degassed prior to use. To reequilibrate the column again to a pH of 7.3, nearly 2 l of imidazole buffer is necessary. Chromatofocusing is carried out at 4° c. Protein fractions are pooled according to the elution pattern and pH and concentrated again.

Binding assay procedures

(a) Ammonium sulphate preticipation method (ASP)

Binding assays are performed according to Wardrop and Polya (1977), modified by Jacobsen (1982). Proteins (20-50 µg = 50-150 µl) are incubated with a constant concentration of labelled IAA (2.5×10^{-9} M) and increasing concentrations of unlabelled hormone ($0\text{-}5 \times 10^{-7}$ M, in 50 mM Tris-HCL, pH 7.8) according to Oostrom et al.(1980). Usually incubation time takes 5 min in ice. Incubation is stopped by adding ice-cold, saturated $(NH_4)_2SO_4$ - solution (4.2 M) in extraction buffer to preticipate the proteins. Total volume is 1 ml. Incubationas well as preticipation of the proteins is done in caps in 10 sec intervals to keep in time exactly. After preticipation, samples are left in ice for some minutes and centrifuged at 25000 x g for 30 min. Supernatants are decanted from the pellets, and the caps are placed upside down on filter-paper. The drops assembling at the cap wall and the remaining liquidity near the pellet are carefully removed with wadding sticks. When the supernatant is removed completly, pellets are redissolved in 0.1 % SDS and whole caps are given into scintillation vials. The vials are shaken vigorously, radioactivity is determined after cooling down. Experiments are done with 2-3 replications / IAA concentration in one experiment.

(b) Nitrocellulose filters (NC)

Binding assays are performed using nitrocellulose filters (Millipore VMWP02500, pore size 0.05 µm) to bind the proteins. Filters are placed on the filter-holder and rinsed with 1 ml of extraction buffer first. The protein-hormone incubation mix is applied to the filters after incubation in ice for 5 min. The incubation mix is sucked through by a vacuum, filters are subsequently twice-rinsed with 1 ml of extraction buffer. The incubation of receptor and hormone is modified: total volume of incubation mix is adjusted to 600 µl with extraction buffer, only 500 µl are applied to the filters and data obtained are corrected by mathematical means. Incubation is done in 20 sec intervals, because it takes more time to change pipet-tips between different incubation steps. Experiments are done with 2 replications/ IAA concentration in one experiment.

(c) Polyethylenimine - treated filters (PEI)

The use of PEI-treated filters was reported first for binding assays with steroid receptors by Bruns et al. (1983). Thick glass filters (Whatman GF/B) are soaked in fresh 0.3 % PEI, pH 7.8 for at least 1/2 h. After soaking, filters are placed on the filter-holder without draining. Incubation was performed in the same way as described under (b). After the assay, filters are left in scintillation cocktail overnight, until they had become transparent. Experiments are done in 2 replications/IAA concentration in one experiment.

Separation of sABP in native PAGE and binding assay

Protein fractions obtained after chromatofocusing crude cytosol, are dialysed against 10 mM diethylbarbituric acid, 50 mM Na - diethylbarbiturate, pH 8.6 for 1 h. After dialysis, proteins are applied to a 1 mm slab gel, prepared according to Altland et al. (1980):

I : 31.25 g Polyacrylamide
1.0 g Bisacrylamide
ad 100 ml a.dest.

II: gel buffer
2.23 g Tris
ad 100 ml a. dest. pH 6.7 (adjusted with H_3PO_4)

III : 1.2 g Ammoniumperoxydisulphate
ad 100 ml a. dest.

Stacking gel:

I 10 %
II 10 %
III 5 %
75 % a.dest.
0.125 % TEMED

Resolving gel:

I 32 %
II 32 %
III 5 %
30 % a.dest.
0.16 % TEMED

Running buffer (" Veronal ")

10 mM Diethylbarbituric acid
50 mM Na - diethylbarbiturate
ph 8.6

Electrophoresis lasts 14 h at 20 mA. For the PAGE binding assays, 2 cm wide lanes are cut of the gel with a stencil, and sliced in 4 mm pieces. Gel pieces are transferred into caps and covered with 200 µl buffer (100 mM Tris-HCL, pH 7.8). In 15 sec intervals 50 µl of 2.5 x 10^{-8} M ^{3}H-IAA are added. After 10 min of incubation, the labelled IAA is removed and 500 µl of unlabelled IAA (5 x 10^{-5} M) are added for the following 10 min. This washing procedure is repeated once. The whole procedure is carried out at 4° C. After the washings, cap walls are dried with wadding sticks and whole caps are given into scintillation vials. Vials are shaken vigorously for 1 min and radioactivity is measured. In parallel, control lanes with protein are stained with Coomasie Brilliant Blue.

References

Altland,K. (1980) Double one-dimensional electrophoresis of human serum transferrin: a new high-resolution screening method for genetically determined variation Hum.Genet. 54, 221-231

Bruns,R.,Lawson-Wendling,K. and Pugsley,T. (1983) A rapid filtration assay for soluble receptors using Polyethylenimine-treated filters, Anal.Biochem. 132, 74-81

Jacobsen,H.J. (1982) Soluble auxin-binding proteins in pea epicotyls, Physiol.Plant. 56, 161-167

Oostrom, H.,Kulescha,Z.van Vliet,Th.B. and Libbenga,K.R. (1980) Characterization of a cytoplasmatic auxin receptor from tobacco-pith callus,Planta 149,44-47

Wardrop,A.J. and Polya,G.M. (1977) Properties of a soluble auxin-binding protein from dwarf bean seedlings, Plant Sci.Lett. 8, 155-163

EXTRACTION OF AUXIN-BINDING PROTEINS FROM TOBACCO

H.J. van Telgen, A.M. Mennes, C. Nakamura, P.C.G. van der Linde, E.J. van der Zaal, A. Quint and K.R. Libbenga

Department of Plant Molecular Biology
Botanical Laboratory
Nonnensteeg 3 2311 VJ Leiden
The Netherlands

1. INTRODUCTION

In our lab several procedures are in use for the isolation of auxin-binding proteins (ABPs) from tobacco tissues. Only for preparation of microsomal fractions containing membrane-bound ABPs one common procedure can be used for all kinds of tissues. However, the procedures used for isolation of soluble ABPs are strongly dependent on the kind of tissue from which extraction is done. For instance, callus tissues are high in polyphenol content and contain a high phosphatase activity. Thus, to reduce contamination of crude preparations different buffer conditions are needed and partial purification is necessary before further experiments can be carried out. Cell suspensions usually pose the same problems. By first isolating nuclei these problems can be greatly overcome. However, when using cell suspensions as source, investments in cell- and tissueculture facilities are mandatory. Shoot tips are usually lower in polyphenol and phosphatase content, but limited availibility of enough plants can seriously confine the possibility of large-scale purification.

The following paragraph will briefly describe the different protocols in use in our lab for isolation of membrane-bound and soluble ABPs from the tissues mentioned above.

2. METHODS

All procedures are carried out at 0 - 4°C.

NATO ASI Series, Vol. H10
Plant Hormone Receptors. Edited by D. Klämbt

2.1. MEMBRANE-BOUND ABP

2.1.1. Extraction procedure

Tissue: callus, cell-suspensions, leaves, whole plant.

Isolation buffer: 50 mM MES-KOH pH 5.0
10 mM $MgCl_2.6H_2O$
.5 M sucrose

- x g material is homogenized in 2-3x volumes of buffer, either directly in a motor-driven Potter homogenizer (15 strokes, 1500 rpm) or, for tough tissues first in a blender (2 times 1' full speed) followed by Potter homogenization.
- filter homogenate through 30 mesh nylon cloth
- centrifuge filtrate 20' 3,600 g
- centrifuge supernatant 60' 150,000 g
- resuspend pellet in buffer
- centrifuge 20' 3,600 g
- measure protein content of supernatant and use directly for binding assay

2.1.2. Binding

NAA binding activity is determined, essentially as described by Vreugdenhil et al (1979) and Maan et al (1983). Total assay volume 1 ml; ABP volume 667 µl. Incubation time 1 hr at 25°C. Binding is 'stopped' by adding 5 volumes ice-cold 5x diluted isolation buffer. Complete sample is poured over GF/C filter, rinsed twice with ice-cold diluted buffer and filter is counted.

Incubation scheme

Tube number	^{3}H-NAA (M)	'cold' NAA (M)	Total NAA (M)
1	5.10^{-8}	-	5.10^{-8}
2	"	5.10^{-8}	1.10^{-7}
3	"	1.10^{-7}	$1.5.10^{-7}$
4	"	2.10^{-7}	$2.5.10^{-7}$
5	"	5.10^{-7}	$5.5.10^{-7}$
6	"	1.10^{-6}	$10.5.10^{-7}$
7	"	2.10^{-6}	$20.5.10^{-7}$
8	"	5.10^{-6}	$50.5.10^{-7}$
9	5.10^{-8}	1.10^{-5}	$100.5.10^{-7}$

2.2. SOLUBLE ABP

2.2.1. Callus tissue

Isolation buffer: 0.2 M boric acid pH 6.8
2 mM sodiumtetraborate
0.25 M KCl
20 mM Tris.HCl
4 mM diethyldithiocarbamate (added just before use)

Dialysis buffer: 20 mM Tris.HCl pH 7.5
5 mM $MgCl_2.6H_2O$
1 mM EDTA
1 mM DTT (dithiothreitol)

- homogenize x g callus in 1.5 - 2x volumes isolation buffer using a blender and/or Potter homogenizer
- centrifuge homogenate 1h 140,000 g
- concentrate supernatant in ultrafiltration cell with PM10 filter to a final volume of 10 ml
- remove any precipitate (10' 3,600 g)
- dialyse to dialysisbuffer and concentrate again to 10 ml
- measure protein content and use either directly or after partial purification for binding assay and/or further experiments
- partial purification can be done using Sephadex G-200 or Phosphoultrogel A6R(LKB)

2.2.2. Cell suspensions

Nuclear isolation buffer: 25 mM MES-KOH pH 6.0
20 mM KCl
20 mM $MgCl_2.6H_2O$
0.6 M sucrose
40% glycerol
10 mM 2-mercaptoethanol

Extraction buffer: 50 mM Tris-HCl pH 7.5
0.75 M KCl
2 mM EDTA
2 mM DTT

Dialysis buffer: see section 2.2.1.

- collect cells and wash with cold distilled water
- homogenize in 60-80 ml nuclear isolation buffer in a Potter homogenizer
- filter through Miracloth
- add Triton X-100 to a final concentration of 0.5%
- swirl gently for 1'; centrifuge 20' 3,600 g
- wash nuclear pellet with cold isolation buffer
- centrifuge 20' 3,600 g
- extract nuclear proteins by lysing nuclei in 10 ml extraction buffer
- centrifuge lysate 1 hr 150,000 g
- dialyse sup to dialysis buffer
- use directly or after partial purification for further experiments

2.2.3. Shoot tips

Extraction buffer: see section 2.2.2.

Dialysis buffer: see section 2.2.1.

- homogenize 30-40 shoot tips in 20 ml extraction buffer in a mortar using acid-purified sand as abrasive with a pestle
- centrifuge homogenate 1 hr 150,000 g
- concentrate supernatant to 6-10 ml in an ultrafiltration cell with PM10 filter
- dilute with 100 ml dialysisbuffer and concentrate again to 10 ml
- use directly or after partial purification over Phosphoultrogel A6R for further experiments

2.2.4. Binding assay

IAA binding activity can be determined essentially according to the scheme described by Oostrom et al. (1980). Total assay volume 1 ml. Binding assay buffer is identical to dialysis buffer (see section 2.2.1.) ABP volume 500 µl. Incubation for 30' at 26°C. We routinely include MgATP and p-nitrophenylphosphate (final concentrations 5 mM each) in the assay mixture.

Number	^{3}H-IAA(M)	'cold' IAA(M)	Total IAA (M)
1	$2.5.10^{-9}$	-	$2.5.10^{-9}$
2	"	$2.5.10^{-9}$	$5.0.10^{-9}$
3	"	$5.0.10^{-9}$	$7.5.10^{-9}$
4	"	$7.5.10^{-9}$	$10.0.10^{-9}$
5	"	$10.0.10^{-9}$	$12.5.10^{-9}$
6	"	$15.0.10^{-9}$	$17.5.15^{-9}$
7	"	$32.5.10^{-9}$	$35.0.10^{-9}$
8	"	$67.5.10^{-9}$	$70.0.10^{-9}$
9	"	$117.5.10^{-9}$	$120.0.10^{-9}$
10	"	$197.5.10^{-9}$	$200.0.10^{-9}$

Separation of bound and free ligand can be achieved in several ways.

a) Dextran-coated charcoal (DCC): After incubation of the binding assay, the test tubes are cooled in a ice-batch for 15 min. After cooling, 0.5 ml of a 3% DCC-suspension (prepared by dissolving 300 mg Dextron T70 in 100 ml dialysisbuffer and adding 3g washed Norit A, stirred for 1h) is added. Samples are vigourously shaken for 15' at 4°C and DCC is pelletted by centrifugation for 20' at 3000 g. 1 ml samples are carefully taken and used for liquid scintillation counting.

This method has two major disadvantages. Firstly: DCC will also absorb protein, so if protein content is low, relatively much protein (including hormone-receptor complexes) is adsorbed. Secondly, if there are many other small molecules in the assay mixture, (for instance phenoles, ATP, p-NPP etc) these will compete with free IAA for the (limited) number of available binding sites in the charcoal. This can result in an increased background.

b) Filtration using polyethyleneimine-coated glassfiber filters (PEI-filter method).
For details about this procedure refer to the procedure described by Hajek et al (1986) in these proceedings.

c) Ultrafiltration using the Amicon MPS-1 micropartition system; filter-holders are equipped with YMT-membrane filters (exclusion limit 30,000 D).

2.2.5. Purification of soluble ABP

For further purification we use an affinity column prepared from 5-hydroxy-IAA coupled to epoxy-activated Sepharose 6B according to the procedure provided the manufacturer. The crude ABP preparation is loaded onto the column in the presence of 5 mM MgATP, 5 mM PNPP and .1 mM $CaCl_2$. As gel filtration buffer we use the dialysis buffer described in section 2.2.1. Proteins are eluted with dialysisbuffer containing 1 mM IAA/0.3 M NaCl. After removal of IAA and NaCl the purified ABP fraction can be used for further experimenting.

3. REFERENCES

Hajek, K., Hess, D. and Jacobsen, H.J. (1986). Comparison of different ligand assays with regard to their reliability and handling ease. In: Proceedings NAW 'Plant Hormone Receptors', D. Klämbt, ed. Springer Verlag, Berlin, Heidelberg, New York.

Maan, A.C., Vreugdenhil, D., Bogers, R., Libbenga, K.R.(1983). The complex kinetics of auxin-binding to a particulate fraction from tobacco pith callus, Planta 158, 10-15.

Mennes, A.M., Nakamura, C., Van der Linde, P.C.G., Van der Zaal, E.J., Van Telgen, H.J., Quint, A., Libbenga, K.R. (1986). Cytosolic and membrane-bound high affinity auxin-binding proteins in tobacco. In: Proceedings NAW 'Plant Hormone Receptors', D. Klämbt, ed., Springer Verlag, Berlin, Heidelberg, New York

Oostrom, H., Kulescha, Z., Van Vliet, Th., Libbenga, K.R. (1980). Characterization of a cytoplasmic auxin receptor from tobacco pith callus. Planta 149, 44-47.

Van der Linde, P.C.G., Bouman, H., Mennes, A.M., Libbenga, K.R. (1984). A soluble auxin-binding protein from cultured tobacco tissues stimulates RNA synthesis in vitro. Planta 160, 102-108.

Vreugdenhil, D., Burgers, A., Libbenga, K.R. (1979). A particle-bound auxin receptor from tobacco pith callus. Plant Sci. Lett. 16, 115-121.

METHODS FOR MEMBRANE PREPARATION, AUXIN AND PHYTOTROPIN BINDING, IN VITRO AUXIN TRANSPORT.

R. Hertel
Institut Biologie III, Universität Freiburg, Schänzlestr.1, D-78 Freiburg i.Br., F.R.G.

Radiochemicals , Special Chemicals.

3-Indol-(1-^{14}C)acetic acid (^{14}C-IAA), 1.78 GBq/mmol, was purchased from Amersham, Braunschweig FRG, ^{3}H-1-N-naphthylphthalamic acid (^{3}H-NPA), 2.03 TBq/mmol, from CEA, Gif-sur-Yvette, France.

Unlabeled 1-N-naphthylphthalamic acid (NPA) was synthesized by Thomson et al (1973). Other phytotropins were obtained from G.F. Katekar, Division of Plant Industry, CSIRO, Canberra, Australia.

Most auxins, phytotropins and ionophores are stored in ethanol at concentrations such that final ethanol concentrations never exceed 0.1% to avoid membrane damage.

Plant Material and Preparation of Membrane Vesicles.

Coleoptiles of maize (*Zea mays* L., many varieties) are grown and harvested as described e.g. by Ray et al (1977); membranes from maize are prepared like those from *Cucurbita*.

Zucchini squash (*Cucurbita pepo* L.; many varieties, e.g. Zucchini "Cocozelle von Tripolis", Wagner, Heidelberg, FRG) is especially used for the in vitro transport tests. Seeds are planted in moist vermiculite and grown in plastic boxes at 25°C and 95% relative humidity in total darkness for 5 to 6 days.

The procedure of Hertel et al (1983) with some modifications can be used to prepare vesicles. 3 cm long hypocotyl segments are cut just below the hook . All subsequent steps are carried out at 2-4°C in daylight. The segments are homogenized in a herb-grinder (Moulinex) for 2 x 3 s, with two weight equivalents of extraction buffer containing e.g. 50 mM HEPES-BTP pH 7.9, 3 mM EGTA (EDTA in case of material to be fractionated on gradients), 1 mM DTE, 250 mM sorbitol or sucrose, and additional salts when desired.

The homogenate is squeezed through a narrow nylon mesh, and then fractionated in a SS34-rotor of a Sorvall R5-C refrigerated centrifuge for 10 min at 6,000g (or at 10,000g when mitochondria are to be removed). The pellet is discarded, and the supernatant is

NATO ASI Series, Vol. H10
Plant Hormone Receptors. Edited by D. Klämbt

centrifuged at 44,000g for 30 min using the same rotor. The pelleted vesicles are resuspended in extraction buffer (1ml/g fresh weight equivalents) and centrifuged again under the same conditions ("wash").

These pellets are surface rinsed with 1 ml test buffer (e.g. 10 mM MOPS-BTP pH 6.0, 250 mM sorbitol and varying salts as desired) followed by gentle homogenization with a glass potter in adequate volumes of the same buffer. With addition of this test medium, ion and pH-gradients can be applied depending on the contents of extraction medium and test buffer. The final vesicle concentration may vary from 0.75g fresh weight equivalents per ml to 2 g per ml depending on the type of experiment performed.

The buffers described here,were developed for in vitro transport studies (see also Clark and Goldsmith, 1986, 1987); microsomes for binding studies can be obtained with similar procedures but with different media - see e.g. Ray et al, 1977.)

Fractionation of Subcellular Particles in Order to Localize Different Auxin Binding Sites or Transport Activities.

To localize possible receptors, carriers or functions on the different cellular membranes, particulate material is fractionated on density gradients, and subsequently tested for different binding properties, for transport activity and for marker enzymes. (For a general survey on separation of subcellular particles see e.g. Reid, 1979).

Precentrifuged homogenates, or else, resuspended 44,000g or 100,000g microsomes are layered on gradients pepared with sucrose, metrizamide, renografin, dextran etc. E.g. linear sucrose gradients are prepared, over a 2 ml 45% sucrose cushion, using 45% and 15% sucrose (w/w) solutions (20 ml) containing 10 mM Tris pH 7, 1 mM EDTA, 1 mM KCl and 0.1 mM $MgCl_2$. After overlayering of 2 or more ml of 15% sucrose followed by the homogenate or microsomes, the tubes are centrifuged at 100,000g for 120 to 180 min. After centrifugation, fractions of definite volume are collected and are tested for different auxin binding and enzyme activities according to standard procedures (Ray, 1977; Dohrmann et al, 1978; Lomax, 1986).

Alternatively, and especially to distinguish presumed plasma membrane metrizamide (Jacobs and Hertel, 1978) or the very similar renografin (Leong and Briggs, 1981) can be used in place of sucrose. Solutions of gradient basal medium containing 28% metrizamide and a mixture of 5% metrizamide and 5% sucrose are employed. Linear gradients (e.g. 20 ml) are constructed over cushions of 2 ml 28% metrizamide in the bottom of the centrifuge tubes. A layer of 1.5 ml 5% metrizamide + 5% sucrose is pipetted onto the top followed by precentrifuged homogenate or resuspended microsomes. Centrifugation time is shorter than with sucrose gradients; otherwise, assays and evaluation of such gradients are as described above.

A gradient medium which is osmotically less active, must be used if the - osmotically sensitive - vesicles are to be tested for in vitro transport after the centrifugation. Lomax (1986) describes linear density gradients prepared with 1-12% Dextran T-70 in 250 mM sucrose, 50 mM Tris-HCl at pH 7.6, overlayed with *Cucurbita* microsomes and spun for 2 h at 75,000g.

A further, quite different method of fractionation has been used with homogenates from maize (Michalke, 1982) and *Cucurbita*. Polyethylenglycol (PEG) is added in the presence of a high amount of salt, and material precipitating between 1.5 and 12% PEG is recovered in fractions and tested.

Binding Tests.

The centrifugation assay for particle-associated ligand-binding sites is an equilibrium test (Ray et al, 1977). Radioactive hormone at a low concentration, e.g. 5×10^{-8}M ^{14}C-NAA, is mixed with the test particles. The particles in two equal volumes of this mixture are pelleted, one sample (A) without further additions, the other (B) after addition of a relatively high concentration of non-radioactive hormone, e.g. 10^{-4} M NAA. The radioactivity in pellets A and B is determined: the difference (A - B) is called saturable or specific binding. Non-specific binding - in presence of high ligand concentration - comprises not only any actual low-affinity or simple partition binding of labelled hormone to membranes in the pellet, but also any free labelled hormone that is carried in solution within and between the membrane vesicles.

Technically, the assay can be performed in small samples (0.2-5 ml) in any high-speed centrifuge (40,000g, or more; 30 min). As a standard assay for e.g. site I or II, radioactive auxins +/- unlabelled auxins or analogs are added to the microsomes or to the gradient-fractioned particles. 1 or 2 ml samples are prepared in test medium (e.g. 0.25 M sucrose; 5 mM $MgSO_4$, 10 mM Na citrate at pH 5.5). The samples are centrifuged at 44,000g or 133,000g for 45-

20 min at 4^{0}; pellets are cut off together with the bottoms of the tubes and radioactivity is counted.

For kinetic resolution, especially with NPA binding, membrane material can be exposed to radioactive ligand and subjected to nitrocellulose membrane filtration (see e.g. Michalke, 1982). For association kinetics, ^{3}H-NPA is added to the particle suspension in binding medium, and sampling is started immediately. With an automatic pipette, 100 or 200 μl are withdrawn and transferred into 10 ml of an ice-cold binding medium, and the total suspension is immediately filtered through a membrane filter (0.45 μm pore-size, 47 or 25 mm diam., Sartorius Membranfilter) into a vacuum flask. Usually the filter is washed with an additional 10 ml of medium. When all the liquid has filtered through, the vacuum is shut off. The total filtration and washing procedure lasts less than 30 s. The filters are counted in a scintillation counter. To observe the dissociation of ^{3}H-NPA from the binding sites, excess non-radioactive NPA (10 uM) is added, or the samples are simply diluted. Sampling is done as described for association.

NPA binding sites - presumably connected with the efflux carrier - can be solubilized from Cucurbita microsomes or plama membrane fractions with 0.3% CHAPS or 1.5% octylglucoside and stabilized with up to 50% glycerol (Thein and Michalke, unpubl.). Binding tests are best performed at pH 5; the material is quickly diluted into pH 7.8 buffer and filtered through Whatman glass filters GF-B precoated for 2 h with 0.3 % polyethylenimine (Bruns et al, 1983; see also Hajek et al, 1987) where the protein-^{3}H-NPA complex is retained.

Accumulation Test. In vitro Auxin Transport. (See also Clark and Goldsmith, 1986, 1987; Hertel et al, 1983; Benning, 1986)

Three different assay procedures, very similar to the binding tests, can be employed. The simple centrifugation assay is used whenever no kinetic analysis is required. Vesicle suspensions are mixed with test buffer containing radioactive probes, ionophores, and additional salts or osmoticum in plastic test tubes (final volume 1 ml). The test tubes are spun in a Sorvall R5-C centrifuge with SM24 rotor at 44,000g for 45 to 60 min. Addition of 2.5 mM $MgSO_4$ to the test buffer improves the compactness of the pellets.

A faster centrifugation assay can be used to measure transient changes in IAA accumulation. Vesicles are resuspended and stored in pH 7.9 low capacity buffer (5 mM HEPES-BTP, 250 mM sorbitol). 0.1 ml of this vesicle suspension are pipetted into polycarbonate tubes containing 0.9 ml pH 5 test buffer (25mM MOPS-BTP, 250 mM sorbitol), added ^{14}C-IAA, and ionophores or phytotropins. The tubes are spun in a Beckman tabletop ultracentrifuge (TLA-100.2 rotor) at ca. 400,000g for 5 min. ^{14}C-IAA is extracted with 0.5 ml methanol each.

Kinetic measurements are also performed using Sartorius nitrocellulose filters of 0.45 μm pore size (4.5 cm in diameter). Resuspended vesicles are incubated with ^{14}C-IAA with or without additional substances. At the appropriate intervals 0.5 ml vesicle solution (1 to 1.25 g fresh weight equivalents) are rapidly pipetted into 2.5 ml ice cold test buffer, mixed, and poured over the prewetted filter under suction. Approximately 5 to 7 s after dilution the filters are clear of liquid and can be transferred to a vial with scintillant and counted.

References

Benning C (1986) Planta, in press

Bruns RF, Lawson-Wendling K, Pugsley TA (1983) Anal Biochem 132:74

Clark KA, Goldsmith MHM (1986) In: Bopp M (ed) Plant Growth Substances 1985. Springer-Verl, Heidelberg, p 203

Clark KA, Goldsmith MHM (1987) This volume

Dohrmann U, Hertel R, Kowalik H (1978) Planta 140:97

Hajek K, Hess D, Jacobsen HJ (1987) This volume

Hertel R, Lomax TL, Briggs WR (1983) Planta 157:193

Jacobs M, Hertel R (1978) Planta 141:1

Leong TY, Briggs WR (1981) Plant Physiol 67:1042

Lomax TL (1986) In: Bopp M (ed) Plant Growth Substances 1985. Springer-Verl, Heidelberg, p 209

Michalke W (1982) In: Marme' D, Marre` E, Hertel R (eds) Plasmalemma and tonoplast. Elsevier, Amsterdam p 129

Ray PM (1977) Plant Physiol 59:594

Ray PM, Dohrmann U, Hertel R (1977a) Plant Physiol 59:357

Reid E (ed) (1979) Plant organelles. Horwood, London

Thomson KS, Hertel R, Müller S, Tavares JE (1973) Planta 109:337

STUDY OF SOLUBILIZED PEA NPA RECEPTORS USING HIGH PERFORMANCE ION EXCHANGE CHROMATOGRAPHY (HPIEC)

Timothy W. Short and Mark Jacobs
Department of Biology, Swarthmore College
Swarthmore, PA, USA 19081

1. Prepare 2 liters of mobile phase: Dissolve 5.88 g Na citrate (10 mM) in 1500 ml distilled H_2O, adjust pH to 6.5, fill to 2 liters with distilled H_2O. To 1 liter of this solution add 28.2g NaCl (0.5 M). This is solution "B", while the remaining citrate solution (no salt) is solution "A". Filter each through a 0.45 micron filter.

2. The HPLC is a Gilson/Rainin with an Apple controller. Attach an Altex DEAE-5PW anion exchange column (Spherogel TSK from Beckman), 7.5 cm x 7.5 mm I.D., between the column valve and a holochrome detector set at 254 nm or 280 nm (280 nm is best for proteins, but most of the preliminary work was done at 254 nm). Equilibrate the HPLC with mobile phase A at 1 ml/min for 3-4 hr. By that point the baseline is relatively stable.

3. A solubilized pea membrane protein preparation is obtained. Third internode tissue (ca. 15 g) from 7 day old etiolated pea seedlings (Pisum sativum, cv. Alaska) is chopped with razor blades and ground for 2 min in a mortar and pestle in 15 ml grinding medium ("GM")(250 mM sucrose, 1 mM EDTA, 1mM DTE, 0.1 mM $MgCl_2$ and 50 mM Tris-Mes, pH 8). The grindate is filtered through a fine nylon mesh,and the tissue remaining on the mesh is reground in a second 15 ml volume of GM and refiltered. The combined filtrates are centrifuged at 15,000 x g for 10 min, then the supernatant of that spin is centrifuged at 40,000 x g for 30 min. The resulting pellet is resuspended in 15 ml resuspension medium ("RM")(250 mM sucrose, 1 mM DTE, 0.5 mM $MgCl_2$, 10 mM Na citrate, pH 6.5) containing 1 % Triton X-100. It is homogenized, incubated for 20 min on ice, and centrifuged at 40,000 x g for 30 min. The supernatant is the "soluble pea membrane protein preparation."

4. 1 ml of soluble protein preparation is loaded into the injection loop. A 60 min gradient profile is then begun with the following points:

Time	% B
0	0
5	0
25	100
50	100
60	0

The injection is performed at 1 min. Fractions are collected either at set times (e.g. every 5 min) or by observing the instantaneous output trace and collecting individual peaks. A lag time is incorporated to allow for the protein to move from the detector to the output collection. The tubes are put on ice as soon as possible after the fractions are obtained, and are subsequently tested for NPA binding using the Amicon micropartition system, the description of which is given in the Jacobs and Short contribution to this NATO workshop.

NATO ASI Series, Vol. H10
Plant Hormone Receptors. Edited by D. Klämbt

EXTRACTION AND ASSAY OF GIBBERELLIN-BINDING PROTEINS FROM CUCUMBER AND PEA

Nasser Yalpani, Zin-H. Liu, and Lalit M. Srivastava
Department of Biological Sciences, Simon Fraser University, Burnaby, B.C., Canada, V5A 1S6

Plant Material

'National Pickling' cucumber seeds are surface sterilized in 50% household bleach, 5 min, and sown in moist, autoclaved vermiculite in flats. Seedlings are grown in darkness at 28°C, 6.5 days, then exposed to natural light for 0.5 day.

'Progress No. 9' or 'Alaska' pea seeds are soaked in lukewarm running water for 24 h and, after surface sterilization, sown in moist, autoclaved vermiculite. Seedlings are grown in darkness at 25°C, 10 days, then exposed to natural light for 0.5 day.

Gibberellin-Binding Protein (GABP) Extraction

All procedures at 0 to 3°C.

Cucumber - Apical 1.5 cm of hypocotyls are cut and pooled into freshly prepared extraction buffer (100 mM Tris, 1 mM EDTA, 50 μM PMSF, 10 mM dithiothreitol, adjusted to pH 7.3 with H_3PO_4), drained and homogenized in an equal volume (1:1, w/v) of extraction buffer using mortar and pestle or blender. The homogenate is passed through 2 layers of nylon cloth and centrifuged at 100,000*g*, 1.5 h. $(NH_4)_2SO_4$ is added to the supernatant to 50%. After equilibration and centrifugation at 24,000*g*, 20 min, the pellet obtained is washed in column buffer (20 mM Tris, 1 mM EDTA, 10 mM dithiothreitol, adjusted to pH 7.0 with H_3PO_4) + 50% $(NH_4)_2SO_4$. The pellet is resuspended in column buffer, particulates removed at 7,000*g*, 5 min, and the supernatant desalted on a column of Sephadex G-25 (Pharmacia). The G-25 eluate is loaded onto a DEAE-cellulose column (DE32,

NATO ASI Series, Vol. H10
Plant Hormone Receptors. Edited by D. Klämbt

Whatman) equilibrated in column buffer and a 0.18 to 0.33 M KCl gradient applied. The GABP fraction elutes with 0.23 to 0.28 M KCl. This fraction is then loaded onto a hydroxylapatite column (High Resolution, Calbiochem) equilibrated with 10 mM K-phosphate, 1 mM EDTA (pH 7.0). After elution with 400 mM K-phosphate, 1 mM EDTA (pH 7.0), the GABP fraction is concentrated and desalted in an ultrafiltration cell (PM 30 membrane, Amicon). Further purification of the GABP can be achieved using Sephacryl S-200 (Pharmacia). The GABP can be stored at -20°C or lower as lyophilized powder from low ionic strength solutions.

Pea - The procedure for GABP extraction is similar to that for cucumber with the following exceptions. Apical 1 cm of epicotyls are homogenized in extraction buffer (100 mM K-phosphate, 1 mM EDTA, 25 μM PMSF [pH 7.4]). To the 100,000*g* cytosol saturated $(NH_4)_2SO_4$ in column buffer (20 mM K-phosphate, 1 mM EDTA [pH 7.0]) is added to 60%. The pellet is resuspended in column buffer and desalted on a column of Sephadex G-50 (Pharmacia). The protein fraction is loaded onto a DEAE-cellulose column (DE32, Whatman) in column buffer. The GABP elutes with 0.1 to 0.2 M KCl.

Assay of GABP

Incubations: High affinity specific binding of $[^3H]GA_4$ by the receptor has to be distinguished from low affinity, nonspecific binding by other macromolecules. The 100,000*g* cytosol, or resuspended $(NH_4)_2SO_4$ precipitate of the cytosol, or protein fractions during purification (> 1.5 mg protein$\cdot$mL^{-1}) are incubated for 1.5 h with $[^3H]GA_4$ (> 1.6 x 10^{12} Bq$\cdot$mmol^{-1}) in the absence and presence of selected, unlabeled GAs. A typical selection of unlabeled GAs is shown in Table 1. In receptor containing fractions, ability to displace radiolabel from GA-binding site correlates with *in vivo* activity of the competitor. Metabolism of ligands has to be inhibited. All solutions, therefore, are kept at 0 to 2°C.

Table 1. Unlabeled GAs used in a typical assay.

Tube	Label	Competitor	*in vivo* activity of competitor
1	50 nM $[^3H]GA_4$	-	-
2	50 nM $[^3H]GA_4$	5.0 μM GA_4	high
3	50 nM $[^3H]GA_4$	5.0 μM GA_3	moderate
4	50 nM $[^3H]GA_4$	5.0 μM GA_4 methyl ester	zero

Filter paper assay: The protocol described below is based on the methodology reported earlier (Keith *et al*., 1982) but incorporates some of the more recent changes. At neutral pH the GABP is negatively charged and binds to DEAE-cellulose filters. Ligands bound by the protein are retained while unbound molecules are washed through. Protein binding to filters is affected by cations, hence low ionic strength buffers are used for incubation as well as wash. Stacks of 2 filter discs (2.4 cm diameter, Whatman DE81) moistened in ice-cold assay buffer (10 mM K-phosphate, 1 mM EDTA [pH 7.0]) are placed on a vaccuum filtration manifold (Hoefer 225 V) capable of handling up to 10 replicates. The filters are equilibrated with 25 mL assay buffer. The vaccum is released and 100 μL aliquots of the incubation mixture are pipetted onto the filters. After exactly 1 min the filters are washed with 100 mL of assay buffer to remove unbound ligand molecules. The filter stacks are allowed to dry for about 5 sec, placed in scintillation vials and 1 mL 100% ETOH added. After 30 min 6 mL of scintillant (Scintiverse 2, Fisher) is added, the vials are shaken and left in the dark for 2 h before counting. The $[^3H]GA_4$ concentration in the incubation mixture is determined by measuring the radioactivity of 10 μL unfiltered aliquots.

While the assay is fast, simple and reliable, several of its characteristics should be noted: To reduce variability a standard assay protocol should be followed. In the absence of protein $[^3H]GA_4$ can bind to the DEAE-filters directly. This binding can be displaced partly by unlabeled GA_4. The resulting 'apparent specific' binding is kept low, 3 to 4 times lower than specific binding in the presence of protein. This is done by adjusting the amount of protein in the incubation mixture and thus loaded on to the filter paper. Many nonreceptor proteins, including BSA, also show specific binding. However, with these proteins, ability to displace [^{3}H]GA_4 from the binding sites does not correlate with *in vivo* bioassay activity of the competing ligands. The capacity to bind proteins and GAs varies with filter paper lot and should be tested for each lot.

Reference

Keith, B., S. Brown, and L.M. Srivastava. 1982. *In vitro* binding of gibberellin A_4 to extracts of cucumber measured by using DEAE-cellulose filters. Proc. Natl. Acad. Sci. (USA) 79: 1515-1519.

ETHYLENE BINDING AND METABOLISM IN *PISUM SATIVUM* L. cv 'ALASKA'

I.O. Sanders, D.R. Robertson, A.R. Smith & M.A. Hall
Department of Botany and Microbiology, University College of Wales, Aberystwyth, Dyfed SY23 3DA, Wales, U.K.

INTRODUCTION

A method will be described which permits the measurement of ethylene binding and metabolism simultaneously in pea epicotyls. It is well established that pea seedlings metabolise ^{14}C-ethylene (Beyer 1975). Three main products of this activity have been identified: ^{14}C-ethylene glycol which has been found in the tissue (Blomstrom & Beyer 1980), and two gaseous products which are absorbed by NaOH and identified as ^{14}C-ethylene oxide and $^{14}CO_2$ (Beyer 1980). In contrast, there is only one brief report on ethylene binding in pea seedlings (Sisler & Yang 1984). As a result of this report, it was necessary to see if some of the ^{14}C measured in the tissue was in fact due to ethylene binding rather than ethylene metabolism. A modification of the ethylene binding assay of Sisler (1979) was used since it permits a distinction between bound ^{14}C-ethylene and ^{14}C-ethylene metabolites in the tissue.

METHODS

EThylene metabolism was measured in five-day-old etiolated pea epicotyl tips using the assay illustrated in Fig. 1. Purification of the ^{14}C-ethylene was required because of the presence of impurities which were rapidly metabolised by pea seedlings to $^{14}CO_2$. ^{14}C-ethylene is incorporated either into the tissue, or into ^{14}C-ethylene oxide and $^{14}CO_2$ which are released from the tissue and absorbed by NaOH to form ^{14}C-ethylene glycol and ^{14}C-bicarbonate (Beyer 1980). The ^{14}C-bicarbonate can be released as $^{14}CO_2$ by the addition of excess acid, and retrapped in a fresh NaOH solution. Thus the two metabolites absorbed in the first NaOH trap can be separated and measured independently of each other.

NATO ASI Series, Vol. H10
Plant Hormone Receptors. Edited by D. Klämbt

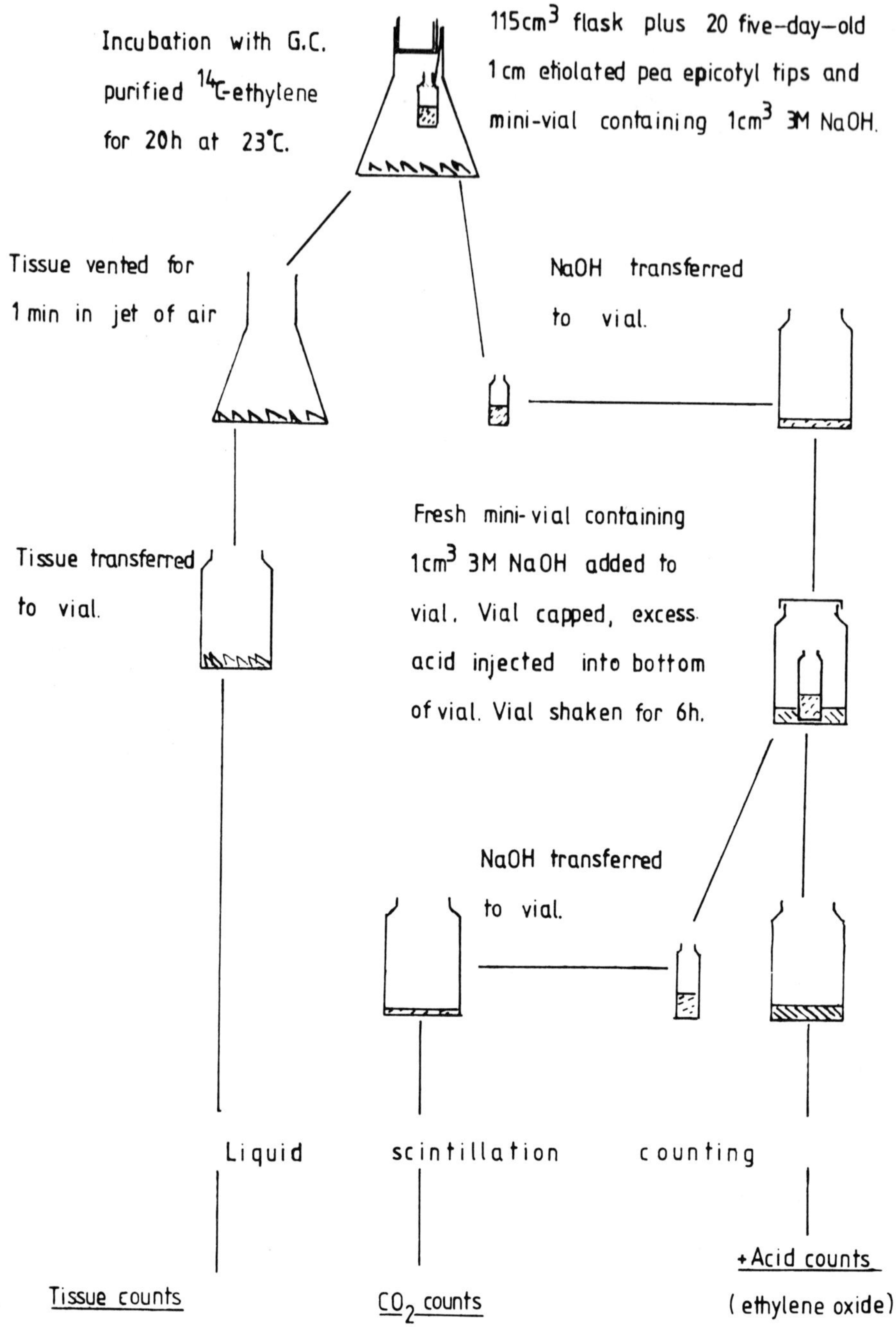

Fig. 1 Scheme illustrating an assay for the measurement of ethylene metabolism in pea seedling tissue.

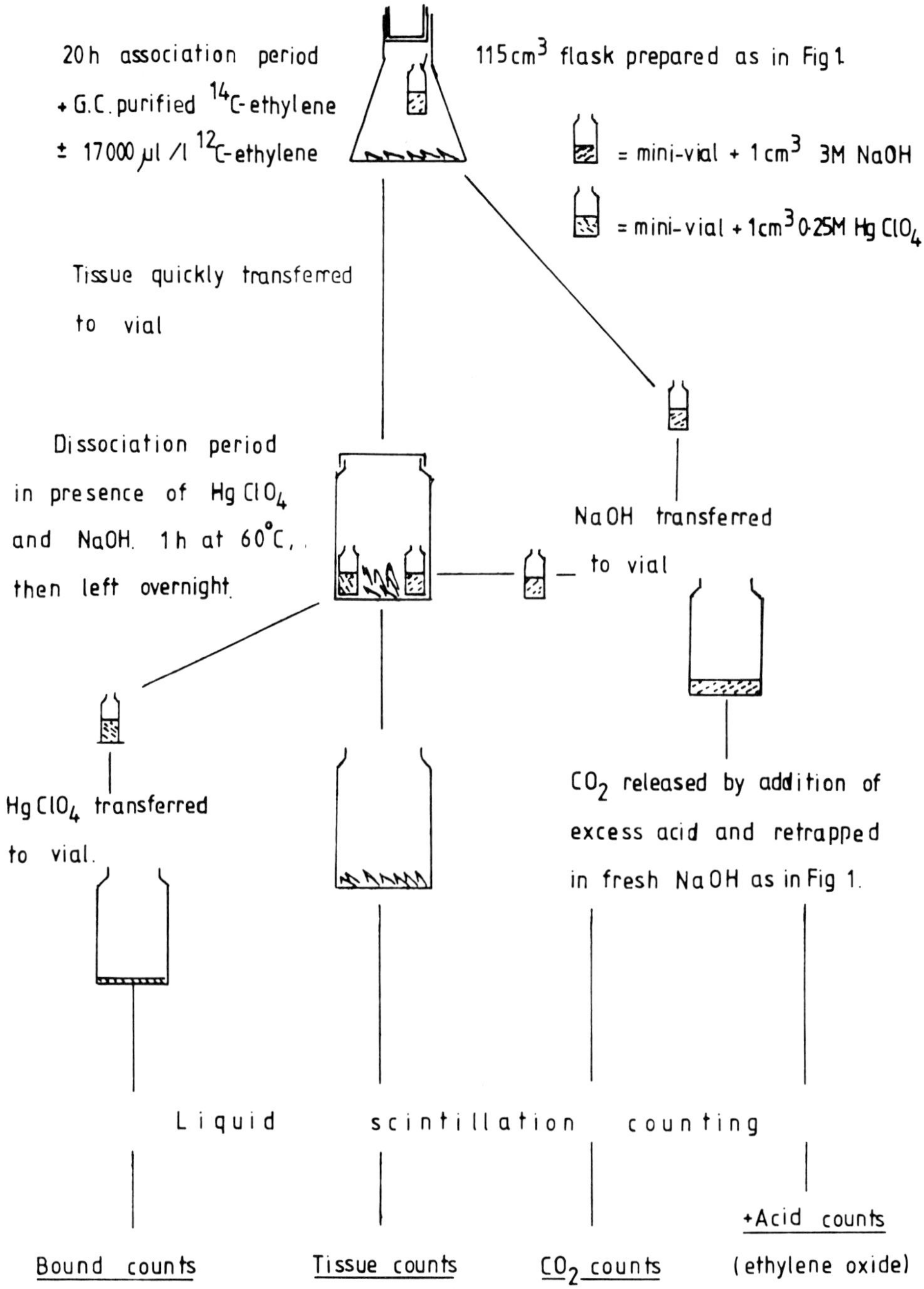

Fig. 2 Scheme illustrating an assay for the simultaneous measurement of ethylene binding and metabolism in pea seedling tissue.

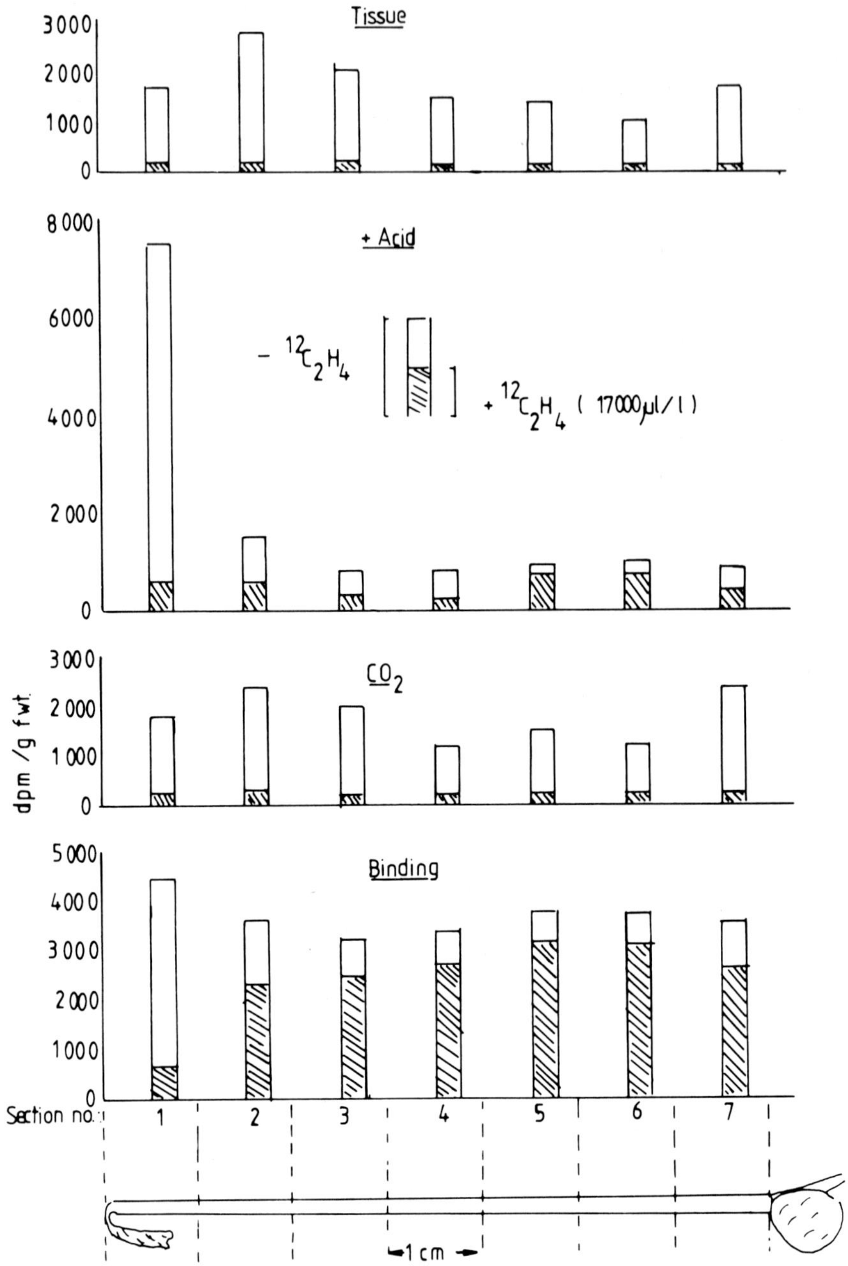

Fig. 3 Ethylene binding and metabolism in serial 1cm sections of five-day-old etiolated pea epicotyls. 20 epicotyl sections were placed in each 115 cm^3 flask and exposed to 1.7 µl/l ^{14}C-ethylene for 20h at 23°C.

The ethylene binding assay of Sisler (1979) was incorporated into the ethylene metabolism assay (Fig 2). After incubation with ^{14}C-ethylene, the tissue is quickly transferred to a vial containing a $HgClO_4$ solution which absorbs any ^{14}C-ethylene released from the tissue. Thus the bound ^{14}C-ethylene is separated from the ^{14}C-metabolites incorporated into the tissue. An NaOH solution is also included in the vial to absorb any additional ^{14}C-ethylene oxide and $^{14}CO_2$ released by the tissue. The quantity of dissolved ^{14}C-ethylene in the tissue was assessed using a displacement assay, where the ^{14}C-ethylene is displaced from the binding sites by an excess of ^{12}C-ethylene (Sisler 1979).

RESULTS AND DISCUSSION

Table 1 compares the data obtained from the metabolism assay of Fig. 1, and the binding and metabolism assay of Fig. 2.

TABLE 1. Ethylene metabolism and binding in five-day-old pea epicotyl tips

The tissue was incubated in 2.1 $\mu l/l$ $^{14}C_2H_4$ $\pm$ 17000 $\mu l/l$ $^{12}C_2H_4$ for 20h, and then assayed for ethylene metabolism as in Fig 1, or ethylene binding and metabolism as in Fig. 2.

Metabolism assay	Tissue (dpm/g fwt)	+Acid	CO_2	Binding
a) $-^{12}C_2H_4$	9377	7153	2418	-
b) $+^{12}C_2H_4$	3207	523	253	-
(a - b)	6170	6630	2166	-
Binding & metabolism assay				
a) $-^{12}C_2H_4$	1900	6998	1730	5400
b) $+^{12}C_2H_4$	173	616	171	1320
(a - b)	1726	6382	1559	4080

The quantity of ^{14}C measured in the tissue is considerably reduced by the inclusion of the binding assay, and this reduction is equivalent to the quantity of binding measured. This suggests that much of the ^{14}C measured in the tissue by the metabolism assay of Fig. 1 is due to ethylene binding rather than ethylene metabolism.

Ethylene metabolism was measured in the top 1cm of five-day-old pea epicotyls because these sections have a high activity per gram

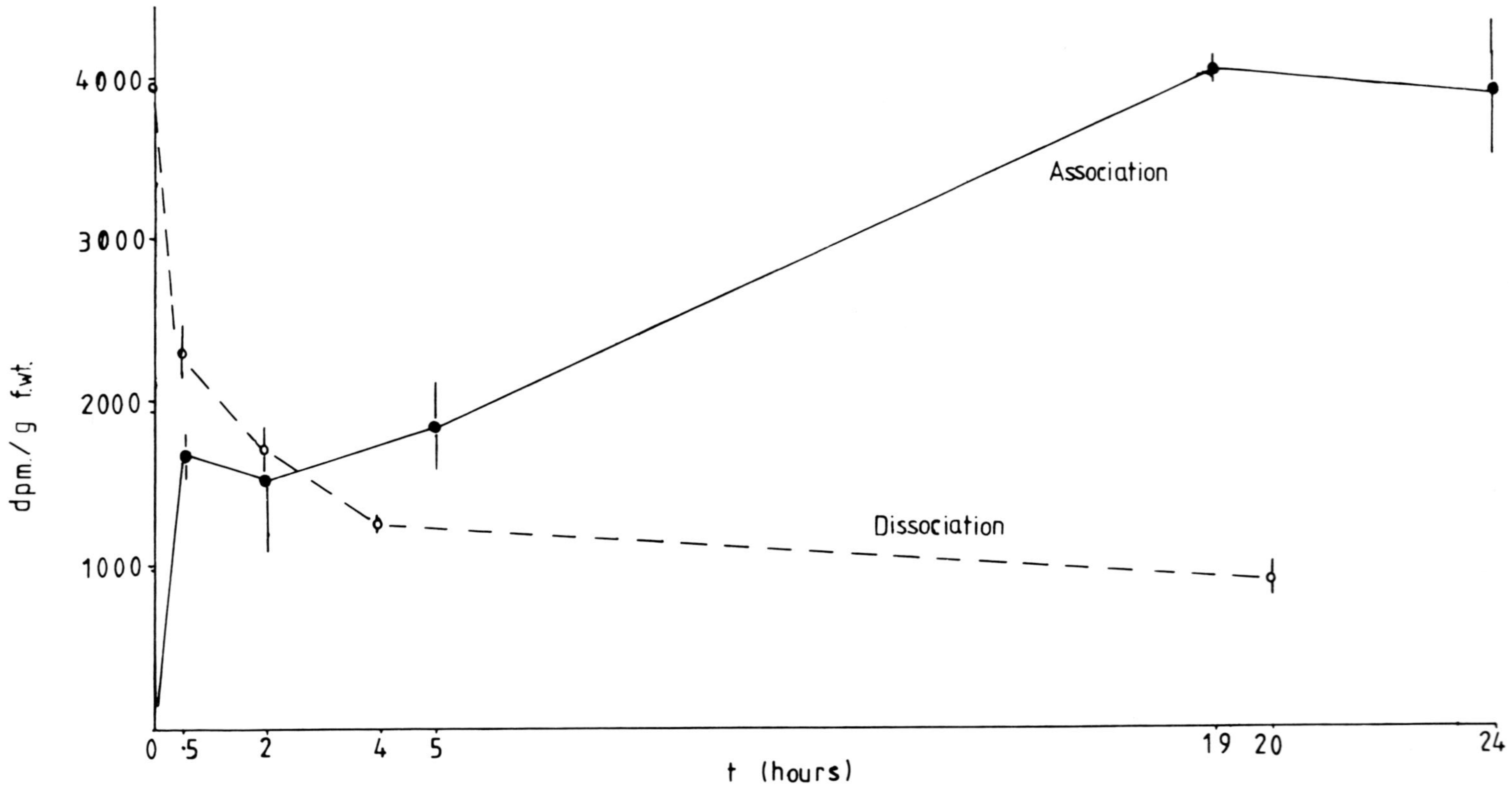

Fig. 4 Association and dissociation curves for ethylene binding in pea epicotyl tips. Five-day-old etiolated pea epicotyl tips were icubated in 2.1 ul/l ^{14}C-ethylene for various time periods to produce an association curve. The dissociation curve was produced by incubating th tissue in 2.1 ul/l ^{14}C-ethylene for 24h, and then venting for various time priods before assaying for ethylene binding.

fresh weight of tissue (Beyer personal communication; Fig. 3). These sections were also found to be suitable for the measurement of ethylene binding because they contain a high binding activity compared to the quantity of ^{14}C-ethylene dissolved in the tissue (Fig. 3). This is probably due to the epicotyl tips containing a large number of immature cells with a low water content, since sections taken from lower down the epicotyl where the cells are more mature exhibit a low binding activity compared to the quantity of dissolved ^{14}C-ethylene (Fig. 3).

Sisler and Yang (1984) measured 13.8 dpm of bound ^{14}C-ethylene per gram fresh weight of three-day-old pea epicotyls. Considerably more bound ^{14}C-ethylene was measured in these experiments (Table 1 & Fig. 3), and this is not entirely due to the higher ^{14}C-ethylene concentrations used in these studies. A 20 hour association period with ^{14}C-ethylene was used instead of the 2 hours used by Sisler & Yang (1984), since the longer period was found to increase the magnitude of binding measured (Fig. 4). It was also found that not all the bound ethylene had dissociated from the binding sites 20 hours after the removal of the ^{14}C-ethylene (Fig. 4), and a 60°C heat treatment was required to ensure full dissociation. The nature and physiological relevance of this binding activity from pea epicotyls awaits discovery.

REFERENCES

Beyer EM (1975) Plant Physiol. 56: 275-278.
Beyer EM (1980) DPGRG/BPGRG Mono. 6: 35-69.
Blomstrom DC, Beyer EM (1980) Nature 283: 66-68.
Sisler EC (1979) Plant Physiol. 64: 538-542.
Sisler EC, Yang SF (1984) Phytochem. 23: 2765-2768.

PURIFICATION OF THE ETHYLENE-BINDING COMPONENT FROM MUNG BEAN SPROUTS AND SEEDS

Edward C. Sisler
Department of Biochemistry
North Carolina State University
Raleigh, North Carolina 27695-7622 USA

The ethylene-binding component has been partially purified from both mung bean sprouts and seeds. The purification procedures and properties differ for sprouts and seeds, so they will be given separately. Whether the different preparations represent different forms of the same component or different components remains to be determined. It is important that means of obtaining these components be given so that they can be studied in more detail and subjected to further purification. Mung bean seeds contain much higher amounts of binding component; however, much greater purification of the component has been achieved from sprouts.

Assay of intact tissue. The plant tissue (usually about 25 g) is placed in desiccators with solid NaOH to absorb CO_2. [^{14}C]Ethylene (110 mc/mM) as the mercury perchlorate complex (7) is introduced in a container with a magnetic stirrer. After sealing, an excess of LiCl is injected into the ethylene-mercury perchlorate complex. The chloride displaces the ethylene from the mercury complex. After 10 minutes of stirring, 99% of the ethylene is in the gas phase. The tissue is exposed for 2-3 hours and then the desiccator is opened, the tissue is shaken in air and quickly (about 30 seconds) placed in a second desiccator containing a scintillation vial in which mercury perchlorate is present. A 2.5-cm^2 piece of fiberglass filter is included to increase the surface area. After about 12 hours, the vial is removed, scintillation fluid is added and the sample is counted. This treatment procedure is carried out both in the presence and absence of unlabeled ethylene. The difference between the values obtained in the presence and absence of unlabeled ethylene is a measure of binding. To assay cell-free systems, the pH is adjusted to pH 6.0 and the samples are made uniform with respect to detergent. Cellulose powder (0.7 g/ml of fluid) is added to facilitate diffusion to the site. The sample is blended, placed in small pans, and placed in desiccators for exposure to ethylene. After about 2 hours' exposure, the samples are removed and, after 3 minutes, placed in small jars containing scintillation vials with

NATO ASI Series, Vol. H10
Plant Hormone Receptors. Edited by D. Klämbt

mercury perchlorate and a 2.5-cm^2 piece of fiberglass filter. The samples are heated to 60 C for 2 hours and allowed to stand for 8 additional hours to collect the [^{14}C]ethylene. The samples are removed, scintillation fluid is added to each vial, and the samples are then counted. This procedure is carried out both in the presence and in the absence of unlabeled ethylene. The difference between labeled ethylene and labeled + unlabeled ethylene is the measure of ethylene binding.

Assay method for ethylene binding in homogenates. The success of isolating the component depends on the use of a successful assay method. The measurement of the ethylene-binding component depends on the competition of ^{14}C-labeled ethylene with unlabeled ethylene (2). When only [^{14}C]ethylene is present, some will be bound to the receptor. If both ^{14}C-labeled ethylene and relatively large quantities of unlabeled ethylene are present, essentially all of the sites will be occupied by unlabeled ethylene. With intermediate levels of unlabeled ethylene, part of the binding sites will be occupied, and by using Scatchard plots (2), the concentration of binding sites can be calculated and the dissociation constant can also be determined.

Binding component from mung bean sprouts. Mung bean sprouts approximately 5-7 days old are obtained commercially or produced by soaking mung beans for 24 hours in water and then placing the beans on cheesecloth in a container in a manner that allows good drainage. About 1 kg of beans/20 liters works well. The beans are covered with cheesecloth and washed twice daily with tap water. The temperature is kept at about 18 C, and it is important to obtain seeds grown in a dry area to minimize microbial growth.

Sprouts (3) are blended with sufficient K_2HPO_4 to make the final pH of the homogenate approximately 7.0. Sufficient water is added to insure good blending. After straining through nylon cloth, the residue is reextracted twice more at pH 7.0. The pH of the total extract is then adjusted to 4.0, and the precipitated material is collected by centrifugation. This precipitate contains the ethylene-binding component. After adjusting the pH of the precipitate to 7.0 and adding water and Triton X-100 to 2%, the mixture is blended. The pH is then adjusted to 4.0 and the mixture is centrifuged. The residue is reextracted twice at pH 7.0 with 2% Triton X-100 and twice re-centrifuged, adjusting the pH to 4.0 each time. The residue is discarded after the third extraction and the extracts are combined for further purification. The binding component may be salted from solution with 313 g/l of $(NH_4)_2SO_4$ and collected by centrifugation. The floating material which contains the binding component is dialyzed and passed over CM-Sephadex using 10 mM potassium phosphate at pH 5.0. About 75% of the binding component is collected on the CM-Sephadex, and it is then eluted with 500 mM potassium phosphate at pH 7.0. This preparation can be concentrated by bringing it to 313 g/l with respect to $(NH_4)_2SO_4$. After dialysis it may be further concentrated in dialysis bags covered with carboxymethyl cellulose. The component can then be subjected to other techniques such as isoelectric focusing (4, 5). The isoelectric point for the major portion of the component appears between 5.5 and 6.5, although a small portion appears lower than pH 4.0 (5). The isoelectric point of the portion not adhering to CM-Sephadex has not been determined. A summary of the purification is given in Table I.

Table I. Purification of Ethylene-Binding Component from Mung Bean Sprouts

Fraction	Binding dpm/1000 g	Binding dpm/g protein	Recovery %
Sprouts	137,000	--	100
Crude	102,700	6,848	74
pH 4.0 Precipitate	79,060	7,186	57
Triton X-100	91,500	183,000	67
CM-Sephadex	54,431	1,088,620	39
Isoelectric focusing	48,000	2,800,000	35

Binding component from mung bean seeds. Mung bean seeds are soaked overnight at 5 C in distilled water. The beans are extracted successively with 5 mg, 2 mg, and 2 mg of Triton X-100 for each gram of seeds, based on the original dry weight. (More Triton X-100 may be used, but these amounts are adequate.) One ml of 0.25 M K_2HPO_4 is added before blending. The extract is centrifuged each time, and the supernatant fluids are combined. After adjusting the extract to 2% Triton X-100, and making to 25% NaCl and 10% $(NH_4)_2SO_4$, the pH is adjusted to 7.5. After separating into two phases and siphoning off most of the lower phase, the extract is centrifuged and the floating material is saved and dialyzed. Much of this material appears to have an isoelectric point below 4.0 (4). This preparation is used for further study. A summary of the purification is given in Table II.

Binding material from an acetone powder of seeds. After soaking seeds in distilled water overnight at 5 C, an acetone powder is prepared by blending

Table II. Extraction of Ethylene-Binding Component from Mung Bean Seeds

Fraction	Binding dpm/1000 g	Binding dpm/g protein
Seeds	16,300,000	67,650
Triton X-100 extract	12,500,000	66,000
$NaCl-(NH_4)_2SO_4$	12,800,000	145,000

with 5 volumes of cold (-15 C) acetone. The powder is collected by filtration and dried at room temperature. It is then stored at -15 C. The binding component may be obtained by soaking overnight in distilled water. Then extraction is made by successively extracting the powder with 5 mg, 2 mg, and 2 mg of Triton X-100 and 1 ml of 0.5 M K_2HPO_4 for each gram of acetone powder. Additional water is added to facilitate blending. After centrifuging, the supernatant solutions are combined and the binding component is salted from solution by adding 250 g NaCl and 100 g $(NH_4)_2SO_4$ for each liter of solution. After adjusting to pH 7.5 and separation into two phases, most of the lower phase is siphoned off. The binding component is then collected by centrifugation. The floating material containing the binding component is dialyzed and stored frozen or at 4 C.

Protein. Protein can be determined by the method of Johnson (1) or by precipitation with 5% trichloroacetic acid followed by successive washing with ethanol, *n*-amyl alcohol, ethanol and ethanol to remove Triton X-100. After dissolving in 0.1 N NaOH, the absorbance is measured at 280 nm (6).

The preparations of the ethylene-binding component represent one case where no physiological function of a receptor is known (mung beans) and one case in which at least part of the component may come from a source in which there is no known function (cotyledons of sprouts) and part from the ethylene-responsive hypocotyl. It would be desirable to obtain a preparation from a source in which all of the tissue responds to ethylene. Unfortunately, the amount of binding component from such tissues is very low (5). Further efforts should be made to find a satisfactory source and to develop methods for obtaining the component from such a source.

LITERATURE CITED

1. Johnson, M. K. Variable sensitivity in the microbiuret assay of protein. Anal. Biochem. 86:320-323. 1978.
2. Sisler, E. C. Measurement of ethylene binding in plant tissue. Plant Physiol. 64:538-542. 1979.
3. Sisler, E. C. Partial purification of an ethylene-binding component from plant tissue. Plant Physiol. 66:404-406. 1980.
4. Sisler, E. C. Isoelectric focusing of the ethylene-binding component using immobilized ampholines. Plant Physiol. 80:s144. (Abstr.) 1986.
5. Sisler, E. C. Distribution and properties of ethylene-binding component from plant tissue. In Y. Fuchs and E. Chalultz (eds), Ethylene: Biochemical, Physiological and Applied Aspects, pp. 45-54. Martinus Nijhoff/Dr. W. Junk Publishers, The Hague. 1984.
6. Warburg, O., and W. Christian. Isolierung und kristallisation des garungsferment Enolase. Biochem. Z. 310:394-421. 1942.
7. Young, R. E., H. K. Pratt, and J. B. Biale. Manometric determination of low concentrations of ethylene. Anal. Chem. 24:551-555. 1952.

CHARACTERISATION AND PURIFICATION OF AN ETHYLENE BINDING COMPONENT FROM DEVELOPING COTYLEDONS OF PHASEOLUS VULGARIS L.

R.A.N. Williams, A.R. Smith and M.A. Hall
Department of Botany and Microbiology, University College of Wales, Aberystwyth, Dyfed SY23 3DA, Wales, U.K.

ABSTRACT

The characteristics of the ethylene binding protein (EBP) found in developing cotyledons of *Phaseolus vulgaris* cv. 'Canadian Wonder' have been determined *in vivo* (1,2) and also *in vitro* after extraction with the non-ionic detergents Triton X-100 and β-octylglucoside (3,4). The EBP has been localised on the endoplasmic reticulum (ER) and protein body membranes (PBM) (5,6), and is a hydrophobic integral membrane protein.

A MW of 60kD has been estimated by gel permeation chromatography and isopycnic sucrose density centrifugation (7).

A procedure for the extraction and purification of the EBP can give purification of up to 600-fold. The drawback is that recoveries are low and the limited capacity of the FPLC mono Q column means that the bulking of many runs is required to produce sufficient purified material.

The tentative identification of the EBP in denaturing polyacrylamide gels is possible by indirect means, but the resolution of the EBP using non-denaturing gels is poor.

One attempt has been made at producing polyclonal rabbit antibodies against a purified preparation but without success.

INTRODUCTION

The ethylene binding activity of the EBP is specific for ethylene ($K_D = 10^{-9}$M) and can be competitively inhibited by physiologically active structural analogues of ethylene such as propylene and acetylene. The EBP does possess some of the requirements suggested by Burg & Burg (8) needed for it to be considered as an ethylene receptor. However, as there is no obvious ethylene mediated response in the tissue, the EBP cannot be considered as a functional receptor. As well as this it possesses a number of anomalous characteristics inconsistent with the molecular requirements for

NATO ASI Series, Vol. H10
Plant Hormone Receptors. Edited by D. Klämbt

an ethylene receptor. The binding of ethylene has a low rate constant of association at 25°C ($k_{+1} = 2.97 \times 10^3\ M^{-1}s^{-1}$) and an even lower rate constant of dissociation ($k_{-1} = 3.18 \times 10^4\ M^{-1}s^{-1}$). The rates of association and dissociation are temperature sensitive and so at low temperatures (4°C) the binding of ethylene is effectively irreversible.

The kinetic characteristics of the EBP are not compatible with the requirements of a putative receptor for ethylene in systems which rapidly perceive and transduce the signal.

The purification of the EBP is undertaken in order to raise EBP-specific antibodies and produce a probe sensitive enough to detect binding proteins in ethylene responsive tissues. The nature of the responsive tissue involved, together with the limited sensitivity of a binding assay using $[^{14}C]$-ethylene means that a more sensitive assay for ethylene binding is required.

The characteristics of the EBP mean that the screening of monoclonal antibodies (McAb) should be straightforward. The antigenic nature of the EBP is unknown so specific polyclonal antibodies (PcAb) are the immediate goal.

Ethylene binding of this type in cotyledons is found throughout a range of P. vulgaris cultivars and other members of the Leguminosae. Sisler (this volume) also finds high levels of binding activity in legume cotyledons and cereal grains. There appears to be a link between ethylene binding and compact seed storage tissues.

During the development of the cotyledons the ethylene binding capacity per cotyledon increases to a maximum of 20 pmol per cotyledon at 28 days after anthesis. The most rapid increase occurs between 20 and 28 days which is at the same time as the most rapid increase in storage protein synthesis (9). Ethylene binding activity does not appear to be directly connected with storage protein synthesis. Isolated ER which could synthesise storage proteins and bind ethylene showed no qualitative or quantitative response in protein synthesis when ethylene was present or not (Cairns, A., unpublished).

BINDING ASSAY FOR EBP

The kinetic behaviour of the EBP and the fact that ethylene is a gas allow for a simple assay. In addition ethylene is not metabolised in the cotyledons so that any radiolabelled ligand applied remains unchanged. The EBP does not modify the bound ethylene in any way. The very slow dissociation rate of specifically bound ethylene means that any non-specifically bound can be removed by venting the sample for 45 sec using a compressed air-stream. The sample is simply incubated for 18h at RT in a gas tight vessel with $[^{14}C]$-ethylene. The free value (F) is obtained by sampling the gas space; the bound value (B) by counting the sample directly after venting. Precise details in ref. (1).

For purification procedures the peeled cotyledons are incubated in $[^{14}C]$-ethylene before extraction. This pre-labelled preparation is used and radiolabel simply followed during subsequent steps, the EBP being $[^{14}C]$-labelled. Assay of samples is possible at any stage (post-labelling). Due to dissociation of pre-labelled samples, any values for purification and recovery will be under-estimates. The difference between pre- and post-labelling can be seen in Tables 1 and 2.

EXTRACTION OF THE EBP FROM MEMBRANES

The EBP is an integral membrane protein and requires a non-ionic detergent to extract it from a crude membrane pellet (96k x g). Binding activity is lost with ionic detergents such as cholate and SDS. Solubilization of the EBP is determined by its presence in the supernatant of a 96k x g centrifugation after 1h. For a 96k x g pellet the requirement is for 20 mg Triton X-100 to each mg of protein for maximum solubilization. After treatment with 0.1M KCl at pH12 the EBP requires 10 mg Triton X-100 per mg protein for maximum solubilization. See Fig. 1.

Less protein is present after the extrinsic membrane proteins have been stripped off using the high salt and high pH treatment. As proteins bind Triton X-100 this fact could account for the reduction in the amount of Triton X-100 required for solubilization. Three lanes on the RH side of Figure 2 show a protein band with a MW of about 60 kD increasing in intensity. The lanes indicated correspond to the points in Figure 1. The detergent is added by weight to the membrane pellet as a 20% solution in 0.02M Tris at pH 7.0. After resuspension the mixture is diluted with 0.02M Tris at pH 7.0 and centrifuged at 96k x g for 1h. The solubilized

EBP is taken from the supernatant fraction.

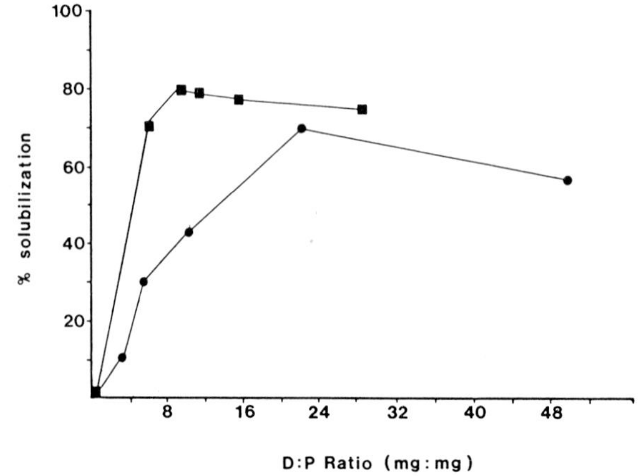

Figure 1. Solubilization curves using Triton X-100 (pH 12, 0.1M KCl treated 96k x g pellet, ■; 96k x g pellet, ●).

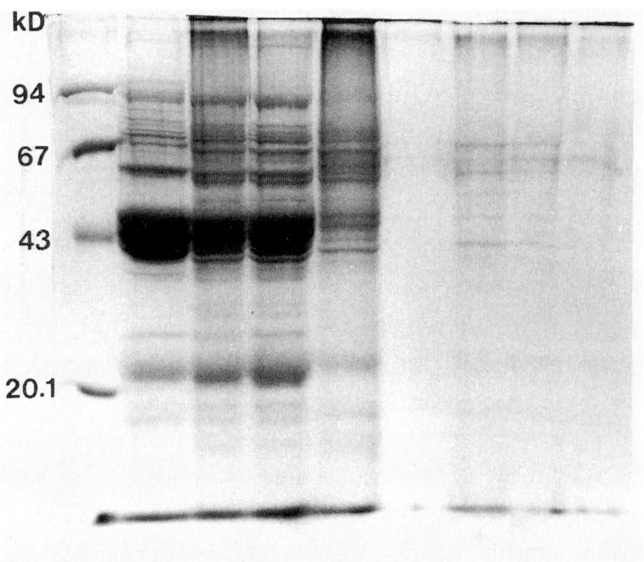

Figure 2. Disc. SDS PAGE (10%) of differential centrifugation steps; 0.1M KCl, pH 12 treatment and increasing Triton X-100 on 100k x g pellet. Coomassie blue stain. Lanes from left to right are: MW standards, crude homogenate, 100k x g supernatant, 100k x g pellet, pH12 treatment 100k x g supernatant, pH12 100k x g pellet followed by 100k x g supernatants of pH12 pellets with increasing D:P ratios.

Triton X-100 is used, although β-octylglucoside is useful for non-denaturing PAGE. Other non-ionic and some ionic detergents remain to be tried for possible advantages they may confer for PAGE, IEF and chromatographic fractionation.

The EBP is sensitive to organic solvents; a range of alcohols, alkanes and ether added to a pre-labelled 96k x g pellet resulted in loss of $[^{14}C]$-ethylene. Only xylene and toluene did not abolish binding activity. The white, pelletable precipitate contained the EBP which, when resuspended in 0.02M Tris at pH9.0, showed binding activity with high affinity ($K_D = 3 \times 10^{-9}$M). Toluene had the effect of reducing the concentration of binding sites from 60 to 40 pmol g^{-1} FW. (see Fig. 3).

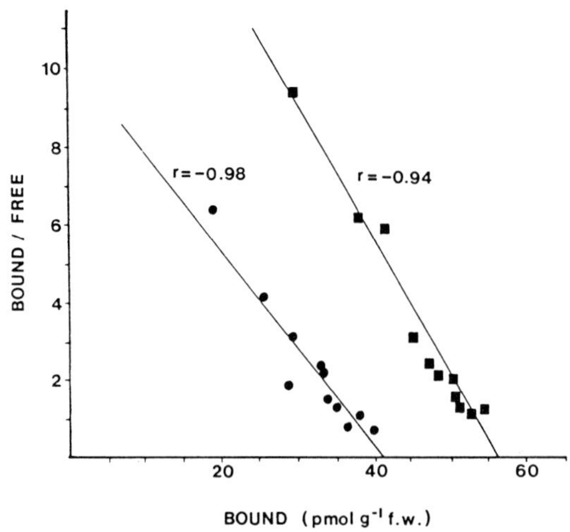

Figure 3. Scatchard plots of Xylene and Toluene precipitated EBP resuspended in 0.02M Tris pH 9.0 (●, Toluene; ■, Xylene).

PURIFICATION OF THE EBP

The steps in the protocol are set out below.

PURIFICATION PROTOCOL FOR THE ETHYLENE BINDING COMPONENT

1. EXTRACT peeled cotyledons with cold 0.02M Tricine, 8% sucrose, pH 9.0 buffer, by chopping finely and using a Polytron homogenizer. Filter through 20μm Miracloth.

2 CENTRIFUGE at 12k x g, 45 min, 4°C. Retain supernatant and discard pellet.

3. CENTRIFUGE at 96k x g, 3.75h, 4°C. Discard supernatant and retain pellet.

4. SOLUBILIZE pellet with 0.02M Tris, 0.05M NaCl, 20% Triton X-114, pH 7.0, using a detergent:protein ratio (mg:mg) of 4:1. Dilute to volume of tube with 0.02M Tris, 0.05M NaCl, pH 7.0.

5. PHASE SEPARATION of the Triton X-114 solubilized extract is achieved by warming to 35°C and centrifuging at 1000 x g for 5 min. to obtain an upper aqueous phase (discarded) and a lower detergent phase (retained) and diluted to volume of tube with 0.02M Tris, pH 7.0.

6. REMOVE DETERGENT by stirring for 2h at R.T. with Amberlite XAD-2 resin beads. A ratio of Amberlite:detergent (mg:mg) of 87.5:1 is used to ensure total removal and precipitation of binding activity.

7. CENTRIFUGE decanted solution at 96k x g for 1h at 4°C. Supernatant discarded, activity-rich pellet retained.

8. PELLET resuspended in 0.02M Tris, 0.1% Triton X-100, pH 8.9. This is used for subsequent anion exchange chromatography.

9. DEAE-Sepharose. Elute applied sample(above) with a gradient of 0.35M NaCl in the same buffer pH 8.9.

10. FPLC MONO Q. Elute sample with a 0-100% gradient of 0.35M NaCl in the same buffer as above, pH 8.9, using a total gradient volume of 20ml. Flow rate of 1ml min^{-1}.

After differential centrifugation the crude membrane (96k x g) pellet is extracted with 20% Triton X-114 in 0.02M Tris at pH7.0. Above 30°C in solution there is a transition by the detergent from the micellar to lamellar state. The lamellar state has a greater density and can be condensed into a heavy detergent phase by centrifugation at 100 x g. More than 90% of the EBP is separated into this phase. The hydrophilic protein separate into the upper aqueous phase. The phase separation of hydrophobic proteins was optimized on a protein basis to ensure maximum recovery of EBP in the detergent phase. A detergent:protein (D:P) weight ratio of 4.0 (mg:mg) was determined. However it is probable that insufficient detergent is present for solubilization (see Fig. 1) at stage 5.

At stage 8 the EBP is not solubilized, the D:P ratio being 1.0. This is probably the reason for the low recoveries and blockage of the FPLC Mono Q anion exchange column. In addition the low capacity of the Mono Q column means that many runs need to be bulked before a sample large enough for generating and screening antibodies can be obtained.

Despite the inadequacies of the protocol, purified preparations of EBP were produced in μg quantities. Tables 1 and 2 show the typical results obtained. Figures 4 and 5 show the FPLC anion exchange fractionation and the protein present in the fractions analysed by 10% disc-SDS PAGE. Unique bands indicated on Fig. 4 in the purified fraction correspond to molecular weights of 18 and 60 KD. Modification of the protocol will ensure the EBP is soluble at all stages.

Table 1. Purification and recovery of ethylene binding activity over a crude homogenate, samples assayed for activity at each step (post-labelling).

STAGE	SPECIFIC ACTIVITY (dpm mg^{-1} protein)	PURIFICATION	RECOVERY (%)
Crude homogenate	502		
96k x g pellet	1463	2.9	32
Triton X-114 detergent phase	3289	6.6	25
96k x g pellet after detergent removal	8.52×10^4	170	10.6
DEAE-Sepharose anion exchange	3.19×10^5	635	3.6
FPLC Mono Q HR(5/5)	nd	nd	0.3

Table 2. Purification and recovery of ethylene binding activity over a crude homogenate, cotyledons inducbated with [^{14}C] -ethylene before extraction.

STAGE	SPECIFIC ACTIVITY (dpm mg^{-1} protein)	PURIFICATION	RECOVERY (%)
Crude homogenate	1140		
96k x g pellet	2853	2.5	95
Triton X-114 detergent phase	5395	4.7	74
96k x g pellet after removal	1.34×10^4	11.8	40
FPLC Mono Q (HR 5/5)	2.23×10^5	196	3.3

NOTE: The % recovery data for FPLC anion exchange has been corrected for the small sample size applied to the column. Protein estimates were made using the BCA protein assay (Pierce Chemical Co.) against a standard curve of BSA.

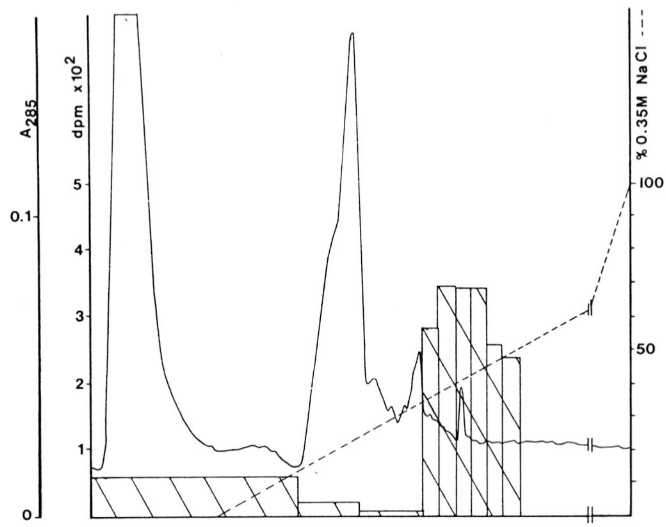

Figure 4. FPLC anion exchange using Mono Q HR(5/5). 0-100% gradient of 0.35M NaCl in 0.02M Tris, 0.1% Triton X-100 pH 8.9. Gradient volume of 20 ml. Trace of UV absorbance and [^{14}C]-ethylene recovered per fraction. Highest peak of activity elutes at 37% of salt gradient. No further activity eluted from column after 50% of total gradient.

LOCALIZATION OF THE EBP ON NON-DENATURING P.A.G.

Non-denaturing P.A. gradient gels have been used to resolve and purify the EBP. It is hoped that excision of a single band exhibiting ethylene binding from the gel could be used directly as an antigen. The native EBP must be used as it is sensitive to SDS and denaturing conditions. A pre-labelled 96k x g pellet was solubilized with either Triton X-100 or β-octyl-glucoside in 0.1M Tris-Glycine at pH 9.0 and loaded onto the gel system in 5% sucrose and 0.01% bromophenol blue. The gel system was that of Cottingham *et al.* (1986) (10) a 3-22% PAG, 1.5mm thick, 0.1% Triton X-100 or 23mM β-octylglycoside in 0.1M Tris-Glycine pH 9.0 throughout. Electrophoresis of samples was carried out at 4°C at 9mA for 16h. The gel was then sliced for

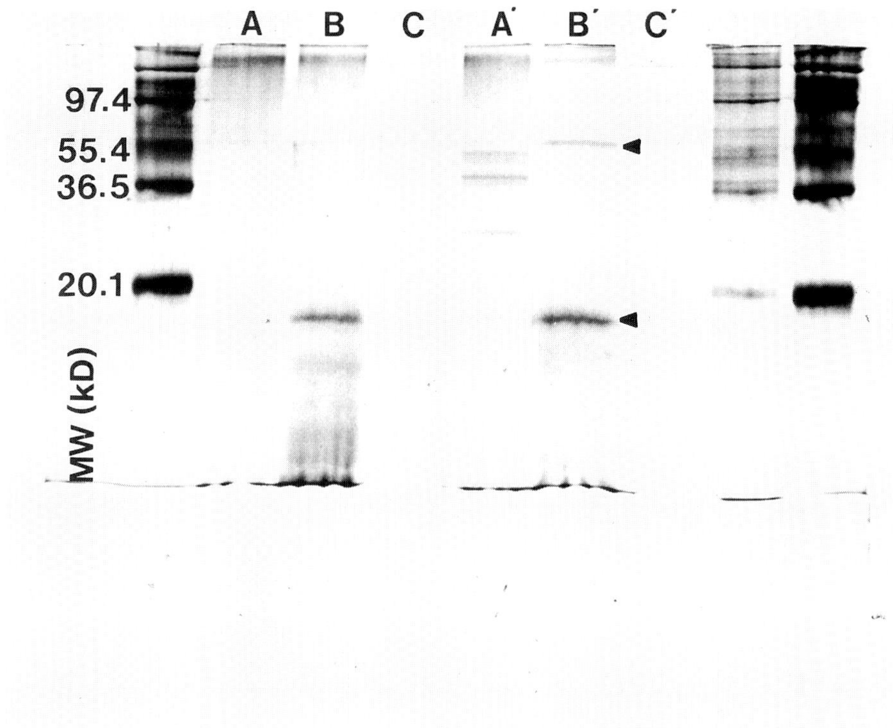

Figure 5. Disc. SDS PAGE (10%) of fractions from FPLC anion exchange. Silver stain. Samples in lanes left to right; MW standards, fractions before elution of radiolabel (A), peak of radiolabel (B), fractions after the peak (C), repeat of previous 3 fractions from another FPLC run (A'B'C'), MW standards.

Coomassie blue staining, autofluorography and scintillation counting of the $[^{14}C]$ in the gel. Autofluorography was carried out by the procedure of Chamberlain et al. (1979) (11) using sodium salicylate as the fluor. Counting of the gel was achieved by squeezing the 1cm gel slices through the barrel of a 5ml syringe and adding distilled water and scintillant. Any procedure which involves digestion or hydrolysis of the gel results in loss of $[^{14}C]$-ethylene.

Fig. 6 shows that no resolution of the EBP is possible. It is expected that better resolution can be obtained using a sample of higher specific activity. Large quantities of protein are electrophorised to ensure enough activity (10-15 x 10^3 dpm) is loaded onto the gel, this may be a reason for the poor resolution achieved so far. With enough purified material there should be fewer problems in resolving the native EBP. Autofluorography is difficult as it involves drying the gel down onto filter paper resulting in substantial loss of [^{14}C]-ethylene due to the temperature sensitivity of the EBP at around 60°C. The technique is also at the limit of its sensitivity (400 dpm cm^{-2}) in the gels run so far.

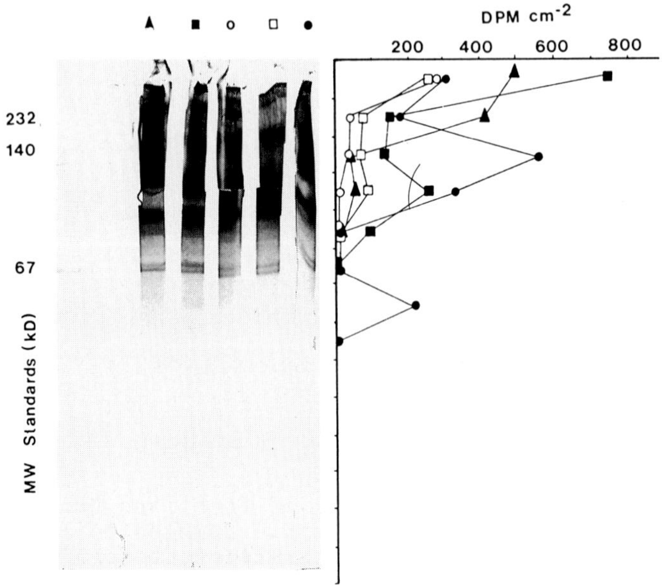

Figure 6. Non-denaturing PAGE of a β-octylglucoside solubilized prelabelled 96k pellet, Lanes are of increasing D:P ratios. (▲, 0; ■ ,3.5; O, 5.3; □ , 10.0; ● , 19.6).

THE ATTEMPTED PRODUCTION OF ANTIBODIES AGAINST A PURIFIED ETHYLENE BINDING PREPARATION

A partially purified preparation from 6 FPLC runs was concentrated to a small volume by ultrafiltration. An initial injection of 60 μg protein of specific activity 1.55 x 10^5 dpm mg^{-1} mixed with an equal volume of Freund's complete adjuvant was administered intramuscularly to a N.Z. white rabbit. Four weeks later the animal was boosted with a similar quantity mixed with an equal volume of incomplete adjuvant, administered subcutaneously at a number of sites on the back. After a further four weeks a sample of serum was taken and tested, the animal was also boosted again with a similar injection. After a further four weeks a second serum test was made.

Two tests were carried out: Ouchterlony double diffusion in 1% agarose containing 0.02M Tricine, 1% Triton X-100, 0.85% NaCl, pH 7.2. No precipitin reaction could be seen when serum dilutions were challenged against partially purified binding protein.

The second test was a double antibody-pelleting assay using SacCel donkey anti-rabbit serum. Serum dilutions were incubated with radiolabelled, partially purified binding protein (a 96k x g pellet after detergent removal). To this was added a donkey anti-rabbit serum coated to cellulose particles. On centrifugation all the rabbit serum should be pelleted together with any immuno-precipitated radiolabel. Using this assay there was no difference from the control (-serum).

CONCLUSIONS

Partial purification of the EBP allows tentative identification of SDS PAGE bands of 18 & 60KD. Electrophoresis of the native EBP on non-denaturing PAG is possible but resolution has not been achieved. Modifications of the purification protocol is required to increase the yield of EBP sufficient to raise antibodies to the EBP. An attempt was made to raise antibodies in a rabbit using a partially purified extract but this was unsuccessful.

The precise electrophoretic behaviour of the EBP on both denaturing and non-denaturing gels is required in order to screen for active sera using Western blotting.

REFERENCES

1. Bengochea T, Dodds JH, Evans DE, Jerie PH, Niepel B, Shaari AR, Hall MA (1980) Planta 148: 397-406.
2. Bengochea T, Acaster MA, Dodds JH, Evans DE, Jerie PH, Hall MA (1980) Planta 148: 407-411.
3. Thomas CJR, Smith AR, Hall MA (1984) Planta 160: 474-479.
4. Howarth CJ, Smith AR, Hall MA (1985) In: Roberts JA, Tucker GA (eds) Ethylene and plant development. Butterworths, London, p.117.
5. Evans DE, Bengochea T, Cairns AJ, Dodds JH, Hall MA (1982) Plant Cell and Environ. 5: 101-107.
6. Evans DE, Dodds JH, Lloyd PC, ApGwynn I, Hall MA '1982) Planta 54: 48-52.

7. Thomas CJR, Smith AR, Hall MA (1985) Planta 164: 272-277.

8. Burg SP, Burg EA (1967) Plant Physiol. 42: 144-152.

9. Opik H (1968) J. exp. Bot. 19: 64-76.

10. Cottingham IA, Kuonen DR, Moore AL (1986) Biochem. Soc. Trans. 14(1): 145-146.

11. Chamberlain A (1979) Anal. Biochem. 98: 132.

SUBJECT INDEX

NATO ASI Series H

Vol. 1: **Biology and Molecular Biology of Plant-Pathogen Interactions.** Edited by J. A. Bailey. 415 pages. 1986.

Vol. 2: **Glial-Neuronal Communication in Development and Regeneration.** Edited by H. H. Althaus and W. Seifert. 865 pages. 1987.

Vol. 3: **Nicotinic Acetylcholine Receptor: Structure and Function.** Edited by A. Maelicke. 489 pages. 1986.

Vol. 4: **Recognition in Microbe-Plant Symbiotic and Pathogenic Interactions.** Edited by B. Lugtenberg. 449 pages. 1986.

Vol. 5: **Mesenchymal-Epithelial Interactions in Neural Development.** Edited by J. R. Wolff, J. Sievers, and M. Berry. 428 pages. 1987.

Vol. 6: **Molecular Mechanisms of Desensitization to Signal Molecules.** Edited by T. M. Konijn, P. J. M. Van Haastert, H. Van der Starre, H. Van der Wel, and M. D. Houslay. 336 pages. 1987.

Vol. 7: **Gangliosides and Modulation of Neuronal Functions.** Edited by H. Rahmann. 647 pages. 1987.

Vol. 8: **Molecular and Cellular Aspects of Erythropoietin and Erythropoiesis.** Edited by I. N. Rich. 460 pages. 1987.

Vol. 9: **Modification of Cell to Cell Signals During Normal and Pathological Aging.** Edited by S. Govoni and F. Battaini. 297 pages. 1987.

Vol. 10: **Plant Hormone Receptors.** Edited by D. Klämbt. 319 pages. 1987.